Elements of Modern Physics

Elements of Modern Physics

(Revised & Updated)

S H Patil
Emeritus Professor
Indian Institute of Technology, Bombay

Ane Books Pvt. Ltd.

New Delhi ♦ Chennai

Elements of Modern Physics (Revised & Updated)
S H Patil

Published by

Ane Books Pvt. Ltd.

4821, Parwana Bhawan, 1st Floor, 24 Ansari Road, Darya Ganj,
New Delhi - 110 002, Tel.: +91(011) 23276843-44, Fax: +91(011) 23276863
e-mail: kapoor@anebooks.com, Website: www.anebooks.com

Branch Office :

- Avantika Niwas, 1st Floor, 19 Doraiswamy Road, T. Nagar,
  Chennai - 600 017, Tel.: +91(044) 28141554, 28141209
  e-mail: anebookschennai@gmail.com, rathinam@anebooks.com

ISBN : 978-93-8116-288-0

Printed at : Trident Enterprises, NOIDA.

Dedicated
to

My Parents

Preface

This book has been thoroughly revised and updated as per the requirement of the students. The book provides a perspective of the important concepts and applications in contemporary physics.

While modern physics developing so rapidly, there is a constant need to revise and update the presentation. The present book tries to do this. Starting with a discussion of special theory of relativity and quantum theory, it describes their applications to atoms, molecules, solids and nuclei. There are two special chapters on the modern description of elementary particles and on general theory of relativity and cosmology. The emphasis is on a logical development of ideas, and historical aspects are referred to mainly as an aid to this. An effort has been made to maintain rigour analytical discussions and precision in descriptions. It is hoped that the book will be useful to an advanced undergraduate student, and as a review to a graduate student.

I am grateful to my colleagues, Dr. S.M. Bharati. Dr. S.M. Chitre, Dr. P.P. Divakaran, Dr. Y.K. Gambhir, Dr. G.V. Dass, Dr. Dipan K. Ghosh, Dr. K.S. Kulkarni, Dr. R.C. Mehrotra, Dr. C.H. Mehta, Dr. G. Mukhopadhyay, Dr. R.S. Patil, Dr. G. Thyagarajan and Prof. Atul Mody, Dept. of Physics, VES, College of Arts, Science and Commerce, Mumbai who ungrudgingly gave me their valuable time in reading parts of the manuscript and made valuable suggestions. I also thank Mr. Sunil Somalwar for going through a part of the manuscript.

Mr. S.B Modak not only provided accurate typing but also executed the entire organization of the book with the help of Mr. D.S. Nakhawa, Mr. Kashipathy and Mr. C.A. Sarmalkar. I owe them gratitude. I acknowledge financial support from the curriculum development programme of IIT, Bombay.

S H Patil

Fundamental Constants

$c = 2.997925 \times 10^{8}$ m/s

$h = 6.6256 \times 10^{-34}$ Js

$m_e = 9.109 \times 10^{-31}$ kg $= 0.511$ MeV/c^2

$e = 1.60206 \times 10^{-19}$ C

$k = 1.38044 \times 10^{-23}$ J/K

$m_p = 938.211$ MeV/c^2

$m_n = 939.505$ MeV/c^2

$\varepsilon_0 = 8.85434 \times 10^{-12}$ F m or C^2/Nm^2

$\mu_0 = 4\pi \times 10^{-7}$ H/m or N/A^2

$N = 6.022 \times 10^{26}$/kmol. number of atoms in 12 kg of ^{12}C

Contents

1

Special Theory of Relativity

Structures of the Chapter

We begin our discussion of modern physics with the theory of relativity which aims at relating the observations made by observers in relative motion with respect to each other. Here *only* the restrictive case of the special theory of relativity is analysed, in which the observers are moving with constant velocity with respect to each other. This will help in choosing appropriate frames of reference and in presenting the later topics in a unified manner. After a brief consideration of the drawbacks of the classical theory, the main results of the special theory of relativity are obtained, and applied to describe some specific physical situations.

1.1 INERTIAL FRAMES OF REFERENCE

Most physical observations describe the behaviour of certain objects in space as a function of time. Since the position of a body can be stated only relative to some other bodies, the description of these observations requires *a frame of reference* which is a technical term for the combination of a set of spatial coordinate axes and a time variable.

It was realised by Galileo and others, that the form of the laws of nature depends on the choice of the frame of reference. Among all the possible frames of reference, there exists a class called the *inertial* frames of reference, in which these laws take a simple form. Inertial frames of reference are those in which a body that is not acted upon by external forces, moves with constant velocity. It is implicit here that if two reference frames move with constant velocity with respect to each other, and one of them is inertial, the other also is an inertial frame. It was found that the laws of mechanics take on the same form in all inertial frames of reference.

1.2 GALILEAN TRANSFORMATIONS

Consider to inertial frames of reference F and F', such that their coordinate axes coincide at $t = 0$, and F' moves with velocity $\mathbf{v}$ along the x-axis with respect to F. Then, it may be expected that the coordinates in the two frames are related by the equations

$$\begin{aligned} t' &= t \\ x' &= x - yt \\ y' &= y \\ z' &= z \end{aligned} \tag{1.1}$$

called *Galilean transformations.* In writing these relations, it is assumed that (*i*) it is possible to define a time t which is the same for all inertial frames of reference, and (*ii*) the distance between two points is independent of the frames of reference.

For Galilean transformations, it is easy to show that the velocities and accelerations in the two frames are related by

$$\mathbf{u}' = \mathbf{u} - \mathbf{v} \tag{1.2}$$

where $\mathbf{v}$ is along the x-direction, and

$$\mathbf{a}' = \mathbf{a} \tag{1.3}$$

respectively. Then, if the interaction potential V is a function of only the distances between particles, Newton's equations in the two frames are:

$$\begin{aligned} m_i\mathbf{a}_i &= \nabla_i V \\ m_i\mathbf{a}_i' &= \nabla_i' V \end{aligned} \tag{1.4}$$

where the subscript i is the particle index. These equations are related by the transformations (1.1) and are of the same form. However, it was observed that the Galilean transformations are not consistent with the dynamical theory of electromagnetic fields as formulated by Maxwell (1865).

1.3 VELOCITY OF LIGHT

It follows from Maxwell's equations for electromagnetic fields that electromagnetic waves travel in vacuum with a speed equal to the ratio of the electromagnetic unit to the electrostatic unit of charge. This ratio is essentially equal to the speed of light so that light itself is taken as a form of electromagnetic radiation.

Now, how does the velocity of light transform from one inertial frame to another? According to Galilean transformations, the velocities are different in different frames and are related by Eq. (1.2). However, Maxwell's equations have no reference to the velocity of the inertial frame and hence imply that the speed of light is independent of the velocity of the inertial frame. Observationally also, the Michelson-Morley experiment (1887) analysed below suggests that the speed of light is independent of the velocity of the inertial frame.

Suppose, the earth is moving with velocity $\mathbf{v}$ in the x-direction with respect to the 'standard' frame in which the velocity of light is c in all directions. Then according to Eq. (1.2), the velocity of light with respect to an observer on Earth is $\mathbf{c}-\mathbf{v}$. The time taken for light to travel along the limb AB of the interferometer (Fig. 1.1), from A to B and back is

$$t_1 = \frac{l_1}{c-v} + \frac{l_1}{c+v} \tag{1.5}$$

While travelling from A to C and back, the velocity $\mathbf{c}-\mathbf{v}$ is parallel to AC and hence perpendicular to $\mathbf{v}$. Therefore

$$\mathbf{c}.\mathbf{v} = v^2 \tag{1.6}$$

and the magnitude of **c**–**v** is $(c^2-v^2)^{1/2}$, so that the time taken for light to travel from A to C and back, is given by

$$t_2 = \frac{2l_2}{(c^2-v^2)^{1/2}} \tag{1.7}$$

Thus the difference in the two times is

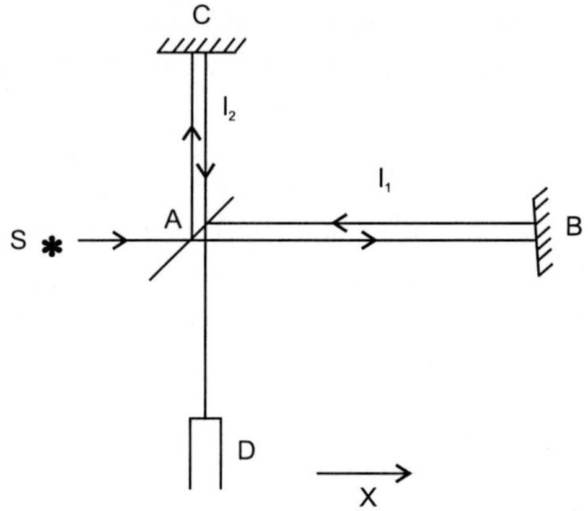

Fig. 1.1 Schematic diagram of the Michelson-Morley experiment.

$$\Delta = t_1 - t_2 = \frac{2l_1 c}{c^2-v^2} - \frac{2l_2}{(c^2-v^2)^{1/2}} \tag{1.8}$$

If the apparatus is turned through 90°, the roles of l_1 and l_2 are interchanged, and the difference in the times becomes

$$\Delta' = t_1' - t_2' = \frac{2l_1}{(c^2-v^2)^{1/2}} - \frac{2l_2 c}{c^2-v^2} \tag{1.9}$$

The expected shift in the interference fringe at D, is

$$\delta = \frac{c(\Delta'-\Delta)}{\lambda}$$

$$= \frac{2(l_1+l_2)}{\lambda}\left[\frac{1}{(1-v^2/c^2)^{1/2}} - \frac{1}{1-v^2/c^2}\right]$$

$$\approx -\frac{(l_1+l_2)}{\lambda}\left(\frac{v^2}{c^2}\right) \text{for } v \gg c \tag{1.10}$$

In the experiment of Michelson and Morley, $l_2 + l_2$ was 22 m, and $\lambda = 5.9 \times 10^{-7}$ m. The value of v is at least of the order $v \approx 30$ km/s corresponding to the velocity of the earth's motion around the sun, even if the motion of the solar system around the galactic centre is ignored. For these values

$$\delta \approx 0.37 \tag{1.11}$$

No such shift was observed in the experiment.

The above result is based on Eq. (1.2) for the transformation of the velocity of light, which was derived from the Galilean transformations (1.2). An attempt was made to salvage the Galilean transformations by postulating that a hypothetical medium called *either*, responsible for the propagation of light, is dragged along by the earth as it moves in space. Then the speed of light with respect to an observer on the earth would remain unaffected by the motion of the earth, analogous to the speed of sound in which case the air is dragged along. This would explain the null result of the Michelson-Morley experiment, but this is in conflict with the observed aberration of starlight received on the earth. It is found that in order to observe a star, the telescope should be titled in the direction of the earth's velocity (Fig. 1.2). This tilt would not be needed if the light-propagating ether was dragged along by the earth. Actually, what is observed is not the absolute tilt but the variation of the tilt as the earth changes it velocity along its orbit around the sun. The either-based explanation became even less tenable after Lodge (1892) showed that the velocity of light is unaffected in the vicinity of rapidly rotating bodies.

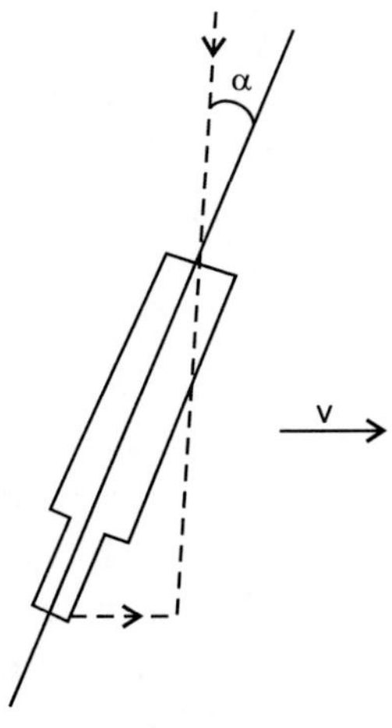

Fig. 1.2 In order that starlight passes along a telescope moving with velocity v, the telescope should be tilted at an angle of $\alpha = v/c$.

1.4 POSTULATES OF SPECIAL RELATIVITY

Einstein (1905) proposed a radically different but, in retrospect, a simple approach to the problem posed by the Michelson-Morley experiment. He started with the principle of relativity but also postulated that the speed of light is the same in all inertial frames of reference, thus giving it the status of a physical law. This immediately explains the Michelson-Morley result but requires that the Galilean transformations (1.2) be discarded. He found that space and time are related in an intimate manner and should be treated on an equal basis. Their relation has a far-reaching influence on the laws of physics. We begin the discussion of Einstein's results with a formal statement of the postulates of the special theory of relativity.

1. The laws of nature are of the same form in all inertial frames of reference.
2. The speed of light is the same in all inertial frames of reference, and is independent of the motion of the source.

It is implicit in the first postulate that, since the coordinates of the different inertial frames are related, the laws of nature written in the various inertial frames can be deduced from one another. It also follows that the Galilean transformations (1.2) relating the coordinates of the inertial frames, cannot be right since they would imply that the speed of light is different in different inertial frames, in contradiction to the second postulate. Hence, a more general relation between the coordinates must be obtained, which incorporates the information that the speed of light is the same in all inertial frames.

1.5 LORENTZ TRANSFORMATIONS

In deriving the transformation equations consistent with the postulates of the special theory of relativity, it was assumed that space is homogeneous, *i.e.*, that all points in space and time are equivalent. This means that the separation between space-time points should remain invariant under translations which implies that the relations between the coordinates of different inertial frames should be linear.

Let us consider again the inertial frames F and F' mentioned in Sec. 1.2, whose axes coincide at time $t = t' = 0$, and F' moves with velocity $\mathbf{v}$ along the x-axis with respect to F. It is assumed that the y-axis is perpendicular to the x'-axis since otherwise the inclinations of the positive and negative y-axis with respect to the x-axis would be different, violating the rotational (or alternatively, left-right) symmetry about the direction of relative velocity. It is also assumed that the y- and z-axes are orthogonal to each other in either of the frames of reference. Finally, since the lengths of two rods, which are at rest in frames F and F' respectively, and which are perpendicular to the x-axis, can be compared while they are passing each other,

$$\begin{aligned} y' &= y \\ z' &= z \end{aligned} \tag{1.12}$$

in order that the relations between F and F' be reciprocal. For the transformation of the x-coordinate, it is noted that the origin of F' travels with velocity $\mathbf{v}$ with respect to frame F, which implies that

$$x' = \alpha(x - vt) \tag{1.13}$$

Taking into account the possibility that time may not be a universal variable,

$$t' = \gamma\,(t - \beta x) \tag{1.14}$$

is for the transformation of the time coordinate.

Let an electromagnetic signal be emitted at $t = 0$, from the origin of F, which also coincides with the origin of F' at that time. Since the speed of light is the same in all inertial frames, the wavefront is described in the two frames by the equations

$$x^2 + y^2 + z^2 = c^2t^2 \tag{1.15}$$

and

$$x'^2 + y'^2 + z'^2 = c^2t'^2 \tag{1.16}$$

respectively. Substituting Eqs. (1.12) - (1.14) in Eq. (1.16)

$$(\alpha^2 - c^2\gamma^2\beta^2)\,x^2 + y^2 + z^2 = (c^2\gamma^2 - v^2\alpha^2)\,t^2 + 2\,(v\alpha^2 + c^2\beta^2)\,xt \tag{1.17}$$

This relation is consistent with Eq. (1.15) provided

$$\alpha^2 - c^2\gamma^2\beta^2 = 1$$

$$\gamma^2 - \frac{v^2}{c^2}\alpha^2 = 1 \tag{1.18}$$

$$v\alpha^2 = c^2\beta\gamma^2 = 0$$

These equations are solved by first eliminating α^2 by using the third equation, and then eliminating γ. The final solutions are:

$$\beta = \frac{v^2}{c^2}$$

$$\alpha = \gamma = \frac{1}{\left(1 - \frac{v^2}{c^2}\right)^{1/2}} \tag{1.19}$$

which lead to the transformations

$$x' = \frac{x - vt}{\left(1 - \frac{v^2}{c^2}\right)^{1/2}}$$

$$y' = y$$

$$z' = z \tag{1.20}$$

$$t' = \frac{t - \frac{v}{c^2}x}{\left(1 - \frac{v^2}{c^2}\right)^{1/2}}$$

These are the celebrated *Lorentz equations*. In obtaining them, the positive root has been chosen, so that $v = 0$ implies $x' = x$, $t' = t$. It is also noted that for small $\frac{v}{c}$ and $x \ll ct$, the Galilean transformations $x' = x - vt$ and $t' = t$ are recovered. The transformations in Eq. (1.20) can be inverted to express x, y, z and t in terms of x', y', z' and t'

$$x = \frac{x' + vt'}{\left(1 - \frac{v^2}{c^2}\right)^{1/2}}; \; y = y'; \; z = z' \tag{1.21}$$

$$t = \frac{t' + \frac{v}{c^2}x'}{\left(1 - \frac{v^2}{c^2}\right)^{1/2}}$$

From these equations, it is seen that frame F moves with velocity $-\mathbf{v}$ with respect to F' so that the relative velocities of the frames are reciprocal.

It should be noted that the deviations of the Lorentz transformations from the Galilean transformations are second order in $\frac{v}{c}$ or $\frac{x}{ct}$ and hence the experiments which can test Lorentz transformations must be accurate enough to detect these second order terms. The Michelson-Morley experiment did have such an accuracy and could prove the inadequacy of Galilean transformations.

Lorentz transformations, though they differ only slightly from Galilean transformations in most physical situations, bring in a profoundly new concept in the kinematics of the universe. They remove the universal character of time and treat it on the same footing as space coordinates. They require that physical space be treated as a 4-dimensional space of space and time coordinates. As might be expected, this mixing of space and time coordinates leads to some unfamiliar consequences. A few of them are discussed here.

1.6 SIMULTANEITY AND TIME DILATION

It follows from Lorentz relations (1.20) that events which are simultaneous in frame F but take place at different positions are not simultaneous in frame F'. For example, if two events take place in frame F at

$$t_1 = t_2 = 0 \tag{1.22}$$

but at positions $x_1 = 0$ and $x_2 = l$ respectively, the corresponding events in frame F' take place at

$$t_1' = 0$$

$$t_2' = -\frac{vl}{c^2\left(1-\frac{v^2}{c^2}\right)^{1/2}} \tag{1.23}$$

i.e., the event at $x_2 = l$ took place earlier in frame F'. In order to understand this result a little better, let the two events correspond to emission of light signals which travel towards the point $x = \frac{l}{2}$ which they will reach at $t = \frac{l}{2c}$ s. However, as observed from frame F', by the time these signals reach the midpoint, the midpoint would have travelled some distance towards the point $x = 0$ and away from the point $x = l$. Therefore, the signal from $x = l$ travels through a longer distance. Because the two signals reach the point $x = \frac{l}{2}$ at the same time, according to an observe in frame F', the event at $x = l$ must have taken place at an earlier time and the events at $x = 0$ and $x = l$ are not simultaneous. He will also observe that since the local clocks at $x = 0$ and $x = l$, stationary in frame F, show time $t = 0$ when the events took place, the F clock at $x = l$ is ahead of the F clock at $x = 0$.

With simultaneity being no longer a universal concept, it is necessary to set up a system of synchronised clocks to measure the time coordinates of events taking place at different positions. Consider now two clocks at rest in frame F, located at a distance l apart. Let a clock at rest in the moving frame F' record time t_0' and t_1' when it passes the clocks of frame F. The corresponding times recorded by the clocks in frame F may be designated by t_0 and t_1. Since these observations correspond to observations taking place at the same place in frame F', one obtains from Eq. (1.21)

$$t_1 - t_0 = \frac{t_1' - t_0'}{\left(1-\frac{v^2}{c^2}\right)^{1/2}} \tag{1.24}$$

Thus, the moving clock appears to run at a slower rate. However, as discussed before, the observer in frame F' will find that the clocks in frame F are not synchronised and the clock he passes later is ahead of the clock he passed earlier by an amount δ. After taking this into account, the reciprocal nature of the two frames then requires that

$$t_1' - t_0' = \frac{(t_1 - \delta) - t_0}{\left(1-\frac{v^2}{c^2}\right)^{1/2}} \tag{1.25}$$

Also, since the relative velocity of the frames is **v**, $t_1 - t_0 = l/v$ which leads to

$$t_1' - t_0' = \frac{l}{v}\left(1 - \frac{v^2}{c^2}\right)^{1/2}$$

$$t_1' - t_0' = \frac{\frac{1}{v} - \delta}{\left(1 - \frac{v^2}{c^2}\right)^{1/2}} \tag{1.26}$$

Therefore, the lag δ is given by

$$\delta = \frac{lv}{c^2} \tag{1.27}$$

The time dilation of moving clocks can be made more physical by considering a clock which consists of a beam of light bouncing back and forth between two mirrors kept at a distance l apart along the y'-direction (Fig. 1.3). In frame F', each round trip takes a time

$$\Delta t' = \frac{2l}{c} \tag{1.28}$$

Viewed from frame F, the beam travels a longer distance along a line making an angle θ with the y-axis, given by $\sin\theta = \frac{v}{c}$. Therefore the corresponding time observed for each trip is

$$\Delta t = \frac{2l}{c\left(1 - \frac{v^2}{c^2}\right)^{1/2}} \tag{1.29}$$

Combining Eqs. (1.28) and (1.29), one has

$$\Delta t = \frac{\Delta t'}{\left(1 - \frac{v^2}{c^2}\right)^{1/2}} \tag{1.30}$$

which tells us that $\Delta t' \leq \Delta t$. Since $\Delta t'$ is the time indicated by the clock at rest in F', this implies that the moving clocks run at a slower rate. This phenomenon is observed for unstable particles which are found to live for a longer time when they are moving (see Example 2, Sec. 1.13).

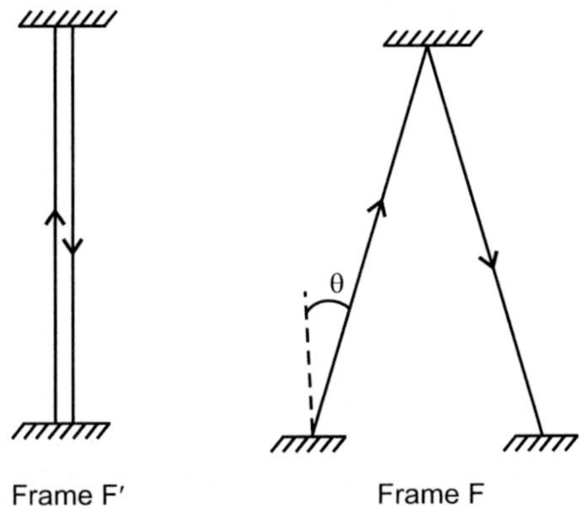

Fig. 1.3 A clock at rest in F' viewed from F.

1.7 LENGTH CONTRACTION

Lorentz transformations lead to unfamiliar results for the measurements of lengths along the direction of motion.

Consider a rod of length l_0, at rest in frame F', parallel to the x'-axis. For measuring the length of this rod in frame F, the positions of the ends of the rod are observed at the same time, *i.e.*, $t_1 = t_2$. Then from Eq. (1.20),

$$x_2' - x_1' = \frac{x_2 - x_1}{\left(1 - \frac{v^2}{c^2}\right)^{1/2}} \tag{1.31}$$

However, since the rod is at rest in frame F', $x_2' - x_1' = l_0$ irrespective of the times of measurements. Therefore the length l of the rod observed from frame F is given by

$$l = l_0 \left(1 - \frac{v^2}{c^2}\right)^{1/2} \tag{1.32}$$

The decrease in the observed length of a rod moving in a direction parallel to itself is called the *Fitzgerald-Lorentz contraction.*

It should again be emphasized that the above observation is reciprocal and an observer in frame F' will find that a rod at rest in frame F, parallel to the x-axis, is shortened by the same factor. Indeed the observer in frame F' can even explain the observation (1.32) from frame F as follows. The observer in frame F' will argue that the scale used by the observer in frame F is shorter by a factor of $\left(1-\frac{v^2}{c^2}\right)^{1/2}$, and therefore the length observed by F should be $l_0\left(1-\frac{v^2}{c^2}\right)$. He will also argue from Eq. (1.21), that $t_2 - t_1 = 0$ corresponds to

$$\begin{aligned} t_2' = t_1' &= -\frac{v}{c^2}(x_2' - x_1') \\ &= -\frac{v}{c^2} l_0 \end{aligned} \tag{1.33}$$

Therefore, the leading end was measured at a time $-\frac{vl_0}{c^2}$ earlier than the trailing end, giving rise to an additional contribution to the measurement of the length of an amount $\frac{v^2 l_0}{c^2}$. Including both of these corrections, the length of the rod should be

$$l' = l_0\left(1-\frac{v^2}{c^2}\right) + \frac{l_0 v^2}{c^2} = l_0 \tag{1.34}$$

in agreement with the measurement from frame F' !

1.8 TRANSFORMATION OF VELOCITIES

In the preceding sections, the implications of the Lorentz transformations for the measurement of position and time intervals in different frames were considered. Here, the relation between velocities of a particle measured in different inertial frames is obtained.

Let $\mathbf{u} = \frac{d\mathbf{r}}{dt}$ be the velocity of a particle in frame F and $\mathbf{u}' = \frac{d\mathbf{r}'}{dt'}$ be the corresponding velocity in frame F' which moves with velocity v. Lorentz transformations (1.20) imply that

$$dx' = \frac{dx - v\,dt}{\left(1-\frac{v^2}{c^2}\right)^{1/2}}, \quad dy' = dy, \; dz' = dz \tag{1.35}$$

and $$dt' = \frac{dt - \frac{v}{c^2}dx}{\left(1 - \frac{v^2}{c^2}\right)^{1/2}}, \tag{1.36}$$

Dividing the position intervals by the time interval,

$$u_x' = \frac{u_x - v}{1 - \frac{vu_x}{c^2}}$$

$$u_y' = \frac{u_y\left(1 - \frac{v^2}{c^2}\right)^{1/2}}{1 - \frac{vu_x}{c^2}} \tag{1.37}$$

$$u_z' = \frac{u_z\left(1 - \frac{v^2}{c^2}\right)^{1/2}}{1 - \frac{vu_x}{c^2}}$$

where the subscript describes the components of the velocities. These expressions relate the velocities of a particle measured in different inertial frames.

As an application of the above formulae, consider the relation between $|\mathbf{u}|$ and $|\mathbf{u}'|$. It follows from Eq. (1.37) that

$$u'^2 \neg c^2 = \frac{(c^2 - v^2)(u^2 - c^2)}{c^2\left(1 - \frac{vux}{c^2}\right)^2} \tag{1.38}$$

For $v^2 < c^2$, which is required for the Lorentz transformations to be physically meaningful, the following important results are obtained:

$$|u'| < c \quad \text{if} \quad |u| < c \tag{1.39}$$

$$|u'| = c \quad \text{if} \quad |u| = c \tag{1.40}$$

$$|u'| > c \quad \text{if} \quad |u| > c \tag{1.41}$$

The first result implies that the relativistic addition of velocities with the speed of each being less than c will again give a velocity with speed less than c.

The statement in Eq. (1.40) is the reappearance of the assumption that the speed of light is the same in all inertial frames of reference. Eq. (1.41) has been considered for particles with speeds greater than the speed of light, known as *tachyons*. It is interesting to note that for tachyons, the equation $u_x v = c^2$, is possible in which case u' tends to infinity. Observation of tachyons would be of great interest since it would imply that information can be transmitted at a speed greater than the speed of light. However, so far, tachyons have not been observed experimentally.

1.9 LORENTZ FOUR-VECTORS

The postulates of the special theory of relativity require that the physical laws retain the same form under Lorentz transformations. This requirement is satisfied if the laws are stated as equalities between terms which have similar transformation properties. For ordinary three-dimensional transformations, these terms are three-dimensional vectors or their generalizations. However, the Lorentz transformations mix the space and time coordinates so that we need to generalize our ideas to four-dimensional vectors.

A relativistic four-vector may be defined to be a set of four quantities $A\mu$ (μ = 1, 2, 3, 0) $\equiv (A_x, A_y, A_z, A_0)$ which transform like the space-time coordinates $x_\mu = (x, y, z, ct)$ under Lorentz transformations (the factor c is included to give the same dimension to all the components), *i.e.*,

$$A_x' = \frac{1}{\left(1-\frac{v^2}{c^2}\right)^{1/2}}\left(A_x - \frac{v}{c}A_0\right) \tag{1.42}$$

$$A_y' = A_y$$

$$A_z' = A_z$$

$$A_0' = \frac{1}{\left(1-\frac{v^2}{c^2}\right)^{1/2}}\left(A_0 - \frac{v}{c}A_x\right)$$

The scalar product between two Lorentz vectors, $A_\mu = (A, A_0)$ and $B_\mu = (B, B_0)$ can be defined as

$$A\cdot B \equiv A_0 B_0 - \mathbf{A}\cdot\mathbf{B} \tag{1.43}$$

which can easily be shown to be equal to $A'\cdot B'$, and hence is called a *Lorentz scalar*. In particular, the 'length' of a vector is defined $(A\cdot A)^{1/2}$ which has the same value in all inertial frames:

$$(A\cdot A)^{1/2} = (A_0 2 - \mathbf{A}\cdot\mathbf{A})^{1/2} \tag{1.44}$$

Here, unlike the three-dimensional case, the length of the vector may be real or imaginary depending on whether $(A_0^2 - \mathbf{A}\cdot\mathbf{A}) \geq 0$ or $(A_0^2 - \mathbf{A}\cdot\mathbf{A}) < 0$, respectively.

Two examples of four-vectors are given which are of particular importance. Consider two events characterized by the vectors $x_\mu = (\mathbf{r}, ct)$, $x_\mu^* = (\mathbf{r}^*, ct^*)$. The separation between these two events,

$$x_\mu^* - x_\mu = (\mathbf{r}^* - \mathbf{r}, c\,(t^* - t)) \tag{1.45}$$

is a 4-vector, and the interval between the two events is defined to be $\Delta\tau$ where

$$(\Delta\tau)^2 = (t^* - t)^2 - \frac{1}{c^2}(\mathbf{r}^* - \mathbf{r})\cdot(\mathbf{r}^* - \mathbf{r}) \tag{1.46}$$

This interval $\Delta\tau$ is a scalar invariant and is called the *proper time interval* between the two events. The proper time interval is said to be timelike if $(\Delta\tau)^2 > 0$, spacelike if $(\Delta\tau) < 0$ and lightlike if $(\Delta\tau)^2 = 0$. Since $(\mathbf{r}^* - \mathbf{r}, c\,(t^* - t)$ transforms as a 4-vector, it may be observed that for a timelike interval, there is an inertial frame in which the events occur at the same place. This is the frame which moves with velocity $\mathbf{v} = (\mathbf{r}^* - \mathbf{r})/(t^* - t)$ with respect to the given frame $(|\mathbf{v}|) < c$ since $|\mathbf{r}^* - \mathbf{r}| < c\,|t^* - t|$. On the other hand, for a spacelike interval, the events occur at the same time in the frame which moves with velocity $v = c^2\,|\mathbf{r}^* - \mathbf{r}|\,|t^* - t|/|\mathbf{r}^* - \mathbf{r}|^2$ with respect to the given frame ($|\mathbf{v}| < c$ since $|\mathbf{r}^* - \mathbf{r}| > c\,|t^* - t|$. Finally, for a lightlike interval, a light pulse starting at $(\mathbf{r}, ct)$ would just reach $(\mathbf{r}^*, ct^*)$.

As a second example of a Lorentz four-vector, consider the set of operators $\left(-\frac{\partial}{\partial x}, -\frac{\partial}{\partial y}, -\frac{\partial}{\partial z}, \frac{1}{c}\frac{\partial}{\partial t}\right)$. The transformations for these operators can be obtained from Eq. (1.21) by using the chain rule and are

$$\left(-\frac{\partial}{\partial x'}\right) = \frac{1}{\left(1 - \frac{v^2}{c^2}\right)^{1/2}}\left[\left(-\frac{\partial}{\partial x}\right) - \frac{v}{c}\left(\frac{1}{c}\frac{\partial}{\partial t}\right)\right]$$

$$\left(-\frac{\partial}{\partial y'}\right) = \left(-\frac{\partial}{\partial y}\right)$$

$$\left(-\frac{\partial}{\partial z'}\right) = \left(-\frac{\partial}{\partial z}\right) \tag{1.47}$$

$$\left(\frac{1}{c}\frac{\partial}{\partial t}\right)=\frac{1}{\left(1-\frac{v^2}{c^2}\right)^{1/2}}\left[\left(\frac{1}{c}\frac{\partial}{\partial t}\right)-\frac{v}{c}\left(-\frac{\partial}{\partial x}\right)\right]$$

Comparing these with Eq. (1.42), it is seen that

$$\left(-\frac{\partial}{\partial x},-\frac{\partial}{\partial y},-\frac{1}{\partial z},\frac{1}{c}\frac{\partial}{\partial t}\right) \tag{1.48}$$

is a 4-vector operator, *i.e.* it is an operator which transforms like a four-vector. The relations (1.47) further imply that the negatives of the scalar product of the operator with itself,

$$\equiv\frac{\partial}{\partial x^2}+\frac{\partial^2}{\partial y^2}+\frac{\partial^2}{\partial z^2}-\frac{1}{c^2}\frac{\partial^2}{\partial t^2} \tag{1.49}$$

is a scalar operator. This operator, which is invariant under Lorentz transformations, is called the *d'Alembertian operator.*

1.10. ENERGY-MOMENTUM FOUR-VECTOR AND RELATIVISTIC DYNAMICS

It may be appreciate that the transformations (1.37) for velocities are involved because the derivatives of position are taken with respect to time t which is not an invariant scalar but the fourth component of a four vector. On the other hand, the derivative of (**r**, ct) with respect to the proper time τ which is a Lorentz scalar, will again give us a 4-vector. It should also be noted that the proper time for the motion of a particle has the physical significance of being the time indicated by a clock moving along with the particle. This can be deduced from Eq. (1.46) by using the fact that $\Delta\tau$ is a scalar invariant, and that the space displacement $\Delta\mathbf{r}$ for a particle is zero in the frame moving with the particle, so that Δt in this frame is equal to $\Delta\tau$.

Consider a particle which is at position $\boldsymbol{r}$ at time t and undergoes a change in position of $\Delta\mathbf{r}$ in time Δt. We then define

$$(\mathbf{p}, p_0)=m_0\lim_{\Delta\tau\to 0}\left(\frac{\Delta\mathbf{r}}{\Delta\tau},\frac{c\Delta t}{\Delta\tau}\right) \tag{1.50}$$

where m_0 is the mass of the particle at rest and $\Delta\tau$ is obtained from Eq. (1.46) as.

$$\Delta\tau=\Delta t\left[1-\frac{1}{c^2}\left(\frac{\Delta\mathbf{r}}{\Delta t}\right)^2\right]^{1/2} \tag{1.51}$$

In terms of the velocity **u** of the particle,

$$\mathbf{p} = \frac{m_0 \mathbf{u}}{\left(1 - \frac{u^2}{c^2}\right)^{1/2}}$$

$$p_0 = \frac{m_0 c}{\left(1 - \frac{u^2}{c^2}\right)^{1/2}} \tag{1.52}$$

Since $\Delta\tau$ is an invariant scalar, it follows that $p_\mu = (\mathbf{p}, p_0)$ is a 4-vector and its transformation relations are

$$p_x' = \frac{p_x - \frac{v}{c} p_0}{\left(1 - \frac{v^2}{c^2}\right)^{1/2}} \quad (a)$$

$$p_y' = p_y;\ p_z' = p_z \quad (b) \tag{1.53}$$

$$p_0' = \frac{p_0 - \frac{v}{c} p_x}{\left(1 - \frac{v^2}{c^2}\right)^{1/2}} \quad (c)$$

The vector **p** is the relativistic generalization of the Newtonian momentum vector $m_0\mathbf{u}$. For the interpretation of p_0, it is noted that for $u \ll c$,

$$cp_0 = m_0 c^2 + \frac{1}{2} m_0 u^2 + \frac{3}{8} m_0 \frac{u^4}{c^2} + \ldots \tag{1.54}$$

where the second term is the Newtonian kinetic energy. Therefore cp_0 may be defined as the energy of the particle, m_0c^2 being the rest energy and the remaining terms being the relativistic generalization of the Newtonian kinetic energy. If T denotes the kinetic energy, then

$$T = cp_0 - m_0 c^2 \tag{1.55}$$

The rest energy does not play a significant role if there is no change of mass in a process, but becomes important if there is a change of mass. Finally, it is noted that the scalar product $p.p$ is

$$p_0^2 - \mathbf{p.p} = m_0^2 c^2 \tag{1.56}$$

an invariant scalar as expected.

The equation of motion for a particle may be written as

$$\frac{d\mathbf{p}}{dt} = \mathbf{f} \tag{1.57}$$

where **f** is the force on the particle. This equation is similar to Newton's equation of motion, with the important difference that the momentum is now given by the relativistic expression (1.52). An equation for $\frac{dp_0}{dt}$ can be deduced using Eq. (1.56) to give

$$\frac{dp_0}{dt} = \frac{\mathbf{p}}{p_0}.\mathbf{f}$$

$$= \frac{\mathbf{u}}{c}.\mathbf{f} \tag{1.58}$$

On multiplying by c, this equation just relates the rate of change of energy to the rate of work done. It should be appreciated that since t is not a scalar, the transformation properties of **f** are rather involved. It is however straight forward to obtain the transformation relations for **f** from those of $\frac{d\mathbf{p}}{dt}$, giving

$$f_x' = \frac{1}{\left(1 - \frac{v^2}{c^2}\right)^{1/2}} \frac{p_0}{p_0'} \left(f_x - \frac{v}{c} \frac{\mathbf{p}}{p_0}.\mathbf{f} \right)$$

$$f_y' = \frac{p_0}{p_0'} f_y;\ f_z' = \frac{p_0}{p_0'} f_z \tag{1.59}$$

It also follows directly from Eqs. (1.57) and (1.58), that if $\mathbf{f} = 0$, both energy and momentum of the particle are constants of motion.

1.11 ELECTROMAGNETIC INTERACTION

Though the theory of electromagnetic fields was formulated before the advent of the special theory of relativity, it is form-invariant under **Lorentz** transformations. In particular, it is consistent with the speed of propagation of electromagnetic radiation being the same in all inertial frames. The transformations of the electromagnetic fields and their interaction with matter are briefly described here.

Electromagnetic fields are described by Maxwell's equations. These equations in rationalised mks units, outside material media, are:

$$\left.\begin{aligned} \nabla \cdot \mathbf{E} &= \frac{\rho}{\varepsilon_0} && (a) \\ \nabla \cdot \mathbf{B} &= 0 && (b) \\ \nabla \times \mathbf{E} + \frac{\partial \mathbf{B}}{\partial t} &= 0 && (c) \\ \nabla \times \mathbf{B} - \frac{1}{c^2}\frac{\partial \mathbf{E}}{\partial t} &= \mu_0 \mathbf{J} && (d) \end{aligned}\right\} \tag{1.60}$$

where **E** is the electric field, **B** is the magnetic field, ρ is the charge density, **J** is the current density, ε_0 is the capacitivity of vacuum and μ_0 is the permeability of vacuum. The motion of a charged particle in the presence of electromagnetic fields is given by

$$\frac{d\mathbf{p}}{dt} = q\,(\mathbf{E} + \mathbf{u} \times \mathbf{B}) \tag{1.61}$$

where q is the charge of the particle, and the expression on the right hand side is called the *Lorentz force.*

It is most convenient to describe the electromagnetic fields in terms of the electromagnetic potentials **A** and ϕ. From Eq. (1.60*b*), it is seen that **B** is of the form

$$\mathbf{B} = \nabla \times \mathbf{A} \tag{1.62}$$

Substitution of this relation in Eq. (1.60*c*) then implies that **E** can be written in the form

$$\mathbf{E} = -\frac{\partial \mathbf{A}}{\partial t} - \nabla\,\phi \tag{1.63}$$

The remaining two Maxwell's equations lead to

$$\nabla^2\,\phi + \frac{\partial}{\partial t}(\nabla \cdot \mathbf{A}) = -\frac{\rho}{\varepsilon_0}$$

$$\nabla^2\,\mathbf{A} - \frac{1}{c^2}\frac{\partial^2 \mathbf{A}}{dt^2} - \nabla\left(\nabla \cdot \mathbf{A} + \frac{1}{c^2}\frac{\partial \phi}{\partial t}\right) = -\,\mu_0 \mathbf{J}$$

It may be observed that Eqs. (1.62) and (1.63) do not determined **A** and ϕ uniquely. A transformation

$$\mathbf{A} \rightarrow \mathbf{A} + \nabla\,\Lambda \tag{1.64}$$

$$\phi \rightarrow \phi - \frac{\partial}{\partial t}\Lambda \tag{1.65}$$

where $\boldsymbol{\Lambda}$ is a scalar function, does not alter **B** and **E.** Therefore some subsidiary conditions can be imposed on **A.** This is done by requiring that

$$\Delta \cdot \mathbf{A} + \frac{1}{c^2}\frac{\partial \phi}{\partial t} = 0 \tag{1.66}$$

a condition known as the *Lorentz condition.* Then Eqs. (1.64) simplify to:

$$\left(\nabla^2 - \frac{1}{c^2}\frac{\partial^2}{\partial t^2}\right)\phi = -\frac{\rho}{\varepsilon_0}$$

$$\left(\nabla^2 - \frac{1}{c^2}\frac{\partial^2}{\partial t^2}\right)\mathbf{A} = -\mu_0 \mathbf{J} \tag{1.67}$$

It is in this form that the Lorentz transformations of Maxwell's equations are most transparent.

It is noted that ($\mathbf{J}$, $c\rho$) transforms as a four-vector. To see this, consider a charge at rest in frame F, and let its charge density in this frame be ρ_0. The charge of the particle is taken to be a universal, Lorentz invariant quantity, an assumption which leads to a correct description of experimental observations. Observed from a frame F' in which the particle moves with velocity $\mathbf{u}$, the particle dimensions appear contracted by the factor $\left(1 - \frac{u^2}{c^2}\right)^{1/2}$ in the direction of $\mathbf{u}$. Since the total charge is an invariant scalar, the charge and current densities in frame F' are

$$\rho' = \frac{\rho_0}{\left(1 - \frac{u^2}{c^2}\right)^{1/2}}. \tag{1.68}$$

$$\mathbf{J} = \frac{\rho_0 \mathbf{v}}{\left(1 - \frac{u^2}{c^2}\right)^{1/2}}$$

Thus, ($\mathbf{J}$, $c\rho$) is proportional to the energy-momentum 4-vector of the particle,

$$\begin{aligned} \mathbf{J}_\mu &= (\mathbf{J}, cp) \\ &= \frac{\rho_0}{m_0}(\mathbf{p}, p_0) \end{aligned} \tag{1.69}$$

and hence transforms as a 4-vector. Furthermore, since $\left(\nabla^2 - \frac{1}{c^2}\frac{\partial^2}{\partial t^2}\right)$ was shown to be a scalar operator (See sec. 1.9), Eq. (1.67) gives the result that

$$A_\mu = \left(\mathbf{A}, \frac{1}{c}\phi \right) \tag{1.70}$$

also transforms as a Lorentz 4-vector. Hence, Maxwell's equations are seen to be consistent with the special theory of relativity.

To show the form invariance of Eq. (1.61) for the motion of a charged particle, the expression for the derivative with respect to the proper time τ is written using Eq. (1.51). Substituting expressions (1.62) and (1.63) for **B** and **E** in the expression for the electromagnetic force, and simplifying, gives

$$\frac{d\mathbf{p}}{d\tau} = -q\left[\frac{1}{m_0}\nabla\left(p_0\frac{\phi}{c} - \mathbf{p}\cdot\mathbf{A}\right) + \frac{d\mathbf{A}}{d\tau}\right] \tag{1.71}$$

The quantity inside the parentheses is a scalar product between the 4-vectors p_μ and A_μ, while ∇ is the space part of the 4-vector operator (1.48). Therefore, both the right hand side and the left hand side of Eq. (1.71) transform as space components of 4-vectors. The corresponding equation for p_0, making use of Eqs. (1.51), (1.56) and (1.61), is

$$\frac{dp_0}{d\tau} = -q\left[\frac{1}{m_0}\frac{\partial}{c\partial\tau}\left(p_0\frac{\phi}{c} - \mathbf{p}\cdot\mathbf{A}\right) + \frac{d}{d\tau}\left(\frac{\phi}{c}\right)\right] \tag{1.72}$$

where the partial derivative applies only to ϕ and **A**; and not to p_0 or **p**. It is now clear that the left hand sides of Eqs. (1.71) and (1.72), as also the right hand sides, transform as 4-vectors and hence the equation of motion of a charged particle in the presence of electromagnetic fields, is form-invariant under Lorentz transformations.

1.12 ZERO-MASS PARTICLES AND DOPPLER SHIFT

It follows from Eq. (1.52) that a particle whose mass is zero, must move with velocity c if its energy and momentum are to be non-zero. It is believed that the following particles have zero mass, and velocity c: the photon designated by γ, which is the quantum of radiation, the neutrinos designated by ν, and the gravitation which is the proposed quantum of gravitational wave. The energy and momentum of these zero-mass particles, and their properties under Lorentz transformations will be considered here.

It was argued by Einstein (See Sec. 2.2 for details) that a radiation of frequency v consists of photons, each carrying a quantum of energy.

$$\begin{aligned} \text{E} &= p_0 c \\ &= hv, \text{ planck's constant } h = 6.67 \times 10^{-34} \text{ J s} \end{aligned} \tag{1.73}$$

Since photons are supposed to have zero mass, one has from Eq. (1.56)

$$|\mathbf{p}| = \frac{h\nu}{c} \tag{1.74}$$

Substituting these expressions in Eq. (1.53*c*) for the transformation of p_0, an expression for the frequency of radiation observed from a moving frame is obtained as:

$$\nu' = \nu \frac{\left(1 - \frac{v}{c}\cos\alpha\right)}{\left(1 - \frac{v^2}{c^2}\right)^{1/2}}, \tag{1.75}$$

where $p_x = p\cos\alpha = \frac{h\nu}{c}\cos\alpha$, α being the angle between **p** and the *x*-axis. This formula is the exact expression for Doppler effect. Similarly, using Eqs. (1.53*a*) and (1.53*c*), the ratio p_x'/p_0' is obtained as

$$\cos\alpha' = \frac{\cos\alpha - \frac{v}{c}}{1 - \frac{v}{c}\cos\alpha} \tag{1.76}$$

This equation relates the directions of propagation in the two frames. In particular, it gives us the relativistic aberration of starlight reaching us. "Unlike the classical Doppler effect, it is observed that the relativistic Doppler shift for radiation depends only on the relative velocity between the source and the observer." For observing the relativistic correction, one may consider transverse Doppler shift for which $\cos\alpha' = 0$, *i.e.* the observer is moving in a direction orthogonal to the direction of propagation. In this case, the transverse Doppler shift is ($\cos\alpha = v/c$)

$$\nu' \approx \nu\left(1 - \frac{1}{2}\frac{v^2}{c^2}\right) \quad \text{for } v \ll c \tag{1.77}$$

in contrast to the classical result of $\nu' = \nu$. The small second order change in the frequency $\left(\frac{\Delta\nu}{\nu} \approx 5.6\times 10^{-16} \text{ for } v \approx 10 \text{ m/s}\right)$ has been observed (1960) by using Mössbauer effect.

In Mössbauer effect, there is recoil-free emission and absorption of photons by atoms embedded in crystals low temperatures (low temperatures are required so that energy is not carried away by the lattice vibrations). The frequencies of the radiation of fairly well-defined except for the uncertainty due to the natural lifetime τ of the excited atom. The radiation has a frequency distribution

$$\rho(\nu) \sim \frac{1}{1+\dfrac{4h^2(\nu-\nu_0)^2}{I'^2}} \tag{1.78}$$

where ν_0 is the central value, and Γ' is the uncertainty in the energy related to the lifetime of the excited state by

$$\tau\Gamma = \hbar \tag{1.79}$$

$\hbar = \dfrac{h}{2\pi}$, $h = 6.67 \times 10^{-34}$ J s being the Planck's constant. In an experiment performed by Hay *et al.* (1960), photons are emitted by excited ^{57}Fe atoms embedded in the crystal, which have energy centred around $h\nu_0$ = 14.4 keV and a linewidth $\Gamma = 4.7 \times 10^{-9}$ eV. The emitter is placed at the centre of a centrifuge. The photons are observed by ^{57}Fe atoms in the ground state, embedded in a crystal and kept at the edege of the centrifuge. When the centrifuge is not rotating, the photons from the emitter are absorbed by the absorber, since the photons have just the right energy for exciting the ^{57}Fe atoms. However, once the centrifuge starts rotating, the absorber sees the photons with shifted frequency given by Eq. (1.77) (the speed is $v = \omega r$, ω being the angular speed and r being the radial distance of the absorber from the centre) and the rate of absorption goes down. The experimental observations for the shifts agree with the shifts given by Eq. (1.77) within experimental errors, thus confirming the predictions of the special theory of relativity for the transverse Doppler shift.

1.13 EXAMPLES

A few examples are now discussed to illustrate some applications, and elaborate the ideas that have been analysed.

Example 1

This example shows that Galilean transformations are not consistent with Maxwell's equations.

Consider an infinitely long, stationary line-charge and a positively charged particle P with charge q, moving away from the line charge with velocity **u**. The only force acting on P is the repulsive force due to the electric field, and it acts in a direction prependicular to the line charge. Now, an observer in a frame moving parallel to the line charge with velocity **v** sees both electric and magnetic fields. The electric field gives rise to a force, again prependicular to the line charge. However, the magnetic field **B** which is prependicular to **u** and **v** gives rise to a force q **(u–v) × B** which has a component parallel to the line charge. This contradicts the result of Galilean transformations that force is invariant.

Example 2

The time-dilation of a moving clock has a dramatic manifestation in terms of the increased lifetime of a moving particle.

In nature, unstable particles are found, whose decay is described in quantum mechanics as a transition from the initial state to the final state. The rate of decay is determined by the transition probability which is defined by λ,

$$|dN(t)| = -\lambda N(t)\, dt \tag{1.80}$$

where $N(t)$ is the number of particles at time t, and $dN(t)$ is the number of particles which decay in time dt. The number of particles remaining at time t is obtained from Eq. (1.80) to be

$$N(t) = N(0)\, e^{-\lambda t} \tag{1.81}$$

The mean lifetime of the particle is then given by

$$\tau_0 = \int t \left|\frac{dN}{N(0)}\right| = 1/\lambda \tag{1.82}$$

Now, if the unstable particles are moving, their dilated lifetime is given

$$\tau = \frac{\tau_0}{\left(1-\frac{v^2}{c^2}\right)^{1/2}} \tag{1.83}$$

where v is the speed of the particles, *i.e.*, the particles live for a longer time.

The dilation of lifetime of moving particles has important implications in the design of experiments. Consider for example, the production of K^+-mesons by fast-moving protons colliding with a target. For a beam of K^+-mesons of momentum 3 GeV/c corresponding to $v \approx 0.98645\, c$, the bubble chamber where the K^+-particles will interact with protons, is kept at a distance of 100 m. Since τ_0 for K^+-mesons is 1.23×10^{-8} s, the value of τ is 7.5×10^{-8} s so that the fraction of K^+-mesons reaching the chamber at $t = d/v$, is

$$\frac{N(t)}{N(0)} = 1.1 \times 10^{-2} \tag{1.84}$$

Without the time-dilation, the fraction would have been about 1.12×10^{-12}, so that with a typical pulse carrying about 10^3 K^+-mesons the experiment would not have been feasible. Therefore, it is time-dilation which makes the experiment feasible.

Example 3

It should be emphasized that it is only the speed of light in vacuum that is invariant. In particular, the speed of light in water is not invariant with respect to different inertial observers, as was shown by Fizeau (1851).

Consider the passage of light through a tube of length l, containing water at rest. The speed of light in water is c/n where n is the refractive index of water. If the water now flows with velocity v parallel to the direction of propagation of light, the observed speed of light can be obtained from Eq. (1.37):

$$u = \frac{\frac{c}{n} + v}{1 + \frac{v}{cn}} \tag{1.85}$$

which is different from $\frac{c}{n}$. This causes a change in the time of passage,

$$\Delta t = \frac{ln}{c} - \frac{ln\left(1 + \frac{v}{cn}\right)}{c\left(1 + \frac{vn}{c}\right)} \tag{1.86}$$

$$\approx \frac{lv}{c^2}(n^2 - 1)$$

and the corresponding phase shift is

$$\Delta\phi \approx 2\pi\ lv\ v_0\ (n^2 - 1)/c^2 \tag{1.87}$$

where v_0 is the frequency of the beam of light.

In the experiment of Fizeau, two parts of a beam traverse the water tube in opposite directions (the set-up is similar to the Michelson-Morley experiment), and interface to produce a fringe shift

$$\Delta N \approx 2\ lv\ v_0\ (n^2 - 1)/c^2 \tag{1.88}$$

This result agrees with experimental observations. It is worth noting that for the classical case, the denominator in Eq. (1.85) is 1 and the corresponding fringe shift is given by Eq. (1.88) but with $n^2 - 1$ replaced by n^2.

Example 4

Whenever energy is extracted from a reaction, chemical or nuclear, it is at the expense of the rest-mass energy. For chemical reactions, the change in the rest mass is no small that it is not possible to observe it directly. On the other hand, for nuclear fusion or fission reactions, the change in the mass is quite substantial and can be measured directly.

In the fusion of hydrogen nuclei into helium nuclei, which is the basic reaction in stars, four protons combine either through the proton-proton cycle or the carbon cycle (see Sec. 9.8) to yield a helium nucleus and two positrons.

$$4p \rightarrow \mathrm{He}^4 + 2e^+ \tag{1.89}$$

The change in the mass for this process can be obtained from $m(p) = 1.007277\ \mu$, $m\ (\mathrm{He}^4) = 4.001506\ \mu$, $m\ (e^+) = 0.000549\ \mu$ ($1\ \mu = 1.6604 \times 10^{-27}$ kg), so that energy available is

$$(\Delta\, m)\, c^2 = 24.7 \text{ MeV} \tag{1.90}$$

The sun today is at a fairly stable stage of its evolution, called the main sequence period. Assuming that the sun will remain on the main sequence till about 10% of the hydrogen is burnt, since its mass is about 2×10^{30} kg and it contains about 75% of hydrogen by mass, the total energy to be released while it is on the main sequence is about 8.9×10^{43} J. The energy emitted by the sun is about 3.9×10^{26} J/s, so that its lifetime on the main sequence is approximately 7×10^9 years. This should be compared with the more detailed estimations of 1.1×10^{10} years for the main-sequence lifetime of the sun, of which it has already spent about 4.5×10^9 years.

Example 5

The transformation properties of the electric field due to a moving charge can in principle be obtained from Eqs. (1.62) and (1.63) where $\left(\mathbf{A}, \frac{1}{c}\phi\right)$ transforms as a four-vector. However, it is simpler to deduce them from the expression for the Lorentz force in Eq. (1.61) and the transformation of force given in Eq. (1.59).

Consider the field due to a charge q at the origin, moving with velocity $\mathbf{u}$ in the x-direction. Then, the force on a unit charge at rest at P is

$$\mathbf{f} = \mathbf{E} \tag{1.91}$$

However, in a frame F' which moves with velocity $\mathbf{u}$, the charge is at rest and therefore, the force on the unit charge is

$$\mathbf{f}' = E'$$

$$= \frac{q}{4\pi\varepsilon_0}\left(\frac{\mathbf{r}'}{r'^3}\right) \tag{1.92}$$

where $\mathbf{r}'$ is the distance between P and the point charge. The two forces are related by Eq. (1.59), so that

$$E_x' = E_x$$

$$E_y' = \left(1 - \frac{u^2}{c^2}\right)^{1/2} E_y \tag{1.93}$$

$$E_z' = \left(1 - \frac{u^2}{c^2}\right)^{1/2} E_z$$

Using the expression for E' given in Eq. (1.92),

$$E_x = \frac{q}{4\pi\varepsilon_0}\frac{x'}{r'^3}$$

$$E_y = \frac{q}{4\pi\varepsilon_0}\frac{y'}{r'^3\left(1 - \frac{u^2}{c^2}\right)^{1/2}}$$

$$E_z = \frac{q}{4\pi\varepsilon_0}\frac{z'}{r'^3\left(1 - \frac{u^2}{c^2}\right)^{1/2}} \tag{1.94}$$

Finally, since $x' = x/(1 - u^2/c^2)^{1/2}$, $y' = y$, $z' = z$, some simplification gives

$$\mathbf{E} = \frac{q\mathbf{r}\left(1 - \frac{u^2}{c^2}\right)}{4\pi\varepsilon_0 r^3\left(1 - \frac{u^2 \sin^2\theta}{c^2}\right)^{3/2}} \tag{1.95}$$

where θ is the angle which the line joining the point P and the charge makes with the direction of the velocity of the charge. The field is seen to be weaker for small angles and angles near π, and has the largest value for $\theta = \pi/2$.

PROBLEMS

1. Show the two successive parallel Lorentz transformations are equivalent to a single Lorentz transformation.

2. A rod of length l_0 in its rest frame, moves with velocity **v** parallel to itself. Obtain the Lorentz contraction of the rod by calculating the time taken by the rod to pass a point (use time dilation) and then multiplying this time by v.

3. Obtain an expression for time dilation considering a clock with light bouncing back and forth along the direction of relative velocity, and using the concept of Lorentz contraction.

4. What is the visually observed rate of a clock which moves with velocity **v** along the line of vision?

5. The incoming primary cosmic rays (mostly protons) create μ-mesons in the upper atmosphere. The lifetime of μ-mesons at rest is 2.15×10^{-6} s. If the mean speed of the meson is 0.998 c, what fraction of the μ-mesons created at a height of 20 km reach the sea level? What is the mean distance travelled by the mesons before they decay?

6. A rod AB parallel to the x-axis, moves along the y-axis with velocity **u**. Show that in a frame F' which moves with velocity v along the x-direction, this rod is inclined to the x'-axis at an angle

$$\tan^{-1} \frac{uv}{c2\left(1-\frac{v^2}{c^2}\right)^{1/2}}.$$

7. A ρ-meson of mass 760 MeV/c^2 decays at rest into two π-mesons of mass 140 MeV/c^2 each. What is the relative velocity of the π-mesons with respect to each other?

8. Show that when force **f** is not parallel to velocity **u**, the acceleration is in general not parallel to either the force or the velocity.

9. A particle of mass M decays at rest into a particle of mass m and a photon. What is the energy of the photon emitted? Apply this to (*a*) Σ^+ (1189.4 MeV) $\rightarrow p$ (938.3 MeV) + γ, (*b*) H (2p) $\rightarrow H$(1s) + γ, the binding energy being 10.2 eV and 13.6 eV respectively.

10. A charged particle emits radiation when subjected to an external field. This is known as *bremsstrahlung*. Show that energy-momentum conservation does not allow a particle in isolation (no external forces) to radiate. The argument is very simple in the centre of mass frame.

11. A proton gains an energy of 1 electron volt (eV) or 1.6×10^{-19} J when it traverses a potential difference of 1 V. If the proton has a mass of 1.67×10^{-27} k.g., what is the velocity of the proton which starts from rest and traverses a potential difference of 10^9 V? What is the velocity of a proton which comes out of the CERN super proton synchrotron with an energy of 270 GeV (1 GeV is equal to 10^9 eV)?

12. A star is observed at the zenith taken to be the z-direction. If the star is moving in the x-direction, and its radiation shows a redshift of 0.003 Å for the H_α Balmer line (λ = 6563Å), what is its velocity with respect to us? If the star were moving towards us with the same speed, what would be the observed wavelength for the H_α line?

13. A charged particle can move in a material medium with a velocity greater than the velocity of light in that medium. It polarizes the nearby atoms which then emit radiation known as Cerenkov radiation. The envelope of the spherical waves is a cone with the vertex at the charged particle and whose surface makes an angle θ with the direction of motion of the particle. Show that $\sin \theta = c/nv$, where v is the speed of the particle and n is the refractive index of the medium ($v \geq c/n$).

2

Introduction to Quantum Ideas

Structures of the Chapter

In this chapter, the background which necessitated the introduction of quantum theory of matter and radiation is discussed. We describe Planck's theory of black-body radiation, photoelectric effect, Compton scattering, matter waves and Davisson-Germer experiment, Rutherford scattering, and the Bohr theory of an atom. The choice of these topics is dictated by their historical importance and the directness with which they lead to the basic rules of quantum mechanics. Though most of these primitive quantum ideas have been replaced by the more universal description in terms of wave mechanics, they still serve a useful purpose in providing a simple picture of quantum phenomena.

Quantum mechanical effects become important in the domain of small distances. To be more precise, the effects are important in measurements which require the knowledge of, say, the momentum p_x and position x of a system to an accuracy such that

$$(\Delta p_x)(\Delta x) \sim h \tag{2.1}$$

where Δp_x and Δx are the errors in the measurement of the x-component of momentum and position of the system, and h is a small number whose value in mks unit is

$$h = 6.67 \times 10^{-34} \text{ Js} \tag{2.2}$$

It is worth pointing out that once again radiation plays an important role in the development of quantum mechanics, through for a different reason: the ideas of wave functions and wave equations already existed for radiation; they only required a reinterpretation in terms of the photon which is the quantum of radiation.

2.1 BLACK-BODY RADIATION

Historically, the first indication of the inadequacy of classical ideas to explain the properties of matter, occurred in what is termed as *black-body radiation*. A black-body is a body which absorbs all the radiation incident on it, and hence is the perfect absorber. Consideration of equilibrium of different bodies at the same temperature implies that it is also the best emitter of radiation energy. A black-body may be idealized by a small hole drilled in a cavity.

If the radiation from a black-body is analysed by a spectrometer (*i.e.* a prism of a grating), it is found (Lummer and Pringsheim, 1900) that the intensity distribution as a function of wavelength, has a well-defined shape (Fig. 2.1). What is most significant is that, for a given temperature, it is a universal curve independent of the properties of the walls of the cavity. In particular, it has a maximum at some wavelength λ_m. As the temperature of the black-body is raised, the intensity of radiation increases at each wavelength, and λ_m shifts to a smaller value such that

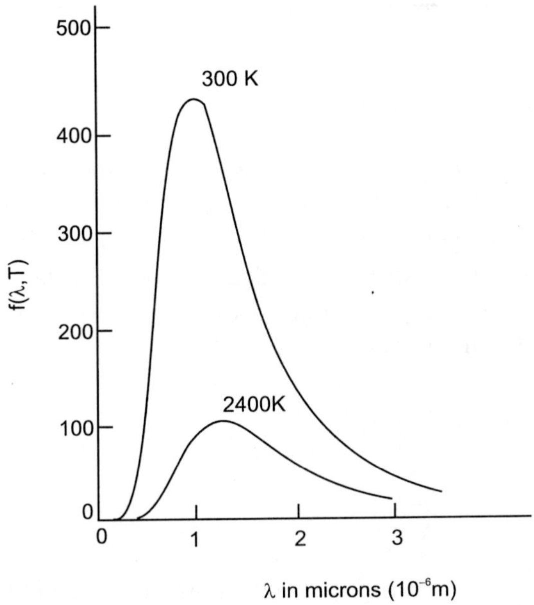

Fig. 2.1 Intensity of radiation from a black-body, $I(\lambda, T)$, in watts per square centimetre per micron.

$$\lambda_m T = \text{constant} \tag{2.3}$$

This result had been obtained by Wien (1893) from a theoretical analysis. The initial approach to understand the nature of black-body radiation was in terms of the standing waves set up in the enclosure and the thermodynamic energy associated with these waves. Consider a plane wave in a cubic box of length l. It may be written in the form

$$\psi = A \sin 2\pi\,(\nu\, t \pm k_x x \pm k_y y \pm k_z z + \alpha) \tag{2.4}$$

where $|\mathbf{k}| = \dfrac{1}{\lambda}$ is the wave number, ν is the frequency of the radiation and the ± signs define the direction in which the wave is travelling. Standing waves are obtained by taking linear combinations of the different terms (with $x = 0$) in Eq. (2.4). Requiring that the waves vanish at the walls passing through the origin, we get

$$\psi_s = A \cos\,(2\pi\, \nu\, t) \sin\,(2\pi\, k_x x) \sin\,(2\pi\, k_y y) \sin\,(2\pi\, k_z z) \tag{2.5}$$

where, in order that the wave should have nodes at the other walls,

$$k_x = \frac{n_x}{2l},\ k_y = \frac{n_y}{2l},\ k_z = \frac{n_z}{2l} \tag{2.6}$$

with n_x n_y and n_z being positive integers (negative integers give the same mode). The wave number and therefore the frequency $\nu = c\,|\,\mathbf{k}\,|$ is obtained from these relations as

$$c = \frac{c}{2l}(n_x^2 + n_y^2 + n_z^2)^{1/2} \tag{2.7}$$

Every possible set of positive integers (n_x, n_y, n_z) gives a possible standing wave, which may be depicted by a point in the 3-dimensional plot of (n_x, n_y, n_z). Since there is one such point per unit volume, the number of states is essentially equal to the volume in this space (provided the volume is large). Therefore, the number of stationary modes with frequency between 0 and ν (which corresponds to the volume in the first octant with $n \leq 2l\,\nu/c$) is

$$N(\nu) = 2\left(\frac{1}{8}\right)\left(\frac{4\pi}{3}\right)\left(\frac{2l\nu}{c}\right)^3$$

$$= \frac{8\pi l^3\,\nu^3}{3c^3} \tag{2.8}$$

where a factor of 2 has been introduced to take into account the fact that for each frequency ν, there are two transverse modes of electromagnetic oscillations.

In the theory of statistical mechanics, the principle of equipartition states that a mean energy of $\frac{1}{2}k\,T$ (k is the Boltzmann constant which has a value of 1.38×10^{-23} J K^{-1}) is associated with each degree of freedom. For example, for an ideal gas with molecules treated as geometric points, the mean energy of each molecule is $\frac{3}{2}kT$ corresponding to its translational motion in three independent directions. However, for a molecule in oscillatory motion. corresponding to each mode of translational motion there is a potential energy term which also contributes a mean energy of $\frac{1}{2}kT$. Now, if a mean energy of kT is assigned to each mode of electromagnetic oscillation, then according to the principle of equipartition, the energy per unit volume, between frequencies ν and $\nu + d\nu$, is given by

$$u\,(\nu)\,d\nu = \frac{dN(\nu)}{l^3}(kT) \tag{2.9}$$

so that energy density per unit volume, per unit frequency is

$$u(\nu) = \frac{8\pi\,\nu^2\,kT}{c^3} \tag{2.10}$$

This is known as Rayleigh-Jeans law and provides a good fit to the experimental results in the region of long wavelengths. However, it is unacceptable since it implies that the energy density increases indefinitely as v increases and therefore the total energy per unit volume is infinite, contrary to physical observations. This is known as the *ultraviolet catastrophe.*

Max Planck re-examined the basic assumptions in the theoretical approach to black-body radiation. He introduced (1900) a new and revolutionary assumption in the basic framework which allowed him to describe the experimental observations with great accuracy. He associated each mode of electromagnetic oscillation with atomic oscillators embedded in the walls of the cavity. However, these oscillators were allowed to have only discrete energies which are integral multiples of hv, and the energy distribution of these oscillators is according to the Maxwell-Boltzmann distribution. Here h is a small number whose value is given in Eq. (2.2), and is called Planck's constant. Since the Maxwell-Boltzmann distribution is given by exp $(-E/kT)$, the average energy for each mode oscillation is

$$\varepsilon = \frac{\sum_{n=0}^{\infty} nh\,v \exp(-nh\,v / kT)}{\sum_{n=0}^{\infty} \exp(-nh\,v / kT)}$$

$$= \frac{hv}{\exp(hv / kT) - 1} \tag{2.11}$$

Using this expression in Eq. (2.10) in place of kT, the following expression is obtained

$$u(v) = \left(\frac{8\pi h\,v^3}{c^3}\right)\frac{1}{\exp(hv / kT) - 1} \tag{2.12}$$

It is easy to see that in the region of long wavelengths, this equation reduces to the Rayleigh-Jeans relation in Eq. (2.10). It gives an exponential damping in the short-wavelength limit. Overall, it is in excellent agreement with experimental observations over a wide range of temperatures. It also leads to many of the special relations for black-body radiation. For example, the energy radiated by a unit area of a black-body, per unit time is given by the Stefan-Boltzmann law:

$$U = \frac{c}{4}\int_0^{\infty} u\,(v)\,d/v \tag{2.13}$$

(the factor of $c/4$ is discussed in Example 1) which, with the substitution $x = h\nu/kT$ and the use of Eq. (2.12), leads to

$$U = \sigma T^4 = \left(\frac{2\pi k^4}{h^3 c^2}\right) T^4 \int_0^\infty \frac{x^3 dx}{e^x - 1} \tag{2.14}$$

The value of the integral is known to be $\pi^4/15$, so that Stefan's constant is given by

$$\sigma = \frac{2\pi^5 k^4}{15 h^3 c^2} \tag{2.15}$$

whose numerical value is in good agreement with the experimental value of

$$\sigma = 5.67 \times 10^{-8} \text{ J/s m}^2 \text{ K}^4 \tag{2.16}$$

Planck's law in Eq. (2.12) can also be used to deduce the properties of λ_m at which the radiation density is maximum. Nothing that $\bar{u}(\lambda) = \bar{u}(\nu)\, d\nu/d\lambda$.

$$\bar{u}(\lambda) = \left(\frac{8\pi hc}{\lambda^5}\right) \frac{1}{\exp(hc/\lambda kT) - 1}$$

$$= \left(\frac{8\pi k^5 T^5}{h^4 c^4}\right) \frac{x^5}{e^x - 1} \tag{2.17}$$

with $x = hc/\lambda kT$. This function has a maximum at a wavelength given by the condition $d\bar{u}(\lambda)/d\lambda = 0$, and a numerical solution (see Example 2) gives

$$x_m = hc/\lambda_m kT = 4.965 \tag{2.18}$$

or $$\lambda_m T = 2.90 \times 10^{-3} \text{ mK} \tag{2.19}$$

This relation, called *Wien's displacement law*, is in very good agreement with the experimental observations. It should be emphasized that Eq. (2.17) implies

$$\frac{\bar{u}(\lambda)}{T^5} = f(\lambda T) \tag{2.20}$$

which had been deduced earlier by Wien (1893) from thermodynamic considerations. It implies that a function of a single variable, namely of λT, gives the complete description of $\bar{u}(\lambda)$ as a function of variables λ and T.

Planck's hypothesis that the energy of the oscillators is quantized, is rather on ad-hoc assumption, though it leads to Planck's law of black-body radiation which is an excellent agreement with the experimental observations. The law of black-body radiation has found a more satisfactory description in terms of later developments of quantum statistical mechanics. However, its historical importance lies in the fact that it was for the first time here that the idea of quantized energy was introduced and used for the description of physical observations.

2.2 PHOTOELECTRIC EFFECT

The quantum hypothesis of Planck postulates that the energies of the atomic oscillator are quantized and that they can change only by an integral multiple or $h\nu$. This does not necessarily imply that the energy of the radiation in the cavity is quantized. It suggests, however, that since the changes in the radiation energy are due to absorption or emission of radiation by the atomic oscillators, the energy of the radiation itself is quantized into multiples of $h\nu$. It is obvious from the previous discussion that this quantized energy would also lead to Planck's law for the black-body radiation. This possibility received strong support from Einstein's explanation of photoelectric effect.

Hertz (1887) found that when a beam of ultraviolet radiation, for example, from a mercury lamp, impinges on the surface of an alkali metal such as Cs, Rb, K or Na, (which have a small work function), electrons are emitted. The number of electrons emitted per second, and their energies can be studied by subjecting the electrons emitted to an electric field as shown in Fig. 2.2 (*a*). The number of electrons that escape from the cathode per second, and are collected by the anode is given by i/e where i is the current, while the maximum kinetic energy of the electrons emitted is given by eV_0 where V_0 is the stopping potential for which the current reduces to zero [Fig. 2.2 (*b*)].

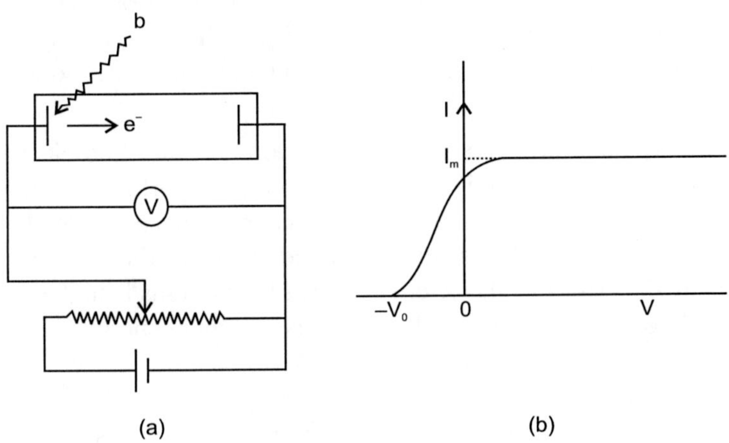

Fig. 2.2 (a) Schematic diagram of the equipment used for studying photoelectric effect. (b) Typical photoelectric current against collector voltage.

The following important points should be noted about the observations:

1. The electrons are emitted without any noticeable time delay (time lag is less tan 10^{-8} s).
2. No electrons are emitted if the frequency of the incident radiation is less than a critical value ν_0.
3. The maximum kinetic energy of the electrons is related to stopping potential by

$$\frac{1}{2} m v_m^{\,2} = eV_0 \tag{2.21}$$

It is independent of the intensity of incident radiation but is proportional to $\nu - \nu_0$, *i.e.*,

$$\frac{1}{2} m v_m^{\,2} \propto (\nu - \nu_0) \tag{2.22}$$

It is very difficult to reconcile the classical wave theory of light with these observations. For example, with an incident radiation of intensity 10^{-10} J/m^2 s, it would require about 5×10^{11} s to absorb an energy of 3 eV by a cross-sectional area of about 10^{-20} m^2 presented by an atom. Actually, a more detailed analysis shows that an atomic oscillator presents an effective area of about λ^2 to light of wavelength λ corresponding to its resonant frequency.

For radiation of $\lambda = 10^{-7}$ m, this means an area of about 10^{-14} m^2, which still implies an accumulation time of about 5×10^5 s in contradiction with the observation that there is no noticeable time delay in the emission of electrons, Nor can the wave theory of radiation explain the existence of the sharp threshold ν_0 for the emission of electrons, if the energy is absorbed continuously. The fact that the maximum kinetic energy of the electrons emitted does not depend on the intensity of incident radiation and that it is proportional to $\nu - \nu_0$, is equally puzzling.

A simple explanation of the various observations of the photoelectric effect, was provided by Einstein (1905). Inspired by Planck's work, Einstein proposed that electromagnetic radiation itself is quantized into quanta of energy $h\nu$ where h is Planck's constant. It is these quanta, called *photons,* that are absorbed as single units by the electrons. If the energy $h\nu$ of the photons is high enough, the electrons are knocked out. From the law of conservation of energy, the maximum kinetic energy of the electron is

$$\frac{1}{2} m v_m^{\,2} = h\nu - e\phi, \quad \textit{for } h\nu > e\phi \tag{2.23}$$

where $e\phi$ is the minimum energy with which the electron is bound within the metal and is called the *work function.* This is the famous Einstein's relation for the photoelectric effect. It explains all the observed features of photoelectric effect. It should be noted that

1. Since photons are absorbed as single units, there is a localization of energy and hence there is no significant time delay in the emission of electrons.

2. Einstein's relation in Eq. (2.23) implies the existence of a critical frequency for the emission of electrons, given by

$$\nu_0 = \frac{e\phi}{h} \tag{2.24}$$

However, since the current increases gradually as V increases from $-V_0$, the effective binding of the electrons inside the metal varies as also the velocity of the emitted electrons. Therefore, the critical frequency in Eq. (2.24) refers to the emission of electrons with minimum binding energy.

3. The maximum kinetic energy of the emitted electrons (having minimum binding energy) is given in terms of the critical frequency, by the relation

$$\frac{1}{2} m v_m{}^2 = h\,(\nu - \nu_0) \tag{2.25}$$

In terms of the stopping potential V_0, one has

$$eV_0 = h\,(\nu - \nu_0) \tag{2.26}$$

Thus, it not only explains all the experimental observations but also gives the ratio of h/e from the slope of the linear plot of V_0 against ν. Using the known value for the charge of the electron, an independent determination of Planck's constant, in good agreement with the value obtained from other considerations such as the black-body radiation, can be obtained.

Some additional observations related to the photoelectric effect are:

1. Only a small fraction (about 5%) of the incident photons, succeeds in ejecting photoelectrons while most of them are absorbed by the system as a whole and generate thermal energy.

2. Photoelectric effect is also possible for isolated atoms in the form of a gas, *e.g.* Na, K vapour, and the process is known as *photoionization.* It is observed by passing a beam of ultraviolet radiation through a chamber containing Na or K vapour, and collecting the electrons ejected by subjecting them to an electric field. It is interesting to not that in photoionization, since the atoms are isolated, there is no collective absorption of photons and every photon absorbed succeeds in ejecting an electron. This can be verified by comparing the number of photons absorbed as deduced from the decrease in the intensity of the beam, with the number of electrons collected by the electric field.

3. The energy required for ejecting the electrons may also be provided by heating the metal, which results in the *thermionic emission* of the electrons. They allows us to calculate, from quantum statistical mechanics, the work function $e\phi$. The value obtained agrees with the one obtained from the photoelectric effect.

4. So far, it has been assumed that an electron receives energy only from a single photon, the process being called a *single-photon process*. The development of lasers has provided light beams of very high intensity which allow us to observe multi-photon processes, in particular the multi-photon photoelectric effect. In this process, an electron ejected from a metal receives energy from N photons. Its kinetic energy is given by

$$\frac{1}{2} m v_m^{\ 2} = Nh\nu - e\phi \tag{2.27}$$

and the critical frequency is $e\phi/Nh$ which is smaller than the corresponding frequency for single-photon processes by a factor of 1/N.

In the analysis of the photoelectric effect, a photon was regarded as a wave packet of energy, with no statements made for the quantization of momentum. In fact since a significant amount of momentum was carried away by the metal, conservation of momentum could not be usefully applied to the photon-electron system. For the photon to acquire the bonafides of a particle, it should have both a quantum of energy as well as a quantum of momentum. This was demonstrated by the discovery of *Compton effect* (1922) in the scattering of x-rays by electrons.

2.3 COMPTON EFFECT

Compton effect is essentially a demonstration of the scattering of a photon by an electron as a particle-particle scattering.

Compton was concerned with the scattering of x-rays of wavelength λ_0 by a thin film of metal. He measured the wavelength distribution of the scattered rays at different scattering angles and found that it had two major components in wavelength. One had essentially the same wavelength as the incident radiation, *i.e.* λ_0, while the other had a slightly longer wavelength [Fig. 2.3 (*a*)], with the separation between the two wavelengths given by $\lambda - \lambda_0 = \lambda_c\ (\lambda - \cos\theta)$. Here λ_c is a constant which is independent of λ_0 or the scattering angle and is called the *Compton wavelength* of the electron. This increase in wavelength or the decrease in frequency is very difficult to understand in terms of the wave description of radiation. However, it can be explained quite accurately by considering the process as a scattering of photons regarded as particles with well-defined energy and momentum by the electrons in the metal.

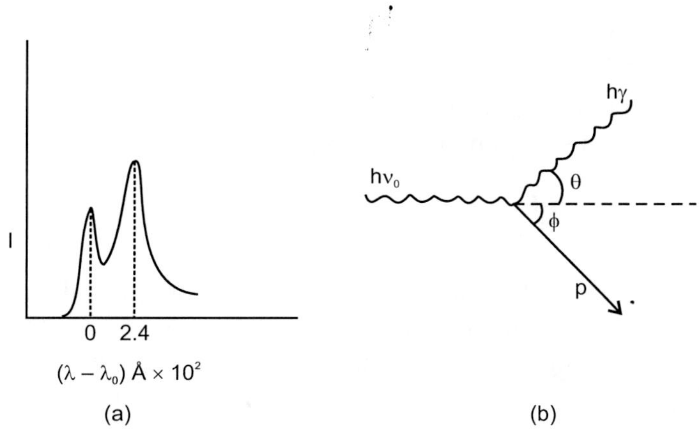

Fig. 2.3 (a) The intensity of the scattered x-rays as a function of $(\lambda - \lambda_0)$ Å × 10^2, for scattering angle $\theta = 90°$. (b) Compton scattering as particle-particle scattering.

If the initial and final photons have energies hv_0 and hv, and momenta $\frac{hv_0}{c}\mathbf{n}_0$ and $\frac{hv}{c}\mathbf{n}$, respectively, where $\mathbf{n}_0$ and $\mathbf{n}$ are unit vectors in the directions of propagation and the momentum of scattered electron is $\mathbf{p}$, then from momentum and energy conservation [Fig. 2.3 (*b*)],

$$\frac{h\nu_0}{c} = \frac{h\nu}{c}\cos\theta + p\cos\phi \tag{2.28}$$

$$0 = \frac{h\nu}{c}\sin\theta - p\sin\phi \tag{2.29}$$

$$hv_0 + mc^2 = h\nu + (p^2c^2 + m^2c^4)^{1/2} \tag{2.30}$$

Eliminating ϕ from the first two equations,

$$p^2 = \frac{h^2}{c^2}(\nu_0^{\ 2} + \nu^2 - 2\nu\,\nu_0\cos\theta) \tag{2.31}$$

Also from Eq. (2.30)

$$p^2 = \frac{h^2}{c^2}(\nu_0 - \nu + mc^2/h)^2 - m^2c^2 \tag{2.32}$$

Equating the two expressions for p^2 leads to

$$\nu_0 - \nu = \frac{h\nu\,\nu_0}{mc^2}(1-\cos\theta) \tag{2.33}$$

or $$\lambda - \lambda_0 = \frac{h}{mc}(1 - \cos\theta) \tag{2.34}$$

This is Compton's expression for the shift in the wavelength of the scattered *x*-rays. It identifies the Compton wavelength as

$$\lambda_c = \frac{h}{mc} \tag{2.35}$$

which depends only on the mass of the scattering particle and has a value of 2.43×10^{-2} Å for the electron. It has an interesting interpretation, that a photon with wavelength λ_c has an energy $h_\nu = mc^2$, *i.e.,* the rest energy of the particle.

In the discussion presented here, it is assumed that the target electron is stationary and free. It is also valid if the electron is weakly bound to the atom with the binding energy of a few eV which is quite small compared with the energies of the *x*-ray photons, which are about 10 keV or greater. However, it may so happen that the electron remains bound in the same state to the atom, even after the collision with the photon (this is more likely to happen if the electrons are strongly bound). In this case, the transfer of energy and momentum is to the atom as a whole, so that the mass of the atom must be used in place of the mass of the electron. Therefore, the corresponding Compton wavelength is much smaller (at least by a factor of 1800) and the resulting Compton shift in the wavelength is negligible. This explains the unshifted component in the spectrum of the scattered *x*-rays, which is called the *Thomson* component.

Some interesting additional features of Compton effect are:

1. The fact that the shift in the wavelength of the radiation is indeed due to the scattering of the radiation by the electron was confirmed by observing the scattered electron (Bothe and Geiger, 1925).
2. The main reason for the spread in the wavelength of the Compton-shifted *x*-rays is that the initial electron is in general not stationary but has a momentum spread even inside the atom. The correction due to the binding of the electron can be taken into account in terms of its momentum distribution (see Example 4).
3. Though the sift $\lambda - \lambda_0$ is independent of λ_0, the intensity of scattering depends on λ_0. It actually increases as $\lambda_0 \to 0$ and hence the effect is more easily observable for *x*-rays than for lower frequency radiation (indeed this is the reason why the sky is blue).
4. The scattering angle ϕ is given by

$$\tan\phi = \frac{\sin\theta}{\nu_0/\nu - \cos\theta}$$

$$= \frac{\cot(\theta/2)}{1 + h\nu_0/mc^2} \tag{2.36}$$

Since in cost cases $h\nu_0$ is of the order of 10 keV, and therefore $h\nu_0 \ll mc^2$, one has the simple relation $\phi \approx \frac{1}{2}(\pi - \theta)$ or the direction in which the electron is scattered bisects the angle which is supplementary to the angle made by the final photon momentum with the initial photon momentum.

2.4 WAVE NATURE OF PARTICLES

The observations of photoelectric effect and Compton scattering firmly establish the particle-like properties of radiation. One could also rephrase this wave-particle duality in the following from: photons are particles with zero mass (since $E^2 - p^2c^2 = 0$ for photons) which have wave-like properties (*e.g.* interference and diffraction). It was suggested by de Broglie (1923) that particles with nonzero mass also possess wave-like properties. This was indeed a daring proposal which went beyond the classical concepts of particles with nonzero mass. Specifically, he proposed that material particles of momentum **p** are associated with a wavelength

$$\lambda = \frac{h}{p} \tag{2.37}$$

called the *de Broglie wavelength*. This idea was an important step in the development of wave mechanics.

The wave properties become easily noticeable only when the obstructing bodies have dimensions comparable with the wavelength. For macroscopic bodies, the de Broglie wavelength is negligibly small. For atomic systems, this wavelength becomes more significant: for an electron with an energy of 100 eV the de Broglie wavelength is about 1 Å, comparable with the wavelength of *x*-rays as also with the size of an atom. Their wave properties may therefore be observed in their scattering by crystals. This was confirmed experimentally by Davisson and Germer (1927) who studied the scattering of electrons by a monocrystal of nickel.

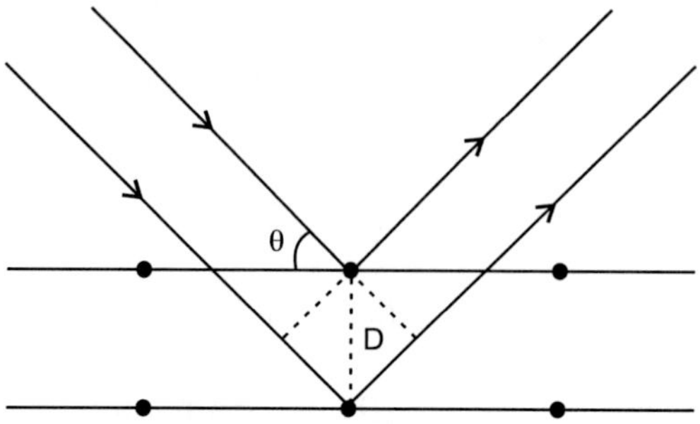

Fig. 2.4 Bragg condition for constructive interference for reflection from different planes.

Consider the scattering of a beam of electrons by a nickel crystal. The intensity of the scattered electrons is measured at various angles, and for different velocities of the incoming electrons. It was found that the scattering is intense when the energy of the incoming electrons was 54 eV and at an angle which was equal to the angle at which a strong reflection by the atomic planes was observed for *x*-rays of wavelength 1.65 Å. If the *x*-rays are incident on a set of atomic planes at distance *d* from each other, at an angle θ (Fig. 2.4), the amplitudes of *x*-rays scattered by the atoms in a given plane will be coherent if the angle of reflection is equal to the angle of incidence. Furthermore, the amplitudes for scattering from different planes will be coherent if the path difference for scattering from two successive planes is an integral multiple of λ, *i.e.,* the Bragg condition is satisfied:

$$2d \sin \theta = n\lambda \tag{2.38}$$

where n is a positive integer. For *x*-rays diffracted by the nickel crystal, maximum intensity occurs at $\lambda = 1.65$ Å (for $n = 1$). On the other hand, the de Broglie wavelength for 54 eV electrons is 1.67 Å, and they too have an intense scattering at the same angle. The ageement between the two wavelengths is striking, indicating that the Bragg equation is satisfied by electrons also, provided de Broglie wavelength is used for the wavelength of the electrons. This should be regarded as a confirmation of de Broglie's idea. The essential point of many of the experiments demonstrating de Broglie's relation is that any experiment of *x*-ray diffraction, in principle can be simulated by an experiment with electrons of corresponding de Broglie wavelength. In particular, Thomson (1927) as also Tartakovsky, obtained a diffraction pattern when an electron beam (of tens of keV) was passed through a thin foil of polycrystalline material. The electrons pick out those crystals whose planes are oriented so as to satisfy Eq. (2.38) and produce a diffraction pattern which is similar to the diffraction pattern produced by *x*-rays of wavelength equal to the de Broglie wavelength for the electrons.

The electron diffraction experiments establish the fact that the electrons possess wave properties exactly as de Broglie had suggested. To clarify that the wave property is not because of the simultaneous participation of a large number of electrons, but is associated with each electron, experiments have been done with very low intensity electron beams so that the electrons pass through the instruments essentially one at a time. With sufficiently long exposure, a diffraction pattern was obtained which differed in no way from the pattern obtained with normal intensity beams, thus suggesting that the wave-like properties are to be associated with individual electrons. The de Broglie wavelength has also been verified for neutral molecules (Estermann and Stern, 1930) and for neutrons; however, in order that they have the same de Broglie wavelength as the x-ray wavelength, their energies ($E = p^2/2m$) have to be of the order of ~ 0.02 eV, *i.e.*, one uses thermal molecules and neutrons.

Electron and neutron diffraction, along with x-ray diffraction have become an indispensable tool for the study of the structures of solids. Their diffraction patterns though qualitatively similar, have important differences which need to be mentioned.

Since the electrons are sensitive to electrostatic forces, electron beams have little penetration. With the development of slow electron beams (10 to 1000 eV) which undergo negligible penetration and for which diffraction occurs essentially at the first atomic layer, electron diffraction has become an important tool for the study of surfaces. Electron beams of fairly high energies, *i.e.,* about 50 keV, have quite small de Broglie wavelengths, and therefore are used in electron microscopes for high resolution studies of small specimens.

On the other hand neutrons, like x-rays, are fairly insensitive to electrostatic forces, and therefore penetrate easily. Neutron diffraction has several advantages:

1. Since neutron scattering is essentially by the nucleus and depends quite distinctively on the structure of the nucleus, neutron diffraction can give more information about crystals formed from different atoms which have nearly equal atomic number and which cannot easily be distinguished by x-rays.

2. The scattering of neutrons by light nuclei such as hydrogen is large, and hence neutron diffraction is an important technique in the study of structures of organic compounds. The scattering of x-rays by light atoms is weak (the coherent cross-section is approximately proportional to Z^2 where Z is the number of electrons).

3. The magnetic ordering of atoms in a crystal, which have nonzero magnetic moment, can be studied by neutron diffraction which therefore is a valuable method of investigating magnetic materials. The main disadvantages of using neutron diffraction are the low intensity of neutron beams available

and the difficulty of detecting neutrons which are electrically neutral. They are usually detected by the α particle emitted through their reaction with boron nuclei.

2.5 ATOMIC SPECTRA

De Broglie's idea allows us to discuss both radiation and particles such as electrons, in a unified manner. They behave like particles in the sense that they have discrete energy and momentum, and yet are diffracted as waves. In this respect, there is no qualitative difference between photons which have vanishing mass and particles such as electrons, neutron and proton which have nonvanishing mass (there are subtle differences, however). Now, for radiation inside a box, one has stationary waves with discrete frequencies, and the corresponding photons have discrete energies. It might then be expected that particles such as electrons also have discrete energies if they are confined to a finite volume. Such a situation is simulated by an electron which is bound inside an atom, and the atomic spectra do indicate that the electron indeed has discrete energies.

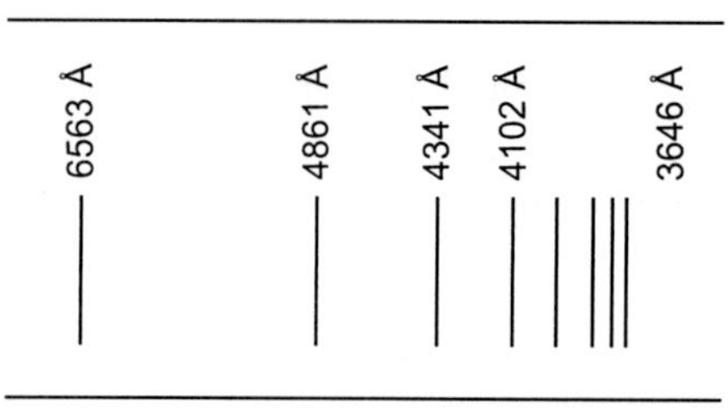

Fig. 2.5 Hydrogen spectrum in the visible and near ultraviolet region.

When a vapour glows, *e.g.* in a flame or due to current discharges, the radiation emitted has well-defined discrete frequencies (unlike thermal emission in a solid, which has a continuous spectrum). Observed from a spectrograph (an instrument for analysing the frequency distribution, using a prism or a grating), sharp lines with a narrow width are observed which are characteristic of the vapour, and which can be used for identifying the components of the vapour. It had been observed that these lines have some order. This is revealed for example in the part of hydrogen spectrum in the visible and near ultraviolet region (Fig. 2.5). Specifically, it was found by J. Balmer (1885) that the hydrogen lines in the visible and near ultraviolet region could be expressed quite accurately by the relation

$$\lambda_n = \lambda_\infty \frac{n^2}{n^2 - 4}, \quad n = 3, 4, \ldots \tag{2.39}$$

where $\lambda_\infty \sim 3646$ Å is the limiting value shown in Fig. 2.5. A simple but important step in the analysis of atomic spectra was taken by Rydberg (1890) who pointed out that the sequence in Eq. (2.39) could be represented in a more suggestive form in terms of the reciprocal of the wavelength called the *wave number,* related to the frequency:

$$\frac{1}{\lambda_n} = R\left(\frac{1}{2^2} - \frac{1}{n^2}\right), n = 3, 4, ... \tag{2.40}$$

$$R = 1.0972 \times 10^7 \text{ m}^{-1}$$

where R is called the *Rydberg constant.* The spectral lines represented by Eq. (2.40) from what is known as the *Balmer series.* Further investigations showed that the hydrogen spectrum has other series, in the ultraviolet and infrared regions. They are represented by formulae similar to Eq. (2.40), *e.g.* Lyman series (1906) in the ultraviolet region by

$$\frac{1}{\lambda_n} = R\left(\frac{1}{1^2} - \frac{1}{n^2}\right), n = 2, 3, ... \tag{2.41}$$

Paschen series (1908) in the infrared region by

$$\frac{1}{\lambda_n} = R\left(\frac{1}{3^2} - \frac{1}{n^2}\right), n = 4, 5, ... \tag{2.42}$$

Brackett series (1922) in the infrared region by

$$\frac{1}{\lambda_n} = R\left(\frac{1}{4^2} - \frac{1}{n^2}\right), n = 5, 6, ... \tag{2.43}$$

The frequencies of the lines in the hydrogen spectrum can be obtained from a single formula

$$\frac{1}{\lambda_{m,n}} = T_1\left(\frac{1}{m^2} - \frac{1}{n^2}\right), m < n \tag{2.44}$$

$m = 1$ giving the Lyman series, $m = 2$ giving the Balmer series, etc.

The spectra of other atoms also show some order, and their frequencies can be represented by

$$\frac{1}{\lambda_{m,n}} = T_1\,(m) - T_2\,(n), m < n \tag{2.45}$$

However, the form of $T(n)$ is generally more complicated than for the hydrogen atom, one of the most useful being $R(n–d)^{-2}$ where δ is a constant known as the quantum defect, $\delta \ll n$.

The discreteness of the spectral lines is a property of the absorption spectrum as well. When radiation passes through a vapour, the vapour absorbs radiation of discrete frequencies which correspond to the frequencies of the emission spectrum, given by Eq. (2.45). This results in the appearance of dark lines corresponding to the frequencies of the radiation absorbed, which coincide with the positions of the spectral lines in the emission spectrum. In fact, helium was first discovered by its absorption lines in the solar spectrum before it could be identified terrestrially.

A photon associated with radiation of frequency ν carries in energy $h\nu$. Therefore, since energy is conserved, it is plausible to conclude that the discrete frequencies of the emission and absorption spectra, given by Eq. (2.45), correspond to transitions of the atom between states characterized by the integers m, n, which have discrete energies

$$E_n = - hcT\,(n) \tag{2.46}$$

In particular, the energy levels of the hydrogen atom are given by

$$E_n\,(H) = -\frac{hcR}{n^2} \tag{2.47}$$

The numerical value of hcR is 13.6 eV, and is called the *ionization potential* of the hydrogen atom.

One of the first attempts to have a model of an atom, which can explain the discrete spectrum of the atom, was due to J.J. Thomson (1903). It had been recognized that the negatively charged electron is one of the fundamental constituents of the atom. Since the atom as a whole is neutral, it should also contain a positively-charged part, called *positive ion* to balance the negative charge of the electron. It was also known that a great majority of the mass is associated with the positive ion. This led Thomson to propose a model of the atom in the form of a sphere of uniform positive charge, in which the small, negatively charged electrons are embedded. The electrons perform simple harmonic motion about their positions of equilibrium, which results in the emission of radiation of characteristic frequencies. Quantitatively, the potential energy of the electron inside the atom is

$$V(r) = \frac{Ze^2}{8\pi\varepsilon_0 R_0^3}\,(r^2 - 3R_0^2) \tag{2.48}$$

where R_0 is the radius of the atom and Z is the atomic number. This potential will cause the electron to oscillate with a frequency

$$\nu = \frac{1}{2\pi}\left(\frac{Ze^2}{4\pi\varepsilon_0 m R_0^3}\right)^{1/2} \tag{2.49}$$

For $Z = 1$ and a frequency $\nu \approx 5 \times 10^{14}\ s^{-1}$ corresponding to a wavelength of $\lambda = 6000$ Å, one gets $R_0 \approx 3$ Å which is comparable to the size of the atom and therefore can be taken as a support for the Thomson model. A suitable arrangement of the electrons in a series of rings, can also explain the existence of homologue series such as Na, K, Rb, etc. which have similar properties. However, the model is totally inconsistent with the experimental observations of the scattering of a particles by thin metal foils, and had to be discarded. Now, it is only of historical interest.

2.6 NUCLEAR MODEL OF THE ATOM

The distribution of charge and mass in an atom can be investigated by the scattering of charged particles by the atom. Such an investigation was carried out by Rutherford and his collaborators.

It was observed by Geiger and Marsden (1909) that in the scattering of α particles (doubly ionised ^{4}He) by the atoms in a thin metal foil, though most of the α particles emerged without much deviation, some of them were scattered through large angles (some close to 180 degrees). The number of such events was at least 10^4 larger than the number expected from the multiple scatterings by the Thomson atoms with uniform positive charge distribution. Nor can these large deflections be caused by the electrons in the atoms, since their mass is very small. From an analysis of the data, Rutherford came to the conclusion that the large deflections are caused by the strong electric field associated with a large mass concentrated in a small volume. On the basis of this conclusion, Rutherford proposed (1911) a nuclear model of the atom, which is the first authentic model of the modern understanding of the atom.

Rutherford's model of the atom consists of a nucleus of positive charge Ze (Z is the atomic number), which carries most of the mass of the atom and is concentrated in a very small region of radius less than 10^{-14} m. Moving around the nucleus are Z electrons at a distance from the nucleus roughly equal to the size of the atom, *i.e.*, 10^{-10} m. The explanation of the α-particle scattering is that most of the α particles go through the atom with only a slight deflection except when they encounter the strong field due to the massive nucleus in which case they undergo a large deflection. The detailed predictions of the calculations based on this model are in very good agreement with the experimental observations.

Consider the scattering of an α particle by a thin metal foil. Let the cross sectional area of the beam be A. Then the total number of atoms which scatter the α particles is

$$n = \rho\, At \tag{2.50}$$

where ρ is the number of atoms per unit volume, and t is the thickness of the foil. For each atom, let the incoming α particles in the impact parameter range (impact parameter is the distance of the nucleus from the initial direction of motion of the α particle) of b to $b + db$ be scattered into angles between θ and $\theta + d\theta$ (Fig. 2.6). Then, the fraction of α particles scattered into angles between θ and $\theta + d\theta$ is

$$\frac{dN}{N} = n\frac{2\pi b db}{A}$$

$$= 2\pi\rho t b\frac{db}{d\theta}d\theta \tag{2.51}$$

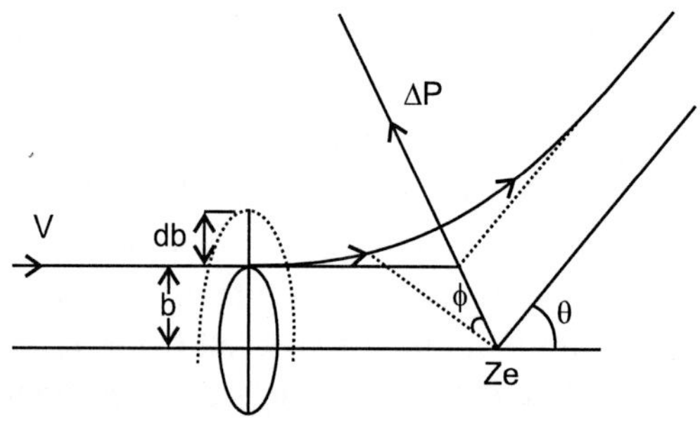

Fig. 2.6 Scattering of α particles by a nucleus of charge *Ze*.

For evaluating this expression, a relation between b and θ is needed, which depends on the force experienced by the α particles. Specifically, the force due to a charge *Ze* of the nucleus is considered. It is shown in Example 6, that for a Coulomb potential $2Ze^2/4\pi\varepsilon_0 r$, the relation between b and θ is

$$b = \frac{Ze^2}{2\pi\varepsilon_0 mv^2}\cot\left(\frac{\theta}{2}\right) \tag{2.52}$$

where m is the mass of α particle and v is its speed. Using this relation in Eq. (2.51), the fraction of particles scattered into solid angle $d\Omega = 2\pi \sin\theta\, d\theta$, is

$$\frac{dN}{N} = \rho t\left(\frac{Ze^2}{4\pi\varepsilon_0 mv^2}\right)^2\frac{d\Omega}{\sin^4\left(\frac{\theta}{2}\right)} \tag{2.53}$$

This is the Rutherford formula for the scattering of α particles by nuclei of change *Ze*.

The important features to note in this formula are that the fraction is (*i*) proportional to the thickness t of the metal foil, (*ii*) proportional to Z^2, (*iii*) inversely proportional to T^2 where $T = \frac{1}{2}mv^2$ is the kinetic energy of the incoming α particles and (*iv*) inversely porportional to $\sin^4 \frac{\theta}{2}$ where θ is the angle of scattering. These properties were tested by Geiger and Marsden by varying the thickness and the composition of the foil, the energy of the incident α particles and the angle of scattering, and were found to be inxcellent agreement with the experimental observations. For examples, they found in an experiment with silver foil, that dN was proportional to 111, 680 and 8800 for $\theta = 150°$, 75° and 37.5°, respectively, other variables remaining the same. For these values, the product $(dN) \sin^4 (\theta/2)$ is proportional to 96.6, 93.4, 93.9, respectively. The near-constancy of the product, though dN itself varies by a large factor, indicates the essential correctness of the θ-dependence of scattering rates.

It may be noted that the Rutherford formula is the same for attractive and repulsive Coulomb potentials. It is not valid for values of the impact parameter b larger than interatomic distances for which an α particle cannot be regarded as being scattered by a single atom in the metal foil. It is also not valid if the α particle approaches the nucleus to a distance (for a head-on collision $r_{min} = Ze^2/\pi\varepsilon_0 mv^2$) less than the size of the nucleus, *i.e.,* about 10^{-14} m, at which the nuclear forces become important.

While the Rutherford model of the atom provides a fairly comprehensive description of the scattering of low-energy α particles by the atoms, there are some implications of the simple model which are in conflict with the classical interpretations of experimental observations. The stability of the atom demands that the electron must revolve around the nucleus. But such an electron, since it is accelerating, must radiate energy continuously according to the classical theory of electromagnetism, and ultimately coalesce with the nucleus. Experimentally, an atom is a highly stable object. Furthermore, it can absorb radiation only of some well-defined frequencies and then emit radiation again of well-defined frequencies. The Rutherford model of the atom is unable to explain these experimental observations. It is with the intention of reconciling the Rutherford model with the observed stability and spectrum of the atom, that Bohr began his search for a model of the atom and came up with what is known as the Bohr model of the atom.

2.7 BOHR MODEL

Bohr (1913) started by assuming that Rutherford's model of the atom is essentially correct but that the classical laws of dynamics need to be modified so as to be applicable to the motion of an electron around the nucleus. The modifications are introdued in the form of the following postulates:

1. Electrons that are bound to the nucleus can move around in only certain discrete orbits. While in these orbits the electrons do not emit electromagnetic radiation, though the motion is accelerated.
2. For circular motion, the allowed orbits are determined by the quantum condition that the angular momentum is $n\hbar$ where $\hbar = h/2\pi$, h being Planck's constant, and n can take positive integral values, $n = 1, 2, \ldots$.
3. Emission or absorption of radiation occurs only when an electron undergoes transition from one allowed orbit to another. The frequency of the radiation emitted or absorbed is given by the relations $h\nu = E_i - E_f$ for emission and $h\nu = E_f - E_i$ for absorption E_f and E_t being the energies associated with the final and initial orbits.

These postulates are quite ad hoc but are justified by the impressive agreement of the predictions with the experimental observations.

Consider an electron with mass m_e moving around a nucleus of mass m_n and charge Ze. Let the distance of the electron and the nucleus, from the centre of mass be r_e and r_n respectively and ω be the angular speed of circular motion. If r is the inter-particle distance, $\mathbf{r} = \mathbf{r}_e - \mathbf{r}_n$, then in the centre of mass frame

$$m_e\mathbf{r}_e + m_n\mathbf{r}_n = 0 \tag{2.54}$$

so that

$$\mathbf{r}_e = \frac{m_n}{m_n + m_e}\mathbf{r}$$

$$\mathbf{r}_n = \frac{-m_n}{m_n + m_e}\mathbf{r} \tag{2.55}$$

It is seen that in the centre of mass frame, the electron and the nucleus are on opposite sides of the centre of mass.

The total energy is

$$E = \frac{1}{2}(m_e r_e^2 + m_n r_n^2)\,\omega^2 - \frac{Ze^2}{4\pi\varepsilon_0 r}$$

$$= \frac{1}{2}m_r r^2\omega^2 - \frac{Ze}{5\pi\varepsilon_0 r} \tag{2.56}$$

where $\qquad m_r = m_e m_n/(m_e + m_n) \qquad (2.57)$

is the reduced mass (it is only slightly smaller than the electron mass), and the angular momentum is

$$L = (m_e r_e^2 + m_n r_n^2)\omega$$

$$= m_r r^2\,\omega \tag{2.58}$$

Now, the dynamical condition for circular orbits is

$$m_e r_e \omega^2 = m_r r \omega^2 = \frac{Ze^2}{4\pi\varepsilon_0 r^2} \tag{2.59}$$

while the quantum condition for the angular momentum is

$$L = m_r r^2 \omega = n\hbar \tag{2.60}$$

Solving for r, Eqs. (2.59) and (2.60) give

$$r = \frac{4\pi\varepsilon_0 \hbar^2 n^2}{m_r Ze^2} \tag{2.61}$$

This is the radius of the nth Bohr orbit. Furthermore, the equilibrium condition in Eq. (2.59) allows us to write the total energy as

$$E = -\frac{Ze^2}{8\pi\,\varepsilon_0 r} \tag{2.62}$$

Using the value of r given in Eq. (2.61), the allowed energies of the atom are

$$E_n = -\frac{hcR}{n^2},\ n = 1, 2, \ldots \tag{2.63}$$

where $$R = \frac{m_r}{4\pi c\hbar^3}\left(\frac{Ze^2}{4\pi\varepsilon_0}\right)^2 \tag{2.64}$$

Substituting the values of the constants, for the hydrogen atom

$$R_H = 1.09678 \times 10^7 \text{ m}^{-1} \tag{2.65}$$

which is very close to Rydberg's original value in Eq. (2.40).

The atom can undergo transitions only between the orbits with the discrete energies E_n given in Eq. (2.63). The frequency of the radiation emitted or absorbed when the atom undergoes a transition from a state with energy E_n to a state with energy E_m, is given by

$$\nu = cR\left(\frac{1}{m^2} - \frac{1}{n^2}\right) \text{ for emission,} \tag{2.66}$$

$$\nu = cR\left(\frac{1}{n^2} - \frac{1}{m^2}\right) \text{ for absorption}$$

It is also implied that if the atom is in the ground state, *i.e.*, the state with the lowest energy, $n = 1$, it continues to remain in that state unless an external radiation of a suitable frequency is incident on it. If the final state has $n = 1, 2, 3, 4$ one gets Lyman, Balmer, Paschen, Brackett series, respectively (Fig. 2.7), of the emission spectra and the corresponding absorption spectra for $n = 1, 2, 3, 4$ in the initial state. Thus, the Bohr model provides a quantitatively satisfactory explanation of the spectrum of the hydrogen atom.

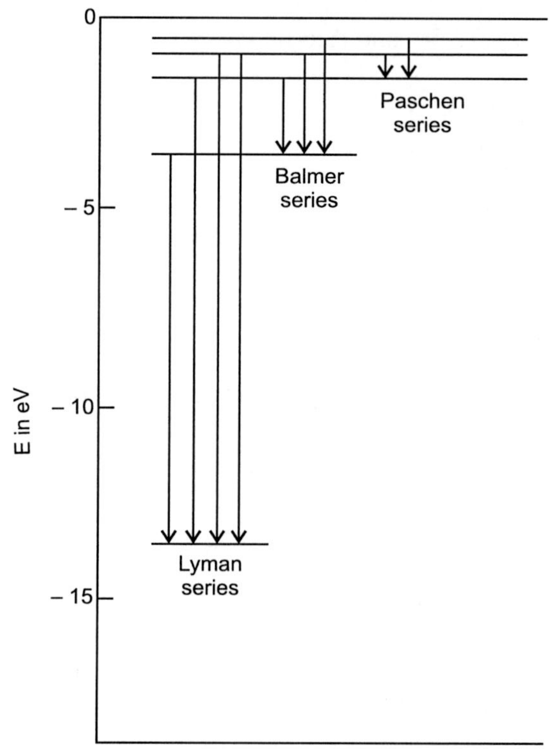

Fig. 2.7 Energy levels of the hydrogen atom, in the Bohr model.

The result in Eq. (2.63) can also be applied to other one-electron atoms such as the deuterium, singly ionized helium atom He^+ or doubly ionized lithium atom Li^{++}, etc. For the deuterium, only the reduced mass is slightly larger than for the hydrogen atom and the energies (being negative) are slightly lower. The existence of the corresponding lines can be used for the detection of the presence of deuterium. For He^+, Li^{++}, etc. the energy levels differ by an additional factor of Z^2. However, since Z is an integer, some of their energy levels will be close to those of the hydrogen atom, the small differences being due to the different reduced masses. For example, the energy levels of He^+ for even n, $n = 2p$, are related to the hydrogen energy levels, by

$$E_{2p}\,(\text{He}^+) = \frac{m_r\,(\text{He}^+)}{m_r\,(\text{H})}\,E_p\,(\text{H}) \tag{2.67}$$

The ratio of the reduced masses is approximately

$$\frac{m_r\,(\text{He}^+)}{m_r\,(\text{H})} \approx 1 + m_e\left(\frac{1}{m(\text{H})} - \frac{1}{m(\text{He}^+)}\right) \tag{2.68}$$

i.e., about 1.000408. Therefore, the transitions between these He^+ levels correspond to frequencies which are slightly higher than those of the hydrogen atom. Indeed, measurements of these small differences give a fairly accurate determination of the ration of $m_e/m(\text{H})$.

Bohr's ideas can be extended to noncircular orbits also. This leads to the conclusion (see Example 7) that the angular momentum does not uniquely determine the energy of the atom, and that there are several angular momentum states which correspond to the same energy. This is an example of what is known as the *degeneracy* of an energy level. However, inclusion of the relativistic corrections shows that these different angular momentum states have slightly different energies. This results in the *multiplicity* of the corresponding spectral lines. Such a fine structure of the lines (the structure is narrower for larger n values), is indeed observed experimentally, but the quantitative predictions of the simple model are not in agreement with the experimental observations.

The Bohr theory of the atom is essentially a theory of single-electron atom. It does not allow a simple generalization to many-electron atoms, not even to helium, and being ad hoc, it is not logically consistent. But its picture of an atom with quantized orbits for the electrons, has retained its utility till today, especially for qualitative arguments.

The existence of discrete atomic energy levels can be observed experimentally from an analysis of collisions between atoms and electrons with known energy. In these collisions, since the mass of the atom is much larger than that of an electron, very little energy is carried away as kinetic energy of the atom. However, if the energy of the electron is sufficient to raise a bound electron to a higher energy orbit, the electron may transfer most of its energy to the atom. This phenomenon was demonstrated by Franck and Hertz (1914). Electrons from a filament are gradually accelerated through a vapour in a tube [Fig. 2.8 (*a*)], towards a grid G and are subjected to a small retarding potential V_0 between the grid and the plate P. When the accelerating potential is sufficiently large to excite an atom, the electron may undergo a collision near G and transfer most of its energy to the atom.

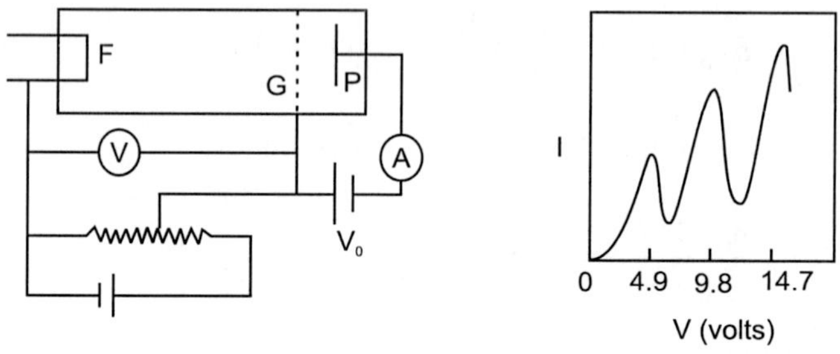

(a) (b)

Fig. 2.8 (a) A schematic diagram of Franck-Hertz experiment.
(b) Variation of current *I* (A) with voltage *V* in volts, for mercury vapour.

It will then be unable to reach the plate P. Thus, as the accelerating potential V is raised from zero, the current arriving at P will increase. When it is just greater than the excitation potential for the atoms, there is a sharp decrease in the current [Fig. 2.8 (*b*)]. It begins to increase again till the next energy level can be excited, and falls again. That the atoms are indeed excited is confirmed by the appearance of the corresponding spectral lines in the radiation emitted as the electrons fall back to the lower energy levels.

2.8 EXAMPLES

A few examples that provided some details and extensions of the ideas discussed are now given.

Example 1

For obtaining the expression for the energy radiated by a unit area per unit time, given in Eq. (2.13), it is noted that the energy crossing a unit area, per unit time, per unit frequency is

$$\frac{dU}{d\nu} = \int v_z \, dE \tag{2.69}$$

where the direction perpendicular to the area is taken as the z direction. Now

$dE = \frac{1}{2} u(\nu)\, d \cos\theta$, where $u(\nu)$ is the energy density, v_z is $c \cos\theta$, and the range of integration is from $\theta = 0$ to $\frac{1}{2}\pi$, so that

$$\frac{dU}{d\nu} = \frac{1}{2} c\, u(\nu) \int_0 \cos\theta \, d\cos\theta = \frac{1}{4} c\, u(\nu) \tag{2.70}$$

Example 2

The position of the maximum in Eq. (2.17) is determined numerically by iteration. Equating the derivative of $u(\lambda)$ to zero, one gets the condition

$$x = 5\,(1 - e^{-x}) \tag{2.71}$$

where $x = hc/\lambda kT$. Inspection suggests that the solution to this is close to $x \approx 5$. First iteration gives

$$\begin{aligned} x &\approx 5\,(1 - e^{-5}) \\ &= 4.9663 \end{aligned} \tag{2.72}$$

while the second iteration gives

$$\begin{aligned} x &\approx 5(1-e^{-4.9663}) \\ &= 4.96516 \end{aligned} \tag{2.73}$$

which agrees with Eq. (2.18).

Example 3

An experiment on the photoelectric effect of a metal gives stopping potentials of 4.62 V for $\lambda = 1850$ Å, and 0.18 V for $\lambda = 5460$ Å. These results can be used to calculate the Planck's constant and the work function of the metal. From Einstein's relation in Eq. (2.23),

$$\frac{hc}{\lambda} = e\phi + eV_0 \tag{2.74}$$

which on using the experimental values, leads to two linear equations in h and ϕ. Solving them, we get $h = 6.64 \times 10^{-34}$ J s and $\phi = 2.1$eV.

Example 4

The wavelength of x-rays scattered by bound electrons (Sec. 2.3) has a spread, mainly due to the fact that the bound electron has a momentum distribution. For estimating the correction due to the non-zero initial momentum, it is noted that the binding energy of the electrons is usually quite small, about 10 eV, compared to the x-ray energies of about 10 keV.

The momentum and energy conservation relations give

$$(\mathbf{p}_f - \mathbf{p}_i)^2 = \frac{h^2}{c^2}\left(\nu_0^{\,2} + \nu^2 - 2\nu_0\,\nu\cos\theta\right) \tag{2.75}$$

$$h\,\nu_0 + mc^2 - \Delta E = h\nu + (p_f^{\,2}c^2 + m^2c^4)^{1/2} \tag{2.76}$$

Here $\mathbf{p}_i$ and $\mathbf{p}_f$ are the initial and final momenta of the electron, and ΔE is the binding energy. Proceeding as in Sec. 2.3

$$\lambda - \lambda_0 = \frac{h}{mc}(1-\cos\theta) + \frac{\lambda_0\,\lambda}{hmc}\left(p_f \cdot p_i - \frac{1}{2}p_i^{\,2} + \Delta EM\right) \tag{2.77}$$

where ΔE has been neglected compared to mc^2. For x-ray energies of a few tens of keV and binding energies of the order of a few eV, the second term on the right hand side is smaller than the first term. Because of the variation of $\mathbf{p}_i$, this term gives rise to a spread in the frequency of the scattered beam [see Fig. 2.3 (*a*)].

Examples 5

In the Davisson-Germer experiment, the x-ray beam was incident normally on the surface AB (see Fig. 2.9). The condition for coherent, maximum reflection is

$$d' \sin\theta = m\lambda \tag{2.78}$$

where m is a positive integer. It can be shown that the Bragg condition reduces to this condition.

For Bragg reflection, the incident and reflected beams make equal angles with the reflection plane AC, the angles being $(\pi - \theta)/2$. The Bragg condition for coherent reflection is

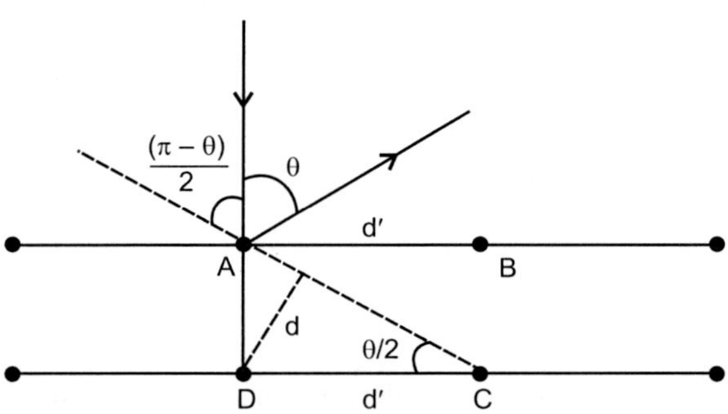

Fig. 2.9 Relation between the Davisson-Germer condition and the Bragg condition

$$2d \sin\frac{1}{2}(\pi - \theta) = n\lambda \tag{2.79}$$

But $d = d' \sin(\theta/2)$, so that the Bragg condition becomes

$$2d' \cos(\theta/2)\sin(\theta/2) = n\lambda \tag{2.80}$$

which is the same as Eq. (2.78) if $n = m$.

Example 6

The orbit for the motion of a particle in a Coulomb potential can be derived by considering the change in the momentum of the particle.

For a particle with impact parameter b (see Fig. 2.6) and scattered at an angle θ, the change in the momentum is

$$|\Delta \mathbf{p}| = 2mv \cos \frac{1}{2} (\pi - \theta)$$

$$= \int_{-\infty}^{\infty} F_p dt$$

$$= \int F \cos \phi \frac{d\phi}{\phi} \tag{2.81}$$

where F_p is the component of the force parallel to $\Delta \mathbf{p}$ and ϕ is the angle between the position vector $\mathbf{r}$ and $\Delta \mathbf{p}$ (see Fig. 2.6). But

$$F = \frac{Ze\,(2e)}{4\pi\,\varepsilon_0\, r^2} \tag{2.82}$$

and

$$mvb = mr^2\phi \tag{2.83}$$

which follows from the conservation of angular momentum.

Therefore

$$2mv \sin (\theta/2) = \frac{Ze^2}{\pi\varepsilon_0 vb} \cos (\theta/2) \tag{2.84}$$

which leads to

$$b = \frac{Ze^2}{2\pi\varepsilon_0 mv^2} \cot (\theta/2) \tag{2.85}$$

used in Eq. (2.52).

Example 7

The Bohr model was generalized by Sommerfeld (1916) to include non-circular elliptic orbits. The generalized quantum conditions are

$$\oint p_\phi \, rd\phi = nh, \; n \text{ is an integer} \tag{2.86}$$

$$\oint p_r \, dr = kh, \; k \text{ is an integer} \tag{2.87}$$

were $p_\perp$ and p_r are components of momentum perpendicular and parallel to the radius, respectively, and the integrals are over the complete orbit. The first condition reduces to the Bohr relation,

$$L = n\hbar \tag{2.88}$$

The total energy is

$$E = \frac{1}{2m_r} p_r^2 + \frac{L^2}{2m_r r^2} - \frac{Ze^2}{4\pi\varepsilon_0 r} \tag{2.89}$$

Using Eqs. (2.89) and (2.88), the quantization condition in Eq. (2.87) becomes

$$2\int_{r_{\min}}^{r_{\max}} dr\,(2m_r E - n^2\hbar^2/r^2 + m_r Ze^2/2\pi\varepsilon_0 r)^{1/2} = kh \tag{2.90}$$

The integral can be evaluated using the theory of complex variables and leads to

$$\frac{Ze^2}{2\varepsilon_0}\left(-\frac{m_r}{2E}\right)^{1/2} - nh = kh \tag{2.91}$$

so that

$$E = -\frac{m^r}{2\hbar^2}\left(\frac{Ze^2}{4\pi\varepsilon_0}\right)^2 \frac{1}{(k+n)^2},\ (k+n) \geq 1 \tag{2.92}$$

Thus, for a given value of the principal quantum number $(k + n)$, states with the same energy exist, for $k = 0, \ldots, k + n - 1$. Thus we have what is called as a degeneracy of order $k + n$ (see. Sec. 3.4).

Example 8

Suppose in addition to the Coulomb attraction, there is a potential energy terms g/r^2. Then the expression for the total energy E is

$$E = \frac{1}{2m_r} p_r^2 + \frac{L^2}{2m_r r^2} + \frac{g}{r^2} - \frac{Ze^2}{4\pi\varepsilon_0 r} \tag{2.93}$$

Since $L^2 = n^2\hbar^2$, this is equivalent to replacing $n\hbar$ by $(n^2\hbar^2 + 2m_r g)^{1/2}$. Therefore, the expression for the quantized energy is

$$E = -\frac{m_r}{2\hbar^2}\left(\frac{Ze^2}{4\pi\varepsilon_0}\right)^2 \frac{1}{(k+(n^2+2m_r g/\hbar^2)^{1/2})^2} \tag{2.94}$$

The energy levels are now different for a given $k + n$ but different k or n values. Thus, the degeneracy due to different angular momentum states is removed by the addition of the potential g/r^2. Indeed, the Coulomb degeneracy is removed by any additional interaction.

Example 9

A quick estimation of the binding energies of the hydrogen atom is obtained by the following simple argument.

A stationary state may be thought of as one for which an integral number of de Broglie wavelengths can be fitted over the orbit, *e.g.* for circular orbits

$$2\pi r = n\ (h/p),\ n = 1, 2, \ldots \tag{2.95}$$

Using this relation, the total energy is

$$E = \frac{1}{2m_r} p^2 - \frac{Ze^2}{4\pi\varepsilon_0 r}$$

$$= \frac{n^2\hbar^2}{2m_r r^2} - \frac{Ze^2}{4\pi\varepsilon_0 r} \tag{2.96}$$

If the state is stable, it corresponds to a minimum of this energy:

$$\frac{dE}{dr} = -\frac{n^2\hbar^2}{m_r r^3} + \frac{Ze^2}{4\pi\varepsilon_0 r^2}$$

$$= 0 \tag{2.97}$$

so that

$$r_{\min} = \frac{n^2\hbar^2}{m_r}\left(\frac{4\pi\varepsilon_0}{Ze^2}\right) \tag{2.98}$$

This leads to the energy levels

$$E_n = -\frac{mr}{2\hbar^2 n^2}\left(\frac{Ze^2}{4\pi\varepsilon_0}\right)^2 \tag{2.99}$$

PROBLEMS

1. If the continuum spectrum of the sun approximates that of a black body, peaking at $\lambda_m \approx 5000$ Å, what is the surface temperature of the sun? What can you infer about the temperature of the material surrounding the sun from the observation of Balmer absorption lines in the spectrum?

2. A spherical object of radius 10 cm and whose surface approximates that of a black body is maintained at a temperature of 1000 K. What is the energy radiated by the body per second? Estimate the energy radiated in the visible range.

3. Show that the number of photons per unit volume of a black body cavity, per unit frequency, is

$$N(\nu) = \frac{8\pi\nu^2}{c^3\,(e^{h\nu/kt} - 1)}.$$

4. A small 10 W source of ultraviolet light of wavelength 1000 Å is held at a distance of 0.1 m from a metal surface. If the radius of an atom is approximately 0.5 Å, how many photons strike an atom per minute? If the efficiency, *i.e.*, the fraction of the photons that succeed in knocking out the photons, is 1%, how many electrons are ejected from a unit area per second?

5. In a photoionization experiment, let ε be the binding energy and let the photon with frequency v be incident. Assuming that the recoil energy of the atom is small, show that the electron comes out with a momentum approximately equal to $[2m(hv - \varepsilon)]^{1/2}$. Obtain an expression for the recoil energy of the atom when the electron comes out at an angle of ϕ with respect to momentum of the photon.

6. Show that the recoil electron in Compton scattering always comes out in the forward hemisphere, *i.e.*, recoil angle is less than or equal to 90°. When is the angle between the scattered photon and the recoil electron, a maximum? What is the corresponding recoil energy of the electron? Assume $hv_0 < mc^2$ for the second part.

7. An x-ray of wavelength $\lambda = 0.1$ Å is scattered by an electron. What are the maximum and minimum wavelengths of the scattered photon? At what angle will the scattered photon have a wavelength of 0.11 Å?

8. An electron has a de Broglie wavelength equal to that of a photon. Compare its kinetic energy with the energy of the photons? What is the limiting value of this ratio for low energy photons? If the photon has an energy of 100 keV, what is the kinetic energy of the electron?

9. What is the de Broglie wavelength of neutrons at room temperature? Can they be used to study crystal structure?

10. A beam of 0.25 eV neutrons incident on a nickel surface is found to produce a maximum at an angle of 70° with the initial beam direction, but no other maxima are produced for angles greater than 70°. Determine the order of the maximum and the interplanar distance for the crystal.

11. When an electron enters a crystal, it is accelerated towards the interior because of an inner potential due to the positive charge of the ions in the crystal. For an electron of speed u, and an inner potential V_i volts show that the bending of the electron beam on entering the crystal is described by a refractive index $\mu = (1 + 2eV_i/mu^2)^{1/2}$. Show also that, in this case, the Bragg relation is modified to read $2d\,(\mu^2 - \cos^2\theta)^{1/2} = n\lambda$.

12. Assume that a particle can be confined to a spherical volume only if its circular orbit can be fitted with an integral multiple of de Broglie wavelengths. Estimate the minimum kinetic energy of a proton confined to a nucleus of diameter 10^{-14} m. What would be the kinetic energy of an electron similarly confined?

13. Use the arguments of example 9 to obtain a rough estimation of the energy levels of a 3-dimensional harmonic oscillator.

14. Calculate the distance of closest approach of a 5 MeV a particle in a head-on collision with a gold nucleus. What is the upper bound of the α-particle energy for which the Rutherford formula is expected to be valid? Take the radius of the gold nucleus to be about 7×10^{-15} m and that of the α-particle to be about 2×10^{-15} m.

15. Show that the fraction of the incident α particles scattered through an angle between θ_1 and θ_2 is given by

$$\pi\rho t\left(\frac{2Ze^2}{4\pi\varepsilon_0 mv^2}\right)^2 [\text{cosec}^2\,(\theta_1/2) - \text{cosec}^2\,(\theta_2/2)]$$

where the terms is defined in Sec. 2.6.

16. The value of R in Eq. (2.63) is found to be 1.0967758×10^7 m^{-1} for the hydrogen and 1.0972227×10^7 m^{-1} for $^4He^+$. From the value of $m_H/m_{He} \approx 0.2517$, estimate the value of m_e/m_H.

17. One of the spectral lines of hydrogen with wavelength 4861.320 Å is accompanied by another line of wavelength 4859.975 Å. Assuming that this is due to the presence of the heavier isotope deuterium, obtain the ratio of the deuterium mass to the proton mass.

18. Apply the Bohr quantization condition to obtain the energy levels of (*i*) a 3-dimensional harmonic oscillator, (*ii*) a particle in a potential $V(r) = -g/r^s$, $s > 0$.

19. A charged particle, moving in the presence of a magnetic field in the z-direction, has circular orbits in the xy plane. Apply Bohr's quantization condition to obtain the energy levels of the particle.

20. Use the Sommerfeld quantization condition $\oint p_x d_x = nh$ to obtain the energy levels of a one-dimensional harmonic oscillator.

21. A free atom undergoes a transition from an energy level E_1 to E_2 by absorbing a photon. Show that the frequency of the photon absorbed is

$$\nu = \frac{(E_2 - E_1)}{h}\left(1 + \frac{E_2 - E_1}{2m_1c^2}\right)$$

where m_1 is the mass of the initial atom. The recoil correction given by the second term is of the order of 10^{-8}, or smaller, for the hydrogen atom. In Mössbauer effect, the atom is embedded in a crystal so that its effective mass is that of the crystal. As a result, the second term is negligible and one has recoil-less absorption of photons.

22. In a Franck-Hertz experiment, hydrogen atoms are bombarded with electrons. What are the wavelengths of the emission lines observed when the electrons are accelerated through a potential difference of 12.5V?

23. Assuming that the earth is a black body in equilibrium at a temperature of 300 K, estimate the temperature of the sun.

24. Suppose that man's power production reaches 20% of the power received from sunlight (this would happen in about 250 years if the present exponential growth continues). What is the expected approximate increase in the surface temperature of the earth?

3

Elements of Quantum Theory

Structures of the Chapter

In Chapter 2, we have discussed some experiments which indicate that radiation should be regarded as being made up of zero-mass particles called photons each of which carries a quantum of energy $h\mathbf{v}$ and momentum $\frac{h\mathbf{v}}{c}\mathbf{n}$. It was also observed that particles with nonzero mass, such as the electron, have wave properties associated with them, and produce diffraction patterns similar to those produced by radiation. Thus, a situation emerges in which radiation and matter exhibit both particles properties, *i.e.,* they carry quantum of energy and momentum, and wave properties, *i.e.,* they produce diffraction patterns.

Here, the basic laws governing the wave-particle behaviour of matter will be analysed and applied to some simple cases. In deducing the laws, we will borrow heavily from the wave properties of radiation and particle properties of matter particles, with which one is familiar.

3.1 A THOUGHT EXPERIMENT

Consider a thought experiment in which a beam of particles is incident on two closely-spaced narrow slits S_1 and S_2, as shown in Fig. 3.1 These slits act as two coherent sources and produce interference fringes on a screen S placed at a distance so far away that the separation between the slits is negligible compared to their distance from the screen. This experiment is Young's double slit experiment for radiation, but here it is reanalysed in terms of the particle which constitute the beam.

The intensity variation on the screen is given by the familiar interference fringes. The interpretation of the interference fringes in terms of the particles is that the intensity is proportional to the number of particles arriving at various points, with no particles coming to regions with zero intensity. It is to be noted that the interference fringes are not due to any interaction between the particles coming from different slits. Indeed, the fringe pattern is independent of the intensity of the initial beam, so that if the particles come only one at a time, they will still avoid the dark spots and the frequency of their arrival is proportional to the intensity (diffraction experiments with electrons coming essentially one at a time have demonstrated this). If on the other hand, one of the slits is closed, the fringes disappear, and one has more or less a uniform intensity. The question that comes up is, how do the particles that are coming one at a time, know the existence of both the slits, which persuades them to come more frequently at the bright spots and avoid the dark spots on the screen.

The wave theory explanation of the interference pattern is that the amplitude of the wave at any point on the screen is a superposition of the amplitude of two coherent waves coming from S_1 and S_2, and the intensity at a point on the screen is given by the modulus squared of the resultant amplitude:

$$I = |\psi(S_1) + \psi(S_2)|_2 \quad (3.1)$$

$$= |Ae^{2\pi i(vt - l_1/\lambda)} + Ae^{2\pi i(vt - l_2/\lambda)}|^2 \quad (3.2)$$

where v is the frequency, λ is the wavelength, and l_1 and l_2 are the distances of the point on the screen from the two slits. Therefore, the intensity of the interference pattern at any point is given by

$$I = 4|A|^2 \cos^2\left(\frac{\pi a x}{\lambda d}\right) \quad (3.3)$$

where a, x and d are as shown in Fig. 3.1 and the path difference is approximately (ax/d). If either of the slits is closed, the interference fringes disappear and a uniform intensity distribution of $I = |A|^2$ results.

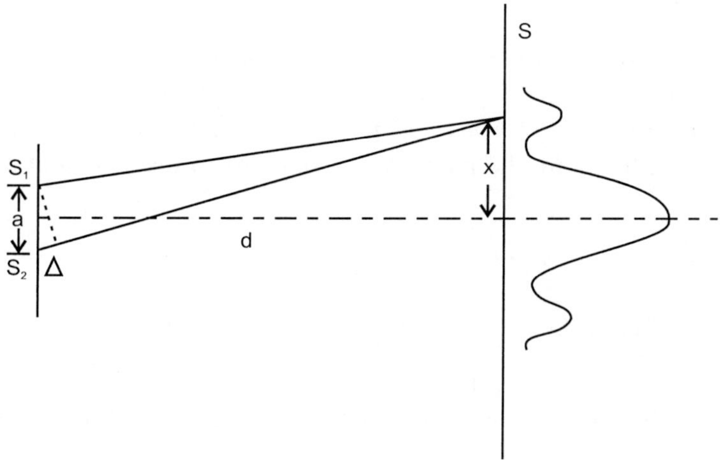

Fig. 3.1 Two slit interference experiment for particles.

3.2 THE WAVE FUNCTION

The observation of the interference fringes, even when particles are coming one at a time, forces us to associate a wave function ψ with the particle. A superposition of allowed wave functions is also a possible wave function of the particle, *e.g.*

$$\psi(S_1 + S_2) = \psi(S_1) + \psi(S_2) \quad (3.4)$$

where $\psi(S_1)$, $\psi(S_2)$ and $\psi(S_1 + S_2)$ are the wave functions when slits S_1, S_2 and $S_1 + S_2$ are open, respectively. The frequency of arrival at the screen is then explained if it is postulated that the probability for a particle to be found in a volume dV is proportional to $|\psi|^2\, dV$. Implicit in this association of a wave function with the particle, is the implication that the position of the particle cannot be predicted precisely, in a single observation. Only a probabilistic prediction can be made that if the experiment is repeated a large number of times, the frequency of a particle being found in different regions is proportional to the square of the wave function associated with each particle. The inability, in general, to predict the results of individual events is inherent in quantum mechanics and is called the *principle of indeterminacy*.

The main problem in quantum mechanics is the determination of the wave function. For obtaining the rules which govern the determination of the wave function, we again invoke our experience with wave functions which describe radiation and therefore, photons. It was shown in Chapter 1 that the scalar and vector fields of electromagnetic radiation in a region where there are no sources, satisfy (see Eq. 1.67) equations of the type

$$\left(\nabla^2 - \frac{1}{c^2}\frac{\partial^2}{\partial t^2}\right)\psi = 0 \tag{3.5}$$

where ψ stands for one of the fields. Such an equation is also satisfied by the electric and magnetic fields as can be seen by applying the curl operator to equations (1.60*c*) and (1.60*d*) and using the other Maxwell equations in Eq. (1.60). Furthermore, the plane-wave solutions to this equation are of the form

$$\psi(\mathbf{r}, t) = Ae^{-2\pi i(\nu t - \mathbf{k}.\mathbf{r})} \tag{3.6}$$

where ν is the frequency and $\mathbf{k} = \frac{\nu}{c}\mathbf{n}$ with $\mathbf{n}$ being a unit vector in the direction of propagation. It is noted that $h\nu$ is the energy of the photon and $h\mathbf{k}$ is its momentum. Therefore, the wave function of the photon is of the form

$$\psi(\mathbf{r}, t) = Ae^{-i(Et - \mathbf{p}.\mathbf{r})/\hbar} \tag{3.7}$$

Actually, for radiation, the polarization vector has also to be considered, which for the present discussion is not relevant. Substituting this function in Eq. (3.5),

$$\left(\frac{1}{c^2}\left(\frac{E^2}{\hbar^2}\right) - \frac{\mathbf{p}^2}{\hbar^2}\right)\psi = 0 \tag{3.8}$$

which is obviously true since for a photon (mass = 0), $E^2 - p^2c^2 = 0$ (using the notation $\mathbf{p}.\mathbf{p} = \mathbf{p}^2 = p^2$).

Suppose we knew first about the photon and wanted the equation which would determine its wave function. Then one could start with the valid relation

$$\left[\frac{1}{c^2}\left(\frac{E}{\hbar}\right)^2 - \left(\frac{\mathbf{p}}{\hbar}\right)^2\right]\psi = 0 \tag{3.9}$$

It is now postulated that the wave equation would be obtained by the substitution

$$E = i\hbar \frac{\partial}{\partial t} \tag{3.10}$$

$$\mathbf{p} = -i\hbar \nabla \tag{3.11}$$

in Eq. (3.9). The choice of the coefficients is dictated by the requirement that the operation of these operators on $\psi(\mathbf{r}, t)$ in Eq. (3.6) should give $h\nu$ and $\frac{h\nu}{c}\mathbf{n}$ which are the energy and momentum of the photon. In any case, the relative coefficients are determined by the requirement of relativistic covariance since $\left(p, \frac{1}{c}E\right)$ and $\left(-\nabla, \frac{1}{c}\frac{\partial}{\partial t}\right)$ both transform as a relativistic 4-vector [see Eq. (1.48)].

The prescription for getting the wave equation is now clear. One starts with the classical expression for the total energy in terms of momentum and position, multiplies it by the wave function and converts it into a wave equation by the substitutions in Eqs. (3.10) and (3.11). It is worth noting that the operator relation in Eq. (3.10) may be used for energy, which either includes or does not include the rest energy term, without changing the essential results since adding a constant to energy in this case only redefines the zero of the energy.

For a nonrelativistic free particle,

$$\left(E - \frac{1}{2m}\mathbf{p}^2\right)\psi = 0 \tag{3.12}$$

which with the substitutions in Eqs. (3.10) and (3.11) leads to

$$i\hbar \frac{\partial \psi(\mathbf{r}, t)}{\partial t} = -\frac{\hbar^2}{2m}\nabla^2 \psi(\mathbf{r}, t) \tag{3.13}$$

This is the celebrated Schrödinger equation for the free particle. It is equally suggestive that for a particle in the presence of an interaction potential, the relation is

$$\left(E - \frac{1}{2m}\mathbf{p}^2 - V(r)\right)\psi(\mathbf{r}, t) = 0 \tag{3.14}$$

Substitution of Eqs. (3.10) and (3.11) then gives

$$i\hbar \frac{\partial \psi(\mathbf{r}, t)}{\partial t} = \left(-\frac{\hbar^2}{2m}\nabla^2 + V(r)\right)\psi(\mathbf{r}, t) \tag{3.15}$$

which is the Schrödinger equation for a nonrelativistic particle in the presence of an interaction potential.

3.3 POSTULATES OF QUANTUM MECHANICS

The discussion so far has been for the purpose of introducing the ideas of wave function and wave equation, and making them plausible. These ideas are now formalized and generalized in terms of some postulates of quantum mechanics.

Postulate 1: Every state of n particles is described by a wave function $\psi(\mathbf{r}_i, t)$, $i = 1, ..., n$ where $\mathbf{r}_i$ are the coordinates of the n particles, such that the probability at time t, of finding the particles in respective volumes d^3r_i about the points $\mathbf{r}_i$, is

$$dP = |\psi(\mathbf{r}_i, t)|^2 \, d^3r_1 ... d^3r_n \tag{3.16}$$

It is relevant to comment here that:

1. Since the particles must be found somewhere, the total probability must be 1,

$$\begin{aligned} \int dP &= \int |\psi(\mathbf{r}_i, t)|^2 \, d^3r_1 ... d^3r_n \\ &= 1 \end{aligned} \tag{3.17}$$

This defines the normalization of the wave function. It is also clear from the definition of probability that the average value of a function $f(\mathbf{r}_i)$ is

$$\langle f(r_i) \rangle = \int |\psi(\mathbf{r}_i, t)|^2 f(\mathbf{r}_i) \, d^3r_i \, ... \, d^3r_n \tag{3.18}$$

2. For a single particle system, the results simplify to

$$dP = |\psi(\mathbf{r}_1, t)|^2 \, d^3r_1$$

$$1 = \int |\psi(\mathbf{r}_1, t)|^2 \, d^3r_1 \tag{3.19}$$

$$\langle f(\mathbf{r}_1) \rangle = \int |\psi(\mathbf{r}_1, t)|^2 f(\mathbf{r}_1) \, d^3r_1 \tag{3.20}$$

In particular, the average position is given by

$$\langle r_1 \rangle = \int |\psi(\mathbf{r}_1, t)|^2 \, \mathbf{r}_1 d^3r_1 \tag{3.21}$$

Postulate 2: The wave function $\psi(r_i, t)$ satisfies the Schrödinger equation

$$i\hbar \frac{\partial \psi(r_i, t)}{\partial t} = H(\mathbf{r}_i, -i\hbar \nabla_i) \, \psi(\mathbf{r}_i, t) \tag{3.22}$$

where H is the Hamiltonian or the energy operator obtained from the classical expression for the total energy by replacing $\mathbf{p}_i$ by-$i\hbar \nabla_i$. For a single article in the presence of an interaction potential, the total energy is $E = \frac{1}{2m} p^2 + V(r)$ which leads to the equation

$$i\hbar \frac{\partial \psi(\mathbf{r}, t)}{\partial t} = -\frac{\hbar^2}{2m} \nabla^2 \psi(\mathbf{r}, t) + V(r) \, \psi(\mathbf{r}, t) \tag{3.23}$$

This equation is valid if the system is nonrelativistic. It requires some modifications if the particle has additional variables such as intrinsic angular momentum called *spin*. If the system has a well-defined energy, its Schrödinger equation reduces to its simpler time-independent form

$$-\frac{\hbar^2}{2m}\nabla^2\psi(\mathbf{r},t)+V(r)\,\psi(r,t) = i\hbar\frac{\partial\psi(\mathbf{r},t)}{\partial t} = E\psi(\mathbf{r},t) \quad (3.24)$$

Postulates 1 and 2 allow us to deduce average values of general dynamical variables. Equation (3.24) on being multiplied by $\psi^*(r, t)$ and integrated over the entire space and on rearrangement of terms leads to

$$\int\psi^*(\mathbf{r},t)\,V(r)\,\psi(\mathbf{r},t)\,d^3r = \int\psi^*(\mathbf{r},t)\left(i\hbar\frac{\partial}{\partial t}\right)\psi(\mathbf{r},t)\,d^3r$$

$$-\int\psi^*(\mathbf{r},t)\left(-\frac{\hbar^2}{2m}\nabla^2\right)\psi(\mathbf{r},t)\,d^3r \quad (3.25)$$

The term on the left hand side, as seen from Eq. (3.21), is the average potential energy. It is therefore reasonable to identify the terms on the right hand side as the average total energy and the average kinetic energy:

$$\langle E\rangle = \int\psi^*(\mathbf{r},t)\left(i\hbar\frac{\partial}{\partial t}\right)\psi(\mathbf{r},t)\,d^3r \quad (3.26)$$

$$\langle\frac{1}{2m}p^2\rangle = \int\psi^*(\mathbf{r},t)\left(-\frac{\hbar^2}{2m}\nabla^2\right)\psi(\mathbf{r},t)\,d^3r \quad (3.27)$$

Now, since $\left(\mathbf{p}, \frac{E}{c}\right)$ transforms as a 4-vector, the relationships in Eqs. (3.10) and (3.11) suggest that, in addition to Eq. (3.26)

$$\langle\mathbf{p}\rangle = \int\psi^*(\mathbf{r},t)\,(-i\hbar\nabla)\,\psi(\mathbf{r},t)\,d^3r \quad (3.28)$$

These results are generalized in the following postulate.

Postulate 3: The average values of E and the dynamical variable $F(\mathbf{p}, \mathbf{r})$ are given by

$$\langle E\rangle = \int\psi^*(\mathbf{r},t)\left(i\hbar\frac{\partial}{\partial t}\right)\psi(\mathbf{r},t)\,d^3r \quad (3.29)$$

and $$\langle F(\mathbf{p}, r)\rangle = \int\psi^*(\mathbf{r},t)\,F(-i\hbar\nabla, r)\,\psi(\mathbf{r},t)\,d^3r \quad (3.30)$$

In quantum mechanics, the choice of dynamical variables which are dynamical observables is not obvious. A minimal requirement is imposed on the dynamical observables that their average values must be real. It is easy show from Eq. (3.30) that both **r** and **p** have real average values and are acceptable as dynamical observables. On the other hand, xp_x is not an observable though the angular momentum $\mathbf{r} \times \mathbf{p}$ can be shown to have a real average value and hence is acceptable as a dynamical observable.

It is often convenient to work with the fourier transform of the wave function rather than with the wave function itself. Writing $\psi(\mathbf{r}, t)$ as

$$\psi(\mathbf{r}, t) = \frac{1}{h^{3/2}} \int f(\mathbf{k}, t) \exp(\mathbf{ik}.\mathbf{r}/\hbar)\, d^3k \tag{3.31}$$

the inverse fourier transform is

$$f(\mathbf{k}, t) = \frac{1}{h^{3/2}} \int \psi(\mathbf{r}, t) \exp(-\mathbf{ik}.\mathbf{r}/\hbar) d^3r \tag{3.32}$$

The fourier transform $f(\mathbf{k}, t)$ is called the wave function in the momentum space. It is easy to show that

$$\int |\psi(\mathbf{r}, t)|^2\, d^3r = \int |f(\mathbf{k}, t)|^2\, d^3k \tag{3.33}$$

and that the average value of momentum given in Eq. (3.28) reduces to

$$\langle \mathbf{p} \rangle = \int |f(\mathbf{k}, t)|^2\, \mathbf{k}\, d^3k \tag{3.34}$$

which justifies the definition of $f(\mathbf{k}, t)$ as the wave function in the momentum space.

3.4 SOME PROPERTIES OF OBSERVABLES AND WAVE FUNCTIONS

In this section, some important properties of quantum mechanical wave functions and observables are described.

It was noted that the average values of dynamical observables must be real. This requirement is satisfied if the operator A corresponding to the observable (the operator is obtained from the appropriate classical variable by the replacement of **p** by $-i\hbar(\nabla)$ satisfies the condition

$$\int \phi^* A\psi d\tau = \left(\int \psi^* A\phi d\tau\right)^* \tag{3.35}$$

where $d\tau$ represents a volume element. Taking $\phi = \psi$ gives the result that the average value is real. An operator A which satisfies this condition is said to be *hermitian*.

Hermitian operators have some special properties, which are mentioned here. Consider an operator equation

$$A\phi_n = E_n \phi_n \tag{3.36}$$

where E_n is a constant. This is an example of what are called *eigenvalue equations*, E_n being the *eigenvalue* of operator A and ϕ_n the corresponding *eigen-function*.

It is natural to interpret this equation as implying that ϕ_n describes a state with a well-defined value E_n for the observable corresponding to A. The eigenvalues and the eigenstates of hermitian operators satisfy the following properties:

1. The eigenvalues of a hermitian operator are real. This follows from Eq. (3.35), If ϕ and ψ are taken to be the same eigenstates.
2. Eigenstates with different eigenvalues are orthogonal in the following sense. In Eq. (3.35), if ϕ and ψ are eigenstates ϕ_n and ϕ_m of A with eigenvalues E_n and E_m respectively, then

$$(E_n - E_m) \int \phi_n^* \phi_m \, d\tau = 0 \tag{3.37}$$

or

$$\int \phi_n^* \phi_m \, d\tau = 0 \quad \text{for} \quad E_n \neq E_m \tag{3.38}$$

The states ϕ_n and ϕ_m are said to be orthogonal to each other. It is also possible in the case of discrete eigenvalues to normalize the eigenstates such that

$$\int \phi_n^* \phi_n = 1 \tag{3.39}$$

The states which satisfy the relations (3.38) and (3.39) are said to be *orthonormal*. It may happen that there are more than one states which have the same eigenvalue. These states are said to be *degenerate* and the number of degenerate states is known as the *degree* of *degeneracy* of the eigenvalue. It is possible to normalize these states and to choose a suitable, orthonormal set of degenerate states.

3. The eigenstates of a hermitian operator are complete, and form a complete basis. This means that any state ψ can be expressed as a linear combination of the eigenstates ϕ_n of a hermitian operator,

$$\psi = \sum_n a_n \phi_n \tag{3.40}$$

The summation may include an integration over a set of states with continuum eigenvalues.

4. Two operators A and B are said to commute if

$$[A, B] \equiv AB - BA = 0 \tag{3.41}$$

where $[A, B]$ is called the *commutator* of A and B. It is possible to choose, as a basis, states which are simultaneous eigenstates of commuting hermitian operators. A particularly important case is obtained if one of the operators is the Hamiltonian (*i.e.* energy operator),

$$[H, B] = 0 \tag{3.42}$$

Then the states can be chosen to be simultaneous eigenstates of H and B. Since the eigenstates of a time-independent Hamiltonian do not change with time (except for the phase factor), this means that the eigenvalues of B for these states do not change with time, and therefore B is conserved.

The solutions to the Schrödinger equation in some simple situations are now discussed.

3.5 FREE PARTICLE

A *free particle* is one on which there are no forces acting. The Schrödinger equation for the free particle is

$$i\hbar\frac{\partial\psi}{\partial t} = -\frac{\hbar^2}{2m}\nabla^2\psi \tag{3.43}$$

Separating the variables, the solution to Eq. (3.43) can be written in the form

$$\psi(\mathbf{r}, t) = f(t)\,\phi(\mathbf{r}) \tag{3.44}$$

Substituting this in Eq. (3.43) and dividing the equation by $\psi(\mathbf{r}, t)$ gives

$$i\hbar\frac{1}{f(t)}\frac{\partial f(t)}{\partial t} = -\frac{\hbar^2}{2m}\frac{1}{\phi(r)}\nabla^2\phi(r) \tag{3.45}$$

which can be satisfied only if both the sides are constant, say E. Then

$$i\hbar\frac{\partial f(t)}{\partial t} = Ef(t) \tag{3.46}$$

$$-\frac{\hbar^2}{2m}\nabla^2\phi(\mathbf{r}) = E\phi(\mathbf{r}) \tag{3.47}$$

Equations (3.46) and (3.47) are eigenvalue equations for the energy, E being the energy eigenvalue and $f(t)$, $\phi(\mathbf{r})$ being the corresponding eigenfunctions. They describe a state with a well-defined energy E.

The solution to Eq. (3.46) is

$$f(t) = \exp(-iEt/\hbar) \tag{3.48}$$

except for an overall constant which will be included in $\phi(\mathbf{r})$. For solving Eq. (3.47), once again a separable form is assumed for $\phi(\mathbf{r})$,

$$\phi(r) = A(x)\,B(y)\,C(z) \tag{3.49}$$

Substituting this in Eq. (3.47) and dividing by $\phi(r)$ gives

$$-\frac{\hbar^2}{2m}\left(\frac{1}{A(x)}\frac{d^2A(x)}{dx^2} + \frac{1}{B(y)}\frac{d^2B(y)}{dy^2} + \frac{1}{C(z)}\frac{d^2C(z)}{dz^2}\right) = E \tag{3.50}$$

For this relation to be valid, each of the three terms in Eq. (3.50) should be a constant. Introducing constants k_x, k_y, k_z one gets

$$\frac{d^2A(x)}{dx^2} = -\frac{k_x^2}{\hbar^2}A(x) \tag{3.51}$$

$$\frac{d^2B(y)}{dy^2} = -\frac{k_y^2}{\hbar^2}B(y) \tag{3.52}$$

$$\frac{d^2C(z)}{dz^2} = -\frac{k_z^2}{\hbar^2}C(z) \tag{3.53}$$

with
$$E = \frac{1}{2m}(k_x^2 + k_y^2 + k_z^2) \tag{3.54}$$

Solutions to these equations finally lead to

$$\psi(\mathbf{r}, t) = \beta \exp\left[-\frac{i}{\hbar}(Et - \mathbf{k}\cdot\mathbf{r})\right] \tag{3.55}$$

with the constants E, $\mathbf{k}$ satisfying the condition in Eq. (3.54). The following points should be noted about this solution:

1. Since the operators corresponding to energy and momentum, $i\hbar\frac{\partial}{\partial t}$ and $-i\hbar\nabla$, operating on this solution give the wave function back but multiplied by constants E and $\mathbf{k}$ respectively, the solutions describe a particle with energy E and momentum $\mathbf{k}$.
2. The solution is not normalizable [see Eq. (3.17)] since $|\psi| = |\beta|$ and $\int |\psi|^2\, dV = \infty$. Nevertheless, it can be used for describing relative probabilities, the probability of finding the particle anywhere being the same. The wave function can be interpreted as describing a beam of noninteracting particles with momentum $\mathbf{k}$, and with $|\beta|^2$ number of particles per unit volume.
3. Since Eq. (3.43) is linear, any superposition of solutions in Eq. (3.55) is also a solution of Eq. (3.43), *i.e.* the general solution can be written as

$$\psi(\mathbf{r}, t) = \frac{1}{h^{3/2}}\int \exp\left[-\frac{i}{\hbar}(Et - \mathbf{k}\cdot\mathbf{r})\right] F(\mathbf{k})\, d^3k \tag{3.56}$$

with E given by Eq. (3.54).

3.6 WAVE PACKET AND THE UNCERTAINTY PRINCIPLE

A wave packet is superposition, as in Eq. (3.56), of plane-wave solutions with nearly the same momenta, so as to give a wave function which is localized in space. Such a wave function may be written in the form

$$\psi(\mathbf{r}, t) = \frac{1}{h^{3/2}} \int f(\mathbf{k} - \mathbf{k}_0) \exp\left[-\frac{i}{\hbar}(k^2 t/2m - \mathbf{k}\cdot\mathbf{r})\right] d^3k \tag{3.57}$$

where $f(k - k_0)$ is significantly nonzero only in a small region about $\mathbf{k} \approx \mathbf{k}_0$. There are some general properties of the wave packet which are demonstrated here by taking the Gaussian form for $f(\mathbf{k} - \mathbf{k}_0)$. For

$$f(\mathbf{k} - \mathbf{k}_0) = \frac{1}{\pi^{3/4}(b\hbar)^{3/2}} \exp\left(-\frac{(\mathbf{k} - \mathbf{k}_0)^2}{2\hbar^2 b^2}\right) \tag{3.58}$$

the wave packet $\psi(\mathbf{r}, t)$ is obtained from Eq. (3.57) by changing the variable of integration to $q = \mathbf{k} - \mathbf{k}_0$, and integrating

$$\psi(\mathbf{r}, t) = \exp\left[-\frac{i}{\hbar}\left(\frac{k_0^2}{2m}t - k_0.r\right)\right] u(\mathbf{r}, t) \tag{3.59}$$

$$u(\mathbf{r}, t) = \frac{b^{3/2} \exp\left[-\frac{b^2}{2}\left(\mathbf{r} - \frac{\mathbf{k}_0}{m}t\right)^2 / (1 + it\hbar b^2/m)\right]}{\pi^{3/4}(1 + it\hbar b^2/m)^{3/2}} \tag{3.60}$$

Thus, the wave packet is a product of a plane wave with momentum $\mathbf{k}_0$, and an envelope which is peaked at $\mathbf{r} = \frac{\mathbf{k}_0}{m}t$. The phase moves with velocity $\mathbf{k}_0.2m$, which is called the *phase velocity*, and the envelope moves with velocity $\mathbf{k}_0/m$, which is called the *group velocity*. Since the envelope determines the location of the particle, it is the group velocity which corresponds to the classical velocity of the particle.

The wave packet brings out an important principle regarding the determination of the position and momentum of a particle. It can be seen from Eq. (3.60) that the wave packet at $t = 0$ is significantly nonzero only for $|x| \lesssim \frac{1}{b}$, so the spread in the x-component of position of the particle is

$$\Delta x \approx \frac{1}{b} \tag{3.61}$$

Similarly, $f(\mathbf{k}-\mathbf{k}_0)$, which is essentially the wave function *i* the momentum space, has nonzero values only for $|(\mathbf{k}-\mathbf{k}_0)x| \lesssim b\hbar$, so that the spread in the x-component of momentum of the particle is

$$\Delta p_x \approx b\hbar \tag{3.62}$$

From Eqs. (3.61) and (3.62), one gets

$$(\Delta x)\,(\Delta p_x) \approx \hbar \tag{3.63}$$

Similar results are valid for the measurements of y or z components. Thus, there is an inherent *uncertainty* in the determination of the position and momentum of a particle. The position and momentum of a particle cannot be simultaneously determined with infinite accuracy. The product of the uncertainties or allowed errors in their measurements must satisfy the Heisenberg uncertainty principle whose special case is stated in Eq. (3.63). According to the *Heisenberg uncertainty principle, the product of the uncertainties in the values of two canonically conjugate variables whose operators are hermitian, cannot be less than* $\hbar$ *in the order of magnitude*. Examples of canonically conjugate variables are (x, p_x), (y, p_y) and (z, p_z). A similar relation for time and energy results from the analysis of the response of a state to a time dependent interaction,

$$(\Delta t)\,(\Delta E) \approx \hbar \tag{3.64}$$

The interpretation of this relation is that if it takes time Δt to measure the energy of a system, there is an inherent uncertainty in the measured value of the energy, given by Eq. (3.64). In particular, this relation implies that unstable particles with lifetime τ have an associated uncertainty in their energy, of order $\hbar/\tau$.

The Heisenberg uncertainty principle can be made into a quantitative statement if Δx, Δp_x, etc. are defined as the standard deviations, *i.e.* $\Delta x = \langle (x-\overline{x})^2\rangle^{1/2}$, $\Delta p_x = \langle (p_x - \overline{p}_x)^2\rangle^{1/2}$, etc. It can then be shown rigorously that

$$(\Delta x)\,(\Delta p_x) \geq \hbar/2 \tag{3.65}$$

a result which is valid for pairs of canonically conjugate, hermitian operators (see *e.g.* Ref. 18).

The Heisenberg uncertainty principle can be easily demonstrated by the thought experiment of Sec. 3.1, which may be regarded as an experiment for determining the position and the momentum of the particle. The position of the particle in the experiment has an uncertainty of

$$\Delta x \approx a \tag{3.66}$$

since it is not known whether the particle passed through slit S_1 or S_2. Similarly, the momentum of the particle also is undetermined to the extent

$$\Delta p_x \approx \frac{w}{d} p \tag{3.67}$$

where w is the width of the central fringe. This uncertainty results from the fact that the particle may come to any point within this fringe. Since $w = \frac{\lambda d}{a}$, and $p = h/\lambda$ by the de Broglie relation, we get

$$(\Delta x)\ (\Delta p_x) \approx h \tag{3.68}$$

which is the same as Eq. (3.63) in order of magnitude. This demonstration brings out the fact that the Heisenberg uncertainty principle is essentially a consequence of associating wave properties with the particles.

3.7 STEP POTENTIAL

As the first example of nonzero potentials, the one-dimensional problem of a article which comes across a sudden change in the potential is considered. The potential may be approximated by

$$\begin{aligned} V(x) &= 0 \qquad \text{for } x < 0 \\ &= V \qquad \text{for } x \geq 0 \end{aligned} \tag{3.69}$$

as shown in Fig. 3.2(a).

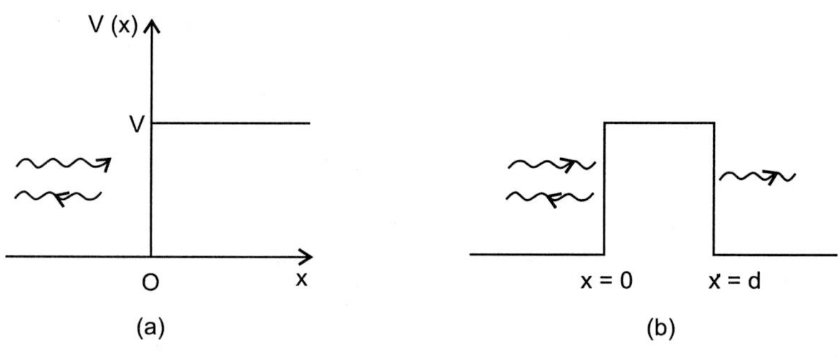

Fig. 3.2 (a) A step potential, (b) A potential barrier.

This Schrödinger equation in one dimension, is

$$i\hbar \frac{\partial \psi (x, t)}{\partial t} = \left(-\frac{\hbar^2}{2m} \frac{\partial^2}{\partial x^2} + V(x) \right) \psi (x, t) \tag{3.70}$$

As before, for states with energy E,

$$y\ (x, t) = \exp\left(-iEt/\hbar\right) \phi\ (x) \tag{3.71}$$

where ϕ (x) satisfies the equation [see Eq. (3.24)]

$$-\frac{\hbar^2}{2m}\frac{d^2\phi(x)}{dx^2} = [E - V(x)]\,\phi(x) \tag{3.72}$$

The solution for $\phi(x)$, for $x < 0$, is

$$\phi(x) = a_+ e^{ipx} + a_- e^{-ipx},\ x < 0 \tag{3.73}$$

with
$$p = \frac{1}{\hbar}(2mE)^{1/2} \tag{3.74}$$

where the first term in the solution corresponds to a particle with momentum $p\hbar$, and the second term to a particle with momentum $-p\hbar$. The solution for $x \geq 0$ is

$$\phi_r(x) = b_+ e^{iqx} + b_- e^{-iqx};\ x \geq 0 \tag{3.75}$$

with
$$q = \frac{1}{\hbar}[2m(E - V)]^{1/2} \tag{3.76}$$

Now, since the potential is piece-wise continuous and finite, it follows from the properties of the differential equation (3.72), that the wave function $\phi(x)$ and its first derivative $\left[\frac{d\phi}{dx}\right]$ are continuous everywhere, in particular at $x = 0$. Therefore, one has

$$a_+ + a_- = b_+ + b_- \tag{3.77}$$

$$a_+ - a_- = \frac{q}{p}(b_+ - b_-) \tag{3.78}$$

The solutions are discussed separately for the two qualitatively different cases, (*i*) $E \geq V$, and (*ii*) $E < V$.

Case (*i*) For $E \geq V$, it is assumed that the particle approaches the barrier from the left and is either transmitted or reflected at $x = 0$. Hence, for $x > 0$, there is only a wave function describing a particle moving to the right which implies that

$$b_- = 0 \tag{3.79}$$

Therefore Eqs. (3.77) and (3.78) give

$$b_+ = \left[\frac{2p}{p+q}\right]a_+$$

$$a_- = \left[\frac{p-q}{p+q}\right]a_+ \tag{3.80}$$

In analogy with the transmission and reflection of classical electromagnetic waves, transmission and reflection coefficients can be defined as:

$$T = \frac{q}{p}\left|\frac{b_+}{a_+}\right|^2 = \frac{4pq}{(p+q)^2}$$

$$R = \left|\frac{a_-}{a_+}\right|^2 = \left[\frac{p-q}{p+q}\right]^2 \tag{3.81}$$

In terms of the refractive index n,

$$n = \frac{p}{q} \tag{3.82}$$

Eqs. (3.81) can be written as

$$T = \frac{4n}{(n+1)^2}$$

$$R = \left[\frac{n-1}{n+1}\right]^2 \tag{3.83}$$

which are the same as the classical transmission and reflection coefficients for electromagnetic waves.

Case (*ii*) For $E < V$, q is imaginary. Writing

$$q = i\alpha \tag{3.84}$$

where
$$\alpha = \frac{1}{\hbar}\,[2m\,(V - E)]^{1/2} \tag{3.85}$$

the solution for $x > 0$, is

$$\phi_r(x) = b_+ e^{-\alpha x} + b_- e^{\alpha x} \tag{3.86}$$

In order to keep the probability finite as $x \to \infty$,

$$b_- = 0 \tag{3.87}$$

The continuity equations (3.77) and (3.78) then imply that

$$b_+ = \left[\frac{2p}{p+i\alpha}\right] a_+$$

$$a_- = \left[\frac{p-i\alpha}{p+i\alpha}\right] a_+ \tag{3.88}$$

This means that $|a_-| = |a_+|$ and

$$\phi(x) = 2a_+ e^{-i\delta} \cos(px + \delta),\ x < 0 \tag{3.89}$$

where
$$\delta = \tan^{-1}(\alpha/p) \tag{3.90}$$

i.e. $\phi(x)$ represents a standing wave.

There are several significant points about the results, that should be noted:

1. Since $\int_{-\infty}^{\infty} |\psi|^2\, dx = \infty$, the wave function is not normalizable.

   However, it could be used to describe a beam of particles with a density of $|a_+|^2$ per unit length, moving to the right, which get either transmitted or reflected at $x = 0$.

2. For $E > V$, *i.e.* case (*i*), the rate of flow of the incoming particles must be equal to the sum of the rates of flow of transmitted and reflected particles. Since the momenta of these particles are p, q and $-p$ respectively, one has the condition

$$p|a_+|^2 = q\,|b_+|^2 + p|a_-|^2 \tag{3.91}$$

   which is equivalent to the requirement that $T + R = 1$ and is easily seen to be satisfied by the relations is Eq. (3.83).

3. For $E < V$, *i.e.* case (*ii*), a standing wave is obtained for $x < 0$, which means that all the incoming particles are reflected. However, the wave function is nonzero for $x > 0$ though it vanishes exponentially as $x \to \infty$. Thus, there is a finite probability of finding the particles in the region $x > 0$ which is a forbidden region in classical mechanics. This is called *barrier penetration* and has no classical analogue in the mechanics of particles (such a phenomenon was observed in optics by Newton). It does not however lead to a paradox since localization or observation of the particle in the classically forbidden region involves a change in the momentum and the energy of the particle which may then have sufficient energy to make this a classically allowed region.

   Electrons in a metal are a case with $E < V$ but when metal is heated it satisfies condition $E > V$ and hence we see thermionic emission. This is used in cathode ray tubes.

4. A special case of interest is the one of $V \to \infty$ in case (*ii*) for which $\alpha \to \infty$ and the wave function is given by

$$\phi(x) = 2ia_+ \sin(px), \qquad \text{for } x < 0 \tag{3.92}$$

$$\phi_r(x) = 0, \qquad \text{for } x \geq 0 \tag{3.93}$$

This wave function could have been obtained from Eqs. (3.73) and (3.86) with $b_- = 0$, by just requiring that the wave function vanishes at $x = 0$, $\phi(0) = \phi_r(0) = 0$, but no further conditions on $d\phi/dx$. Indeed this prescription considerably simplifies that calculations whenever the potentials jump to infinity.

5. The quantum mechanical phenomenon of a particle penetrating classically forbidden barriers gives rise to an interesting observation of trapped particles escaping through classically forbidden barriers. Consider a situation [Fig. 3.2(*b*)] where the barrier exists only for $0 \leq x \leq d$, *i.e.* $V(x) = V$ for $0 \leq x \leq d$ and $V(x) = 0$ elsewhere. In this case, if a beam of particles, is incident from the left with energy $E < V$, the wave function is given by

$$\begin{aligned} \phi(x) &= a_+ e^{ipx} + a_- e^{-ipx} && \text{for } x < 0, \\ &= b_+ e^{-\alpha x} + b_- e^{\alpha x} && \text{for } 0 \leq x \leq d, \\ &= c_+ e^{ipx} && \text{for } x > d \end{aligned} \tag{3.94}$$

where p and α are given in Eqs. (3.74) and (3.85) respectively. Continuity of the wave function and its derivative at $x = 0$ and $x = d$, allows us to determine a_-, b_+, b_- and c_+ in terms of a_+. Since the particles can penetrate the forbidden barrier, in general b_+, b_- and c_+ are nonzero. Thus, the particles can cross a barrier even if classically, the energy is insufficient to pass over the barrier, and the probability of transmission is given by the ratio $|c_+|^2/|a_+|^2$. This effect is termed as *tunnelling* and provides a satisfactory explanation for the decay of unstable particles (*e.g.* U_{235}, etc.) as a tunnelling of trapped particles through a potential barrier.

A **scanning tunneling microscope** (STM) is an instrument for imaging surfaces at the atomic level. It is based on the concept of **quantum tunneling.** When a conducting tip is brought very near to the surface to be examined, a bias (voltage difference) applied between the two can allow electrons to tunnel through the vacuum between them. The resulting *tunneling current* is a function of tip position, applied voltage, and the local density of states (LDOS) of the sample. Information is acquired by monitoring the current as the tip's position scans across the surface. For an STM, good resolution is considered to be 0.1 nm lateral resolution and 0.01 nm depth resolution. With this resolution, individual atoms within materials are routinely imaged and manipulated. (**Source : Wikipedia**)

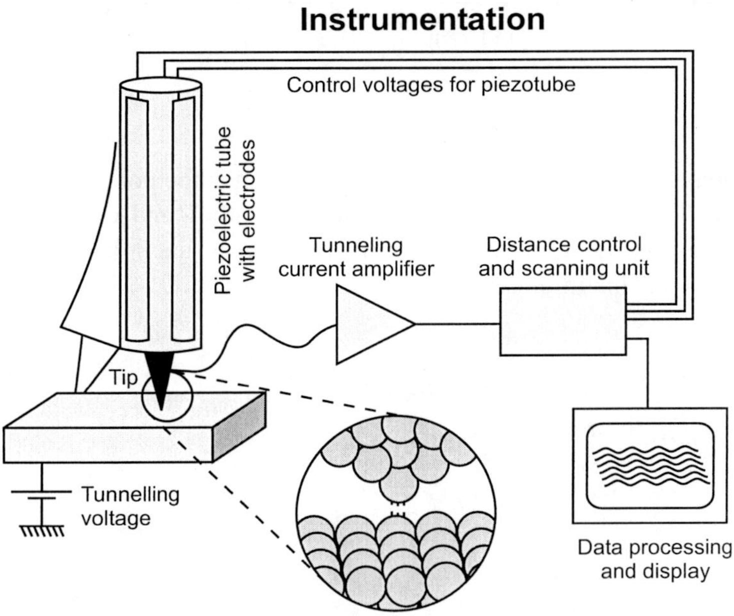

Fig. 3.3

Fig. 3.4 Image of reconstruction on a clean Gold surface

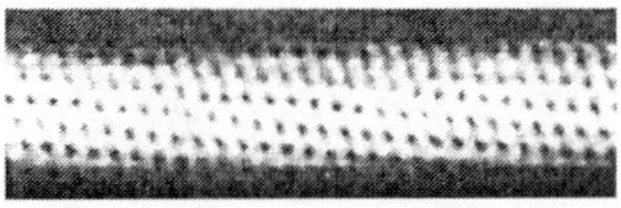

Fig. 3.5 An image of single-walled carbon nanotube

3.8 PARTICLE IN A BOX

An example of a particle in a potential which allows only discrete values for the energy of the particle is now described.

Consider a particle in a one-dimensional potential which is zero for $0 \le x \le l$, and is infinite for $x < 0$ or $x > l$ (see Fig. 3.3). The Schrödinger equation for the particle with well-defined energy is

$$-\frac{\hbar^2}{2m}\frac{d^2\phi(x)}{dx^2} = E\phi(x),\ 0 \le x \le l \tag{3.95}$$

whose solutions can be written as

$$\phi(x) = a \sin(px + \alpha) \tag{3.96}$$

$$p = \frac{1}{\hbar}(2mE)^{1/2} \tag{3.97}$$

The wave function is zero for $x < 0$ or $x > l$ since the potential is infinite in this region. Since the potential jumps to infinity at $x = 0$ and $x = l$, the boundary conditions as discussed in Sec. 3.7 are that the wave function should vanish at $x = 0$ and $x = l$. This implies that

$$\sin \alpha = 0, \tag{3.98}$$

$$pl = n\pi, \; n = 1, 2, \ldots \tag{3.99}$$

Therefore, the solutions are

$$\phi_n(x) = \left(\frac{2}{l}\right)^{1/2} \sin\left(\frac{n\pi}{l}x\right), \; n = 1, 2, \ldots \tag{3.100}$$

$$E_n = \frac{\hbar^2\pi^2}{2ml^2}n^2 \tag{3.101}$$

where $a = (2/l)^{1/2}$ has been used as required by the normalization condition in Eq. (3.19)

$$|a|^2 \int_0^l \sin^2\left(\frac{n\pi}{l}x\right) dx = 1 \tag{3.102}$$

Some of the significant points to note are as follows:

1. A state with $n = 0$ is not acceptable since this would correspond to a state which is zero everywhere. The lowest energy state, called the *ground state*, therefore corresponds to $n = 1$, and has a nonzero energy.

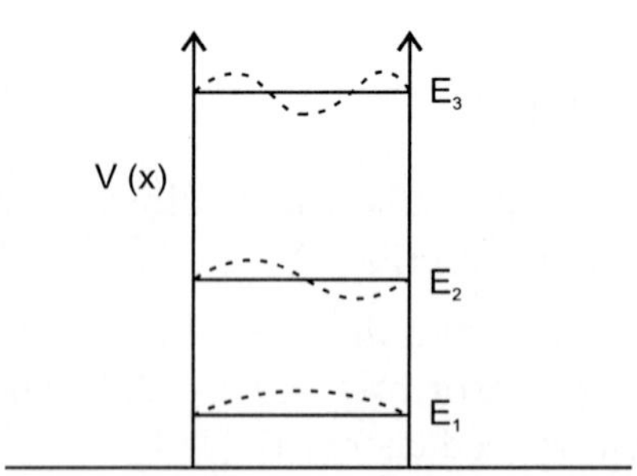

Fig. 3.6 Energy levels and wave functions (dashed lines) for a particle inside a box.

2. The discreteness of the energy levels is significant only for small m and l. For example, if $m \approx 10^{-3}$ kg and $l \approx 0.1$ m, the separation between the energy levels is of the order of 10^{-62} J which is quite negligible. On the other hand, for an electron in an atom, $m \approx 10^{-30}$ kg and $l \approx 10^{-10}$ m, so that $\Delta E_n \sim (60n)$ eV and the discreteness becomes important.

3. It is seen that the states with $n \geq 2$ have nodes inside the box. Since probability density is given by $|\phi_n(x)|^2$, this means that there are some regions where the particle will not be found, which is totally incompatible with the classical ideas of trajectories.

4. If the potential in the region $x < 0$ and $x > l$, is not infinite but finite, the wave function can penetrate into this region. Therefore, it is not forced to be zero at $x = 0$ or $x = l$. Hence, the wave function varies more gently inside the region $0 \leq x \leq l$, and the energies, which are related to the second derivative of the wave function are lower than in the case where the potential is infinite for $x < 0$ and $x > l$.

5. It is observed that

$$\int_0^l \phi_n(x)\, \phi_{n'}(x)\, dx = 0, \quad \text{for } n \neq n' \tag{3.103}$$

$$\int_0^l \phi_n(x)\, \phi_n(x)\, dx = 1 \tag{3.104}$$

which can together be written as

$$\int_0^l \phi_n(x)\, \phi_{n'}(x)\, dx = \delta_{n,n'} \tag{3.105}$$

where the *Kronecker delta* $\delta_{n,n'}$ is 1 for $n = n'$ and zero otherwise. Thus, these states are orthonormal. It can also be shown that any general state of a particle in the box can be written as a linear combination of the energy eigenstates, *i.e.*

$$\psi(x) = \sum_{n=1}^{\infty} a_n \phi_n(x)$$

$$\sum_{n=1}^{\infty} |a_n|^2 = 1 \tag{3.106}$$

which means that the eigenstates $\phi_n(x)$ are complete. The orthonormality and completeness are important properties associated with the eigenstates of any physical observable (see Sec. 3.4).

Due to significant progress in semiconductor technology, now we have systems known as quantum dots and quantum wells which are particles confined in a small volume.

A **quantum dot** is a portion of matter (*e.g.* semiconductor) **whose excitons** (electron-hole pair) are confined in all three spatial dimensions. As a result, such **materials have**-electronic properties intermediate between those of bulk semiconductors and those of discrete molecules.

Quantum dot technology is a good candidate for use in solid-state quantum computation. By applying small voltages to the leads, the flow of electrons through the quantum dot can be controlled and thereby precise measurements of the spin and other properties therein can be made. With several entangled quantum dots or qubits, plus a way of performing operations, quantum calculations and the computers that would perform them might be possible.

A **quantum well** is a potential well with only discrete energy values. One way to create quantization is to confine particles, which were originally free to move in three dimensions, to two dimensions, forcing them to occupy a planar region. The effects of quantum confinement take place when the quantum well thickness becomes comparable to the de Broglie wavelength of the carriers (generally electrons and holes), leading to energy levels called "energy bands", *i.e.,* the carriers can only have discrete energy values.

Fig. 3.7 Researchers at Los Alamos National Laboratory have developed a wireless device that efficiently produces visible light, through energy transfer from thin layers of quantum wells to crystals above the layers.

Quantum wells are in wide use in diode lasers, including red lasers for DVDs and laser pointers, infra-red lasers in fiber optic transmitters, or in blue lasers. They are also used to make **HEMTs** (High Electron Mobility Transistors), which are used in low-noise electronics. Quantum well infra-red photodetectors are also based on quantum wells, and are used for infrared imaging.

A **quantum well laser** is a laser diode in which the active region of the device is so narrow that quantum confinement occurs. The wavelength of the light emitted by a quantum well laser is determined by the width of the active region rather than just the bandgap of the material from which it is constructed. This means that much shorter wavelengths can be obtained from quantum well lasers than from conventional laser diodes using a particular semiconductor material. The efficiency of a quantum well laser is also greater than a conventional laser diode.

(*Source : Wikipedia*)

3.9 SIMPLE HARMONIC OSCILLATOR

A problem of considerable importance is that of the simple harmonic oscillator. For most oscillating systems, it can be used as the first approximation.

For a state of well-defined energy, the time independent Schrödinger equation for a simple harmonic oscillator in 1-dimension, is

$$-\frac{\hbar^2}{2m}\frac{d^2\phi(x)}{dx^2}+\frac{1}{2}kx^2\phi(x) = E\phi(x) \tag{3.107}$$

For obtaining the solutions to this equation, the asymptotic behaviour of $\phi(x) \sim \exp(-\alpha^2x^2/2)$ where $\alpha^4 = mk/\hbar^2$, is first separated out by writing

$$\phi(x) = \exp(-\alpha^2x^2/2)\,\eta(x) \tag{3.108}$$

where $\eta(x)$ satisfies the equation

$$\frac{d^2\eta(x)}{dx^2} - 2\alpha^2x\frac{d\eta(x)}{dx} = \left(\alpha^2 - \frac{2mE}{\hbar^2}\right)\eta(x) \tag{3.109}$$

Substitution of a series solution for $\eta(x)$ into Eq. (3.109) and equating the coefficients of x^k gives

$$\eta(x) = \sum_{k=0}^{\infty} b_k\, x^k$$

$$(k+2)(k+1)\, b_{k+2} = \left(2\alpha^2k + \alpha^2 - \frac{2mE}{\hbar^2}\right)b_k \tag{3.110}$$

For a general value of E, the infinite series for $n(x)$ gives an asymptotically increasing solution $\phi(x) \sim \exp(\alpha^2x^2/2)$ which is not normalizable. However, for some special values of $E = (n + 1/2)\dfrac{\alpha^2\hbar^2}{m}$, n a positive integer, the series in Eq. (3.110) terminates at $k = n$ and we get normalizable wave functions in terms of *Hermite polynomials* H_n of order n. The wave functions and their energies are

$$\phi_n(x) = \pi^{-1/4}\left(\frac{\alpha}{2^n\, n!}\right)^{1/2} \exp(-\alpha^2x^2/2)\, H_n(\alpha x)$$

$$E_n = \left(n + \frac{1}{2}\right)\hbar\omega,\ n = 0, 1, 2, \ldots \tag{3.111}$$

where $\omega = (k/m)^{1/2}$. It is observed that again the ground state energy is not zero. Its value of $\frac{1}{2}\hbar\omega$ is called the *zero-point energy.*

The first three Hermite polynomials are

$$\begin{aligned} H_0(\alpha x) &= 1 \\ H_1(\alpha x) &= 2\alpha x \\ H_2(\alpha x) &= 4\alpha^2x^2 - 2 \end{aligned} \tag{3.112}$$

The harmonic oscillator problem can be solved more elegantly by using operator algebra. Defining

$$a = \left(\frac{m\omega}{2\hbar}\right)^{1/2}\left(x + \frac{\hbar}{h\omega}\frac{\partial}{\partial x}\right)$$

$$a\dagger = \left(\frac{m\omega}{2\hbar}\right)^{1/2}\left(x - \frac{\hbar}{m\omega}\frac{\partial}{\partial x}\right) \tag{3.113}$$

it can easily be shown that

$$\begin{aligned} aa\dagger - a\dagger a &= 1 \\ aH - Ha &= \hbar\omega a \\ a\dagger H - Ha\dagger &= -\hbar\omega a^{+} \end{aligned} \tag{3.114}$$

with the Hamiltonian H (*i.e.* energy) being

$$H = -\frac{\hbar^2}{2m}\frac{d^2}{dx^2} + \frac{1}{2}kx^2 \tag{3.115}$$

Therefore if $\psi_0(x)$ is the ground state with energy E_0, then using Eq. (3.114)

$$Ha\psi_0(x) = (E_0 - \hbar\omega)\, a\psi_0(x) \tag{3.116}$$

$$Ha\dagger\psi_0(x) = (E_0 + \hbar\omega)\, a\dagger\psi_0(x) \tag{3.117}$$

Since $\psi_0(x)$ is the ground state, this implies that

$$a\psi_0(x) = 0 \tag{3.118}$$

and $a\dagger\psi_0(x)$ is an eigenstate with energy $E_0 + \hbar\omega$. It is easy to solve for $\psi_0(x)$ from Eqs. (3.118) and (3.113) which give

$$\psi_0(x) = \pi^{-1/4}\, \alpha^{1/2} \exp(-\alpha^2x^2/2) \tag{3.119}$$

for the normalized ground state wave function, which with the help of Eq. (3.107) can be shown to have an energy $\frac{1}{2}\hbar\omega$. The eigenstates with energies $E_0 + n\hbar\omega$ can be developed with the repeated operation of $a\dagger$ as in Eq. (3.117).

$$\psi_n(x) = \frac{1}{(n!)^{1/2}}(a\dagger)^n\,\psi_0(x),\, n \geq 0$$

$$E_n = \left(n + \frac{1}{2}\right)\hbar\omega \tag{3.120}$$

The normalization is obtained by the repeated use of the first relation in Eq. (3.114) and Eq. (3.118). It is clear from the definition of $a\dagger$, tht $\psi_n(x)$ are alternatively even and odd functions of x.

3.10 SMALL PERTURBATIONS

It is often the situation that a realistic problem we encounter cannot be solved exactly but differs slightly from a solvable problem. If the difference is small, approximate solutions can be obtained by using perturbation theory.

Consider a situation where the solutions to the eigenvalue equation for the energy are known,

$$H_0\phi_n = E_n\phi_n \tag{3.121}$$

where ϕ_n are normalized, but the solution to the following equation is to be found,

$$(H_0 + \lambda V)\psi = E\psi \tag{3.122}$$

where λ is small and E is close to E_n. Multiplying Eq. (3.122) by ϕ_n^* and integrating over the space τ,

$$\int\phi_n^*\,(H_0 + \lambda V)\,\psi\,d\tau = E\int\phi_n^*\,\psi\,d\tau \tag{3.123}$$

One integrating the first term by parts, the Hamiltonian H_0 operates on ϕ_n^* giving $E_n\phi_n^*$ (H_0 is real), which then leads to

$$E = E_n + \lambda\frac{\int\phi_n^*\,V\psi\,d\tau}{\int\phi_n^*\,\psi\,d\tau} \tag{3.124}$$

This is an exact relation. An approximate expression for E can be obtained by noting that the second term is small (λ is small), and therefore ψ can be replaced by ϕ_n in this term, which leads to

$$E \approx E_n + \lambda \int \phi_n^* \, V \, \phi_n \, d\tau \tag{3.125}$$

For obtaining more accurate expressions for E, better approximations for ψ should be used.

The expression in Eq. (3.125) is valid provided ϕ_n is an isolated state. If there are more than one degenerate states corresponding to the energy level E_n, say $\phi_n^{(i)}$ ($i = 1, 2, ...$) which are orthonormal, ψ may be approximated by a linear combination of the degenerate states. Writing

$$\psi = \sum_{i=1} a_i \, \phi_n^{(i)} \tag{3.126}$$

where $\sum_i |a_i|^2 = 1$, Eq. (3.124) gives

$$\sum_i (\lambda \int \phi_n^{(j)*} \, V \phi_n^{(i)} \, d\tau) \, a_j = (E - E_n) \, ai \tag{3.127}$$

Thus, $(E - E_n)$ and a_i are obtained from this set of equations. For example, if the degeneracy is of order two, *i.e.* $i = 1, 2$, the solutions are

$$E = E_n + \lambda \, (V_{11} + xV_{12}) \tag{3.128}$$

where $x = a_2/a_1$ is

$$x = \frac{V_{22} - V_{11} \pm [(V_{22} - V_{11})^2 + 4V_{12} \, V_{21}]^{1/2}}{2V_{12}} \tag{3.129}$$

$$V_{ij} = \int \phi_n^{(i)*} \, V \, \phi_n^{(j)} \, d\tau \tag{3.130}$$

The degenerate states in this case split into two levels corresponding to the two states given by the two values of x.

3.11 ANGULAR MOMENTUM

In the discussion so far, emphasis has been on the energy and momentum of the particle. However, for a particle in the presence of a rotationally invariant 3-dimensional potential, the angular momentum of the particle plays an important role. The operator corresponding to the angular momentum observable has some interesting properties which are discussed briefly.

The angular momentum of a particle is given by

$$\mathbf{L} = \mathbf{r} \times \mathbf{p} \tag{3.131}$$

while its square is given by

$$\mathbf{L}^2 = (\mathbf{r} \times \mathbf{p}) \cdot (\mathbf{r} \times \mathbf{p}) \tag{3.132}$$

Here, since **r** and **p** are operators, their order must be maintained. The expression for $\mathbf{L}^2$ simplifies to

$$\mathbf{L}^2 = \mathbf{r}.(\mathbf{p} \times (\mathbf{r} \times \mathbf{p})) \tag{3.133}$$

$$= \sum_{i,j} [r_i p_j \, r_i p_j - r_i i_j \, r_j p_i]$$

and with $\mathbf{r.p} = -i\hbar r \dfrac{\partial}{\partial r}$ Eq. (3.133) becomes

$$\mathbf{L}^2 = r^2 p^2 + 2\hbar^2 r \frac{\partial}{\partial r} + \hbar^2 r^2 \frac{\partial^2}{\partial r^2} \tag{3.134}$$

Using this relation the kinetic energy T can be written as

$$T = \frac{1}{2m} \mathrm{p}^2$$

$$= -\frac{\hbar^2}{2mr^2} \frac{\partial}{\partial r} r^2 \frac{\partial}{\partial r} + \frac{\mathrm{L}^2}{2mr^2} \tag{3.135}$$

which shows that the angular momentum is an important term in the kinetic energy.

The expressions for the angular momentum operators in terms of spherical coordinates are obtained from Eq. (3.131) as

$$L_x = i\hbar \left(\sin\phi \frac{\partial}{\partial\theta} + \cos\phi \cot\theta \frac{\partial}{\partial\phi} \right)$$

$$L_y = -i\hbar \left(\cos\phi \frac{\partial}{\partial\theta} - \sin\phi \cot\theta \frac{\partial}{\partial\phi} \right) \tag{3.136}$$

$$L_z = -i\hbar \frac{\partial}{\partial\phi}$$

and

$$\mathbf{L}^2 = -i\hbar \left[\frac{1}{\sin\theta} \frac{\partial}{\partial\theta} \left(\sin\theta \frac{\partial}{\partial\theta} \right) + \frac{1}{\sin^2\theta} \frac{\partial^2}{\partial\phi^2} \right] \tag{3.137}$$

The wave functions corresponding to well-defined values of $\mathbf{L}^2$, satisfy the equation

$$-i\hbar \left[\frac{1}{\sin\theta} \frac{\partial}{\partial\theta} \left(\sin\theta \frac{\partial}{\partial\theta} \right) + \frac{1}{\sin^2\theta} \frac{\partial^2}{\partial\phi^2} \right] Y(\theta, \phi) = \lambda Y(\theta, \phi) \tag{3.138}$$

As before, factorizable form for the solution is assumed,

$$Y(\theta, \phi) = P(\theta)\, F(\phi) \tag{3.139}$$

Substituting this in Eq. (3.138) and multiplying by $(\sin^2\theta)/\hbar^2 Y(\theta, \phi)$, gives

$$-\frac{1}{F(\phi)}\frac{\partial^2 F}{\partial \phi^2} = \frac{\sin\theta}{P(\theta)}\frac{\partial}{\partial\theta}\left(\sin\theta\frac{\partial}{\partial\theta}P(\theta)\right) + \frac{\lambda}{\hbar^2}\sin^2\theta \tag{3.140}$$

Since the two sides depend on different variables, each must be a constant, say m^2, so that

$$\frac{d^2F(\phi)}{d\phi^2} + m^2 F(\phi) = 0 \tag{3.141}$$

$$\frac{1}{\sin\theta}\frac{d}{d\theta}\left(\sin\theta\frac{dP(\theta)}{d\theta}\right) - \frac{m^2}{\sin^2\theta}P(\theta) + \frac{\lambda}{\hbar^2}P(\theta) = 0 \tag{3.142}$$

The solutions to the Eq. (3.141) are

$$F(\phi) = e^{im\phi} \tag{3.143}$$

However, if the condition that the wave function at every physical point must be single-valued is imposed, then $F(\phi) = F(\phi + 2\pi)$ which means that the values of m are restricted to $m = 0, \pm 1, \pm 2$ etc. It is easily seen that

$$-i\hbar\frac{\partial}{\partial\phi}F(\phi) = m\hbar\, F(\phi),\ m = 0, \pm 1, \pm 2, \ldots \tag{3.144}$$

which implies that the state is an eigenstate of L_z with eigenvalue $m\hbar$. For obtaining solutions to Eq. (3.142), the substitution, $v = \cos\theta$ is first used. Then, the equation for $m = 0$ reduces to *Legendre's differential equation*. For $m = 0$, substitution of a series solution for $P(\theta)$ into Eq. (3.142) and equating the coefficients of similar terms, gives

$$P(\theta) = \Sigma\, b_k v^k,\ v = \cos\theta \tag{3.145}$$

$$(k+1)(k+2)\, b_{k+2} = \left(k^2 + k - \frac{\lambda}{\hbar^2}\right)b_k$$

For an arbitrary value of λ, the series diverges at $v = \pm 1$. However, for $\lambda = l(l+1)\hbar^2$, l a positive integer, the series terminates at $k = l$ and we get well-behaved solutions:

$$\lambda = l(l+1)\hbar^2, \qquad l = 0, 1, 2, \ldots \tag{3.146}$$

$$P_l(v) = \frac{1}{2^l\, l!}\frac{d^l}{dv^l}(v^2 - 1)^l,\ v = \cos\theta$$

These are *Legendre Polynomials* of order l, the first few of them being

$$\begin{aligned} P_0(\cos\theta) &= 1 \\ P_1(\cos\theta) &= \cos\theta \\ P_2(\cos\theta) &= \frac{1}{2}(3\cos^2\theta - 1) \end{aligned} \tag{3.147}$$

The solutions for $m \neq 0$ are somewhat more complicated, and for $l \geq m \geq 0$, are given by

$$P_l^m(v) = (1-v^2)^{m/2} \frac{d^m}{dv^m} P_i(v), \; v = \cos\theta \tag{3.148}$$

called the *associated Legendre functions*. Combining these solutions with those in Eq. (3.143) gives the solutions to Eq. (3.138) as

$$Y_l^m(\theta, \phi) = \left[\frac{(2l+1)}{4\pi} \frac{(l-m)!}{(l+m)!}\right]^{1/2} (-1)^m e^{im\phi} P_l^m(\cos\theta) \tag{3.149}$$

with $\lambda = l(l+1)\hbar^2$, l and m being integers, and $l \geq m$. $Y_l^m(\theta, \phi)$ are called *spherical harmonics*, and are defined for negative integers m by the relation

$$Y_l^m = (-1)^m (Y_l^{-m})^* \tag{3.150}$$

Their normalization is chosen such that they are orthonormal,

$$\int Y_l^{m*}(\theta, \phi)\, Y_{l'}^{m'}(\theta, \phi)\, d\cos\theta\, d\phi = \delta_{l,l'}\, \delta_{m,m'} \tag{3.151}$$

They are simultaneous eigenfunctions of L_z and $\mathbf{L}^2$ since

$$\begin{aligned} L_z Y_l^m(\theta, \phi) &= m\hbar\, Y_l^m(\theta, \phi), \; m \leq l \\ \mathbf{L}^2 Y_l^m(\theta, \phi) &= l(l+1)\hbar^2 Y_l^m(\theta, \phi) \end{aligned} \tag{3.152}$$

It is easy to show that they satisfy the important property

$$Y_l^m(\pi - \theta, \phi + \pi) = (-1)^l\, Y_l^m(\theta, \phi) \tag{3.153}$$

i.e. for $\mathbf{r} \rightarrow -\mathbf{r}$, they are even for even l and odd for odd l. The first few of these functions are:

$$\begin{aligned} Y_0^0(\theta, \phi) &= \frac{1}{(4\pi)^{1/2}} \\ Y_1^0(\theta, \phi) &= \left(\frac{3}{4\pi}\right)^{1/2} \cos\theta \\ Y_1^{\pm 1}(\theta, \phi) &= \mp\left(\frac{3}{8\pi}\right)^{1/2} \exp(\pm i\phi)\sin\theta \end{aligned} \tag{3.154}$$

Apart from playing an important role in the discussion of the kinetic energy in spherical coordinates [Eq. (3.135)], the angular momentum operator plays a significant role in determining the rotational energy levels of a rigid rotator. For example, the rotational energy levels of a *di*-atomic molecule are given by the Hamiltonian

$$H = \frac{1}{2I}\mathrm{L}^2 \tag{3.155}$$

where I is the moment of inertia, and the corresponding energy levels are given by

$$E = \frac{1}{2I} l\,(l+1)\hbar^2, \quad l = 0, 1, \ldots \tag{3.156}$$

3.12 EXAMPLES

Here some important properties of quantum mechanical system and their applications are discussed.

Example 1

Since indeterminacy is not an essential part of classical mechanics, it is suggetive that classical measurements may be related to the averages of quantum mechanical measurements. This is illustrated by ***Ehrenfest's*** theorem.

Consider the time-derivative of the average position given by

$$\begin{aligned} \frac{d}{dt}\langle \mathrm{r} \rangle &= \frac{d}{dt}\int \psi^* \mathbf{r}\psi\, d^3r \\ &= \int \psi^* \mathbf{r}\frac{\partial \psi}{\partial t} d^3r + \int\left(\frac{\partial \psi}{\partial t}\right)^* \mathbf{r}\,\psi\, d^3r \end{aligned} \tag{3.157}$$

Using the Schrödinger equation, and cancelling the potential energy terms (potential is real),

$$\frac{d}{dt}\langle \mathbf{r} \rangle = \frac{i\hbar}{2m}\left[\int \psi^* \mathbf{r}\nabla^2\,\psi\, d^3r - \int(\nabla^2\,\psi^*)\,\mathbf{r}\,\psi\, d^3r\right] \tag{3.158}$$

Integrating by parts gives

$$\begin{aligned} \frac{d}{dt}\langle \mathbf{r} \rangle &= \frac{1}{m}\int \psi^*(-i\hbar\,\nabla)\,\psi\, d^3r \\ &= \frac{\langle \mathbf{p} \rangle}{m} \end{aligned} \tag{3.159}$$

which is analogous to the classical result that the momentum is the product of mass and velocity. Proceeding in a similar way, it can be shown that

$$\frac{d\langle \mathbf{p}\rangle}{dt} = -i\hbar\left[\int \psi^* \nabla \frac{\partial \psi}{\partial t} d^3r + \int\left(\frac{\partial \psi^*}{\partial t}\right)\nabla\psi d^3r\right] \tag{3.160}$$

which on using the Schrödinger equation once again and integrating by parts, leads to

$$\frac{d\langle \mathbf{p}\rangle}{dt} = -\int \psi^*(\nabla V)\psi \, d^3r$$

$$= \langle -\nabla V\rangle \tag{3.161}$$

This relation is analogous to Newton's second law in classical mechanics.

It is to be noted that if the uncertainties in the values of the various dynamical quantities can be neglected, Eqs. (3.159) and (3.161) represent the classical behaviour of particles in terms of approximate trajectories.

Example 2

The angular momentum operators provide an interesting illustration of the properties of hermitian operators. It follows from Eq. 3.131) or from Eqs. (3.136) and (3.137) that the angular momentum operators satisfy the commutation properties

$$[L_x, \mathbf{L}^2] = [L_y, \mathbf{L}^2] = [L_z, \mathbf{L}^2] = 0 \tag{3.162}$$

but

$$[L_x, L_y] = i\hbar L_z$$

$$[L_y, L_z] = i\hbar L_x$$

$$[L_z, L_x] = i\hbar L_y \tag{3.163}$$

Thus, functions that are simultaneous eigenstates of $\mathbf{L}^2$ and L_x, or $\mathbf{L}^2$ and L_y, or $\mathbf{L}^2$ and L_z, but not of L_x and L_y, and L_z, or L_z and L_x can exist. For example, Y_l^m (θ, ϕ) are simultaneous eigenstates of $\mathbf{L}^2$ and L_z but not of L_x or L_y (except for $l = 0$). However, it is seen that $L_\pm = L_x \pm iL_y$ have the useful property

$$L_z L\pm = L_\pm L_z \pm \hbar L_\pm \tag{3.164}$$

If this equation operates on Y_l^m (θ, ϕ), then

$$L_z\,[L_\pm Y_l^m(\theta, \phi)] = (m \pm 1)\,\hbar\,[L_\pm Y_l^m(\theta, \phi)] \tag{3.165}$$

which means that operation by $L_\pm$ produces eigenstates with eigenvalues $(m \pm 1)\hbar$, or annihilates that state if states with eigenvalues $(m \pm 1)\hbar$ do not exist. Hence $L\pm$ are called *raising* and *lowering* operators.

It may also be noted that for potentials which are functions of the radial distance r only,

$$[H, \mathbf{L}] = 0 \tag{3.166}$$

so that it is possible to obtain eigenstates of H which are simultaneously eigenstates of $\mathbf{L}^2$ and L_z, or $\mathbf{L}^2$ and L_x, or $\mathbf{L}^2$ and L_y.

Example 3

Consider the tunnelling of particles across a barrier potential of height V and with d. Imposing the conditions of continuity of the wave function in Eq. (3.94), and its derivative at $x = 0$ and $x = d$,

$$a_+ + a_- = b_+ + b_-$$

$$a_+ - a_- = \frac{i\alpha}{p}(b_+ - b_-)$$

$$b_+ e^{-\alpha d} + b_- e^{\alpha d} = c_+ e^{ipd} \quad (3.167)$$

$$b_+ e^{-\alpha d} - b_- e^{\alpha d} = -\frac{ip}{\alpha} c + e^{ipd} \tag{3.167}$$

Expressing c_+ in terms of a_+, we get

$$\frac{a_+}{c_+} = \frac{1}{4} e^{ipd}\left[\left(1+\frac{i\alpha}{p}\right)\left(1-\frac{ip}{\alpha}\right)e^{\alpha d} + \left(1-\frac{i\alpha}{p}\right)\left(1+\frac{ip}{\alpha}\right)e^{-\alpha d}\right] \tag{3.168}$$

so that the transmission coefficient T is given by

$$\left|\frac{a_+}{c_+}\right|^2 = \frac{1}{T} = \frac{1}{16}\left|2\left(e^{\alpha d} + e^{-\alpha d}\right) + i\left(\frac{\alpha}{p} - \frac{p}{\alpha}\right)\left(e^{\alpha d} - e^{-\alpha d}\right)\right|^2$$

$$= 1 + \frac{1}{16}\left(\frac{\alpha}{p} + \frac{p}{\alpha}\right)^2 \left(e^{\alpha d} - e^{-\alpha d}\right)^2 \tag{3.169}$$

For a very broad barrier, *i.e.* for large d, this leads to

$$T \approx \frac{16\alpha^2 p^2}{\left(\alpha^2 + p^2\right)} e^{-2\alpha d} \tag{3.170}$$

Example 4

Bohr's correspondence principle states that *a quantum system tends* (in a particular sense) *to its classical analogue, for large quantum numbers*. This is demonstrated for a particle in a box.

For a particle in a one-dimensional box of length l, the probability of finding it in the region $B \le x \le B + b$, (see Sec. 3.8), is

$$P_b = \frac{2}{l}\int_B^{B+b} \sin^2\left(\frac{n\pi}{l}x\right) dx$$

$$= \frac{b}{l} - \frac{1}{2n\pi}\left(\sin\frac{2n\pi x}{l}\right)\Bigg|_B^{B+b} \to \frac{b}{l} \text{ for } n \to \infty \tag{3.171}$$

which is the value expected for a classical system.

Example 5

As an application of perturbation theory, consider a particle of charge q, in the presence of a constant electric field E, inside a 3- dimensional box of dimensions $(l_x) \times (l_y) \times (l_z)$.

In the absence of the electric field, the wave functions and the energies are [see Eqs. (3.100) and (3.101)]

$$\phi_n (x, y, z) = \phi_{nx} (x)\, \phi_{ny} (y)\, \phi_{nz} (z) \tag{3.172}$$

$$E_n = \frac{\hbar^2\pi^2}{2m}\left(\frac{n_x^2}{l_x^2} + \frac{n_y^2}{l_y^2} + \frac{n_z^2}{l_z^2}\right),\ n_x = 1, 2, \text{etc.} \tag{3.173}$$

where $$\phi_{nx} (x) = \left(\frac{2}{l_x}\right)^{1/2} \sin\left(\frac{n_x\pi}{l_x}x\right),\ 0 \le x \le l_x \tag{3.174}$$

$$= 0 \text{ for } x \langle\, 0 \text{ or } x \,\rangle\, l_x$$

and similar expressions for ϕ_{ny} (y) and ϕ_{nz} (z). If a weak constant electric field $\mathbf{E}$ is introduced, the additional potential energy is

$$V = -q\,\mathbf{E}\cdot\mathbf{r} \tag{3.175}$$

The change in the energy due to this term is given by perturbation theory (see Sec. 3.10) as

$$E - E_n \approx -q \int |\phi_n (x, y, z)|^2\, \mathbf{E}\cdot\mathbf{r}\, d\tau \tag{3.176}$$

$$= -\frac{1}{2}\, q\, (E_x l_x + E_y l_y + E_z l_z) \tag{3.177}$$

Thus, to the leading order, all the energy levels are shifted by the same amount.

Example 6

For the 3-dimensional harmonic oscillator, the Hamiltonian

$$H = \frac{1}{2m}\mathbf{p}^2 + \frac{1}{2}kr^2 \tag{3.178}$$

is the sum of 1-dimensional Hamiltonians in the three directions. Hence the wave function is a product of the three wave functions,

$$\psi(x, y, z) = \psi_{nx}(x)\,\psi_{ny}(y)\,\psi_{nz}(z) \tag{3.179}$$

where $\psi_{nx}(x)$, etc. are given in Eq. (3.111), and the energy is

$$E = \left(n_x + n_y + n_z + \frac{3}{2}\right)\hbar\omega \tag{3.180}$$

These solutions can be written in terms of spherical coordinates so as to exhibit the angular momentum content of the states. For example, the ground state is

$$\psi(r, \theta, \phi) = \frac{2\alpha^{3/2}}{\pi^{1/4}} Y_0^0(\theta, \phi) \exp(-\alpha^2 r^2/2)$$

$$E = \frac{3}{2}\hbar\omega \tag{3.181}$$

while the state with $n_x = 1$, $n_y = n_z = 0$ can be written as

$$\psi(r, \theta, \phi) = \frac{2\alpha^{5/2}}{\pi^{1/4}3^{1/2}}[Y_1^{-1}(\theta, \phi) - Y_1^1(\theta, \phi)]\, r \exp(-\alpha^2 r^2/2)$$

$$E = \frac{5}{2}\hbar\omega \tag{3.182}$$

PROBLEMS

1. A one-dimensional wave packet has the form $\psi(x) = 0$ for $|x| > a$, and $\psi(x) = (2a)^{-1/2}$ for $|x| \le a$. What is the wave function in the momentum space? Demonstrate the uncertainty principle for this wave packet.
2. For a particle coming from the left with energy E, and a potential changing from $V(x) = 0$ for $x < 0$ to $V(x) = -V_0$ for $x \ge 0$, obtain the transmission and reflection coefficients T and R respectively. Show that $R + T = 1$.

3. Show that the frequency of radiation emitted when the particle inside a one-dimensional box undergoes a transition from $(n + 1)$ state to n state, tends to the classical frequency of motion inside the box, for $n \to \infty$. This is another illustration of Bohr's correspondence principle.

4. Consider a wave function $Ae^{-r/a}$ for the ground state of the hydrogen atom, r being the separation between the electron and the proton. Determine A, a, and the ground state energy. What are the classical and quantum mechanical probabilities of finding the electron at a separation greater than $2a$?

5. If a three-dimensional harmonic oscillator has a solution of the form $AY_1^0 (\theta, \phi)\, re^{-ar^2}$, determine a, A, and the energy in terms of mass and force-constant of the oscillator.

6. For a particle inside a one-dimensional square well defined by $V(x) = 0$ for $|x| \geq a$ and $V(x) = -V_0$ for $|x| < a$, obtain a relationship between the binding energy, a, and V_0. Show that for $V_0 \to 0$, there is a bound state with energy $E \to -2ma^2V_0^2/\hbar^2$ (this is a shallow bound state in the sense that $E/V_0 \to 0$ as $V_0 \to 0$).

7. For a particle in a one-dimensional box, obtain the standard deviations $\sigma(x)$ and $\sigma(p)$ for position and momentum respectively. Show that $\sigma(x)$ $\sigma(p) = \hbar\left(\dfrac{n^2\pi^2}{12} - \dfrac{1}{2}\right)^{1/2}$, and that it is greater than $\hbar/2$.

8. For a particle in a one-dimensional potential well defined by $V(x) = \infty$ for $x < 0$, $V(x) = -V_0$ for $0 \leq x \leq a$ and $V(x) = 0$ for $x > a$, obtain a relation between the binding energy, a, and V_0. Show that these states are the same as the odd states in Problem 6.

9. Show that p_z and L_z are hermitian operators. Show that the commutator $[p_z, L_z] = 0$ and $[p_y, L_z] = i\hbar p_x$. Also show that the wave function in Problem 5 is an eigenstate of L_z but not of p_z or p_x. Calculate the expectation values of L_z, p_z, p_x and p^2 for this wave function.

10. Obtain the first order perturbation to the energy of the ground state for a one-dimensional harmonic oscillator in the presence of a perturbing potential λx^4.

11. Obtain the first order perturbations to the energy of the ground state and the first excited states for a particle in a cubic box with the centre at the origin and the edges parallel to the coordinate axes, due to a perturbing potential λ_{xy}.

4

The One-Electron Atom

Structures of the Chapter

In this chapter, the one-electron atom is analysed within the framework of wave mechanics. It is the ability of quantum mechanics to describe the detailed properties of the one-electron atom which has, more than any thing else, established the essential validity of quantum mechanical ideas, at least as a calculational tool for describing small-distance phenomena.

The wave functions and the energy levels of the nonrelativistic one-electron atom are first obtained. The corrections due to spin-orbit interaction and other relativistic effects are then introduced perturbatively. Together, these results provide a very satisfactory description of the one-electron energy levels including the fine structure. Finally, the effect of the nuclear spin on the atomic energy levels is discussed and a brief introduction to the formal description of spin-$\frac{1}{2}$ particles is given.

4.1 SOLUTIONS OF THE SCHRÖDINGER EQUATION

The total energy of an electron and a nucleus of charge Ze, is

$$E = \frac{1}{2} m_e \dot{r}_e^{\,2} + \frac{1}{2} m_n \dot{r}_n^{\,2} - \frac{Ze^2}{4\pi\varepsilon_0 |\mathbf{r}_e - \mathbf{r}_n|} \tag{4.1}$$

In the centre of mass frame defined by Eq. (2.54), Eq. (4.1) has the form

$$E = \frac{\mathbf{p}^3}{2m\mathbf{r}} - \frac{Ze^2}{4\pi\varepsilon_0 r} \tag{4.2}$$

where
$$\mathbf{r} = \mathbf{r}_e - \mathbf{r}_n \tag{4.3}$$

$$\mathbf{p} = m_r \dot{\mathbf{r}} \tag{4.4}$$

$$m_r = \frac{m_e m_n}{m_e + m_n} \tag{4.5}$$

The Schrödinger equation follows from Eq. (4.2). For states with well-defined energy E, one can write the wave function in the from

$$\psi(\mathbf{r}, t) = \phi(\mathbf{r}) \exp(-iEt/\hbar) \tag{4.6}$$

with $\phi(\mathbf{r})$ satisfying the time-independent Schrödinger equation

$$-\frac{\hbar^2}{2m_r}\nabla^2\phi(r) - \frac{Ze^2}{4\pi\varepsilon_0 r}\phi(\mathbf{r}) = E\phi(\mathbf{r}) \tag{4.7}$$

where the first term represents the kinetic energy. Because the potential is a function only of r, it is preferable to write the Laplacian operator in terms of spherical coordinates. It is particularly convenient to use the expression in Eq. (3.135) in terms of which the Schrödinger equation becomes

$$-\frac{\hbar^2}{2m_r r^2}\frac{\partial}{\partial r}r^2\frac{\partial}{\partial r}\phi(\mathbf{r}) + \frac{\mathbf{L}^2}{2m_r r^2}\phi(\mathbf{r}) - \frac{Ze^2}{4\pi\varepsilon_0 r}\phi(\mathbf{r}) = E\phi(\mathbf{r}) \tag{4.8}$$

with the angular nominatum term $\mathbf{L}^2$ given by Eq. (3.137).

The solutions to Eq. (4.8) in the factorizable form can be written as

$$\phi(\mathbf{r}) = (r)\, Y(\theta, \phi) \tag{4.9}$$

Dividing Eq. (4.8) to $\phi(\mathbf{r})$ leads to

$$\frac{1}{R(r)}\left[\frac{\hbar^2}{2m_r r^2}\frac{d}{dr}r^2\frac{d}{dr} - \frac{Ze^2}{4\pi\varepsilon_0 r} - E\right]R(r)$$

$$\frac{1}{2m_r r^2 Y(\theta,\phi)}\mathbf{L}^2 Y(\theta,\phi) \tag{4.10}$$

which implies that once the left hand side is independent of θ and ϕ, $Y(\theta, \phi)$ must satisfy the eigenvalue equation

$$\mathbf{L}^2 Y(\theta, \phi) = \lambda Y(\theta, \phi) \tag{4.11}$$

The solutions to this equation were discussed in Sec. 3.11, and are the spherical harmonics $Y_l^m(\theta, \phi)$ given in Eq. (3.149), which satisfy the equations:

$$L_z Y_l^m(\theta, \phi) = m\hbar\, Y_l^m(\theta, \phi), \quad m = 0, \pm 1, \ldots, \pm l$$

$$\mathbf{L}_2 Y_l^m(\theta, \phi) = l(l+1)\,\hbar^2\, Y_l^m(\theta, \phi), \quad l = 0, 1, \ldots \tag{4.12}$$

These solutions $Y_l^m(\theta, \phi)$ reduce the radial equation to

$$\frac{\hbar^2}{2m_r}\left[-\frac{1}{r^2}\frac{d}{dr}r^2\frac{d}{dr} + \frac{l(l+1)}{r^2}\right]R(r) - \frac{Ze^2}{4\pi\varepsilon_0 r}R(r) = ER(r) \tag{4.13}$$

There are two important classes of solutions to the radial equation. It is found that solutions exist for all positive values of E. They exhibit oscillatory behaviour for $r \to \infty$ and are not normalizable. These solutions can be used to describe a beam of particles scattered by the Coulomb potential, and lead to Rutherford scattering. The solutions which are of greater interest are the ones for negative E which correspond to bound state solutions. The steps followed in obtaining the negative energy solutions are as follows:

1. Obtain the asymptotic behaviour of $R(r)$, which is finite for $r \to \infty$. It is

$$R(r) \to \exp\left[-\left(-2m_r E/\hbar^2\right)^{1/2} r\right] \tag{4.14}$$

2. Define

$$u\ (r) = R\ (r)\ \exp\ [(-2m_r E/\hbar^2)^{1/2} r] \tag{4.15}$$

and consider a series solution for $u\ (r)$,

$$u\ (r) = \sum_{k=s}^{\infty} b_k r^k\ ,\ k = s,\ s+1,\ \ldots,\ b_s \neq 0 \tag{4.16}$$

3. Impose the condition that the solution for $u(r)$ does not alter the asymptotic behaviour of $R(r)$. This constraint on the asymptotic behaviour leads to the result that the series in Eq. (4.16) must terminate for a finite value of k, say $k = s + p$, p is an integer, *i.e.* $b_k = 0$ for $k > (s + p)$.

Substituting the above expressions in Eq. (4.13), and equating the coefficients of the same powers of r, in particular of r^{k-2}, gives

$$b_k \frac{\hbar^2}{2m_r}[l\ (l+1) - k\ (k+1)] = b_{k-1}\left[\frac{Ze^2}{4\pi\varepsilon_0} - \frac{\hbar^2 k}{m_r}\left(-\frac{2m_r E}{\hbar^2}\right)^{1/2}\right] \tag{4.17}$$

Since $b_{s-1} = 0$, we have $s = l$ or $s = -l - 1$. For $l \neq 0$, the $s = -l - 1$ solutions are not normalizable and hence are discarded. For $l = 0$, the first term in Eq. (4.16) for the $s = -l -1$ solution, is $b_{-1}\ r^{-1}$. However, Eq. (4.17) for $k = 0$ gives $b_{-1} = 0$ which is inconsistent. Hence, only the $s = l$ solution need be considered. The requirement that the series terminates, *i.e.*

$b_k = 0$ for $k = 1 + p + 1$, $p \geq 0$, then leads to

$$\left(\frac{m_r\ Ze^2}{4\pi\varepsilon_0\hbar^2}\right) = (l + p + 1)\left(-\frac{2m_r E}{\hbar^2}\right)^{1/2} \tag{4.18}$$

Thus the negative-energy solutions exist only for energies

$$E_n = -\frac{m_r}{2\hbar^2}\left(\frac{Ze^2}{4\pi\varepsilon_0}\right)^2 \frac{1}{n^2},\ n = 1, 2, \ldots \tag{4.19}$$

$$n = l + p + 1,\ p = 0, 1, \ldots$$

with $l + p$ being the highest power of r in the series solution for $u(r)$. The wave functions corresponding to the solutions are related to *Laguerre polynomials*, and are given by

$$R_{n,l}\ (r) = \left[\left(\frac{2Z}{na_1}\right)^3 \frac{(n-l-1)!}{2n\left[(n+l)!\right]^3}\right]^{1/2} \rho^l e^{-p/2p}\ L_{n+l}^{2l+1}(\rho) \tag{4.20}$$

$$\rho = \frac{2Zr}{na_1}, a_1 = \frac{\hbar^2}{m_r}\left(\frac{4\pi\varepsilon_0}{e^2}\right)$$

where a_1 is the radius of the first Bohr orbit for $Z = 1$, and L_{n+1}^{2l+1} are the *associated Laguerre polynomials* given by

$$L_j^i(\rho) = \frac{d^i}{d\rho^i} L_j(\rho) \tag{4.21}$$

$L_j(\rho)$ being the Laguerre polynomials,

$$L_j(r) = e^{\rho} \frac{d^j}{d\rho^j}(\rho^j e^{-p}) \tag{4.22}$$

Here $R_{n,l}(r)$ are normalized to satisfy the condition

$$\int [R_{n,l}(r)]^2 r^2 dr = 1 \tag{4.23}$$

Collecting the radial and angular parts, the solutions are

$$\phi_{n,l,m}(\rho, \theta, \phi) = R_{n,l}(r)\, Y_l^m(\theta, \phi) \tag{4.24}$$

with

$$E_n = -\frac{m_r}{2\hbar^2}\left(\frac{Ze^2}{4\pi\varepsilon_0}\right)^2 \frac{1}{n^2} \tag{4.25}$$

The first four solutions are

$$\phi_{1,0,0}(\mathbf{r}) = \frac{1}{\pi^{1/2}}\left(\frac{Z}{a_1}\right)^{3/2} \exp(-Zr/a_1)$$

$$\phi_{2,0,0}(\mathbf{r}) = \frac{1}{(32\pi)^{1/2}}\left(\frac{Z}{a_1}\right)^{3/2}\left(2 - \frac{Zr}{a_1}\right)\exp(-Zr/2a_1)$$

$$\phi_{2,1,0}(\mathbf{r}) = \frac{1}{(32\pi)^{1/2}}\left(\frac{Z}{a_1}\right)^{3/2}\frac{Zr}{a_1}\exp(-Zr/2a_1)\cos\theta \tag{4.26}$$

$$\phi_{2,1,1}(\mathbf{r}) = \frac{1}{(64\pi)^{12}}\left(\frac{Z}{a_1}\right)^{3/2}\frac{Zr}{a_1}\exp(-Zr/2a_1)\sin\theta\exp(i\phi)$$

The wave functions of one-electron atom are characterized by three quantum numbers. These are the principal quantum number n, the angular momentum quantum number l which determines the angular momentum, and the magnetic quantum number m which determines the z-component of the angular momentum. However, the energy E_n depends on only the principal quantum number n. This is a special property of the attractive $1/r$ potential. Thus, there is the accidental degeneracy that there are several states with different l values but the same value of n, which have the same energy. It then follows that since for a given l, there are $2l + 1$ states with $m = 0, \pm 1, ..., \pm l$, and for a given n there are n states with $l = 0, 1, ..., n - 1$ all of which have the same energy, the degeneracy of the states with energy E_n is

$$\sum_{l=0}^{n-1}(2l+1) = n^2 \tag{4.27}$$

Actually, the l-degeneracy gets multiplied by another factor of 2 due to the fact that the electron has an intrinsic angular momentum called spin (this is discussed later) which allows it to be in two independent spin states with the same energy.

In the Schrödinger description of the atom, there are no particle trajectories. The wave functions predict only the probability for finding the electron at various distances from the nucleus. Even then, it may be expected that the average values of radial distance have some correspondence with the distances of Bohr trajectories. Defining the average values as

$$\langle r^n \rangle = \int |\phi(\mathbf{r})|^2 r^n d\tau \tag{4.28}$$

gives, after some involved calculations (for details see Ref. 14)

$$\langle r \rangle = \frac{a_1 n^2}{Z}\left[1 + \frac{1}{2}\left(1 - \frac{l(l+1)}{n^2}\right)\right] \tag{4.29}$$

$$\langle \frac{1}{r} \rangle = \frac{Z}{a_1 n^2} \tag{4.30}$$

$$\langle \frac{1}{r^2} \rangle = \frac{Z^2}{a_1^2 n^3 \left(l + \frac{1}{2}\right)} \tag{4.31}$$

$$\langle \frac{1}{r^3} \rangle = \frac{Z^3}{a_1^3\, n^3\, l\,(l + \frac{1}{2})\,(l+1)}, \text{ for } l > 0 \tag{4.32}$$

Only for $1/r$ is the average value the same as the corresponding value for the Bohr orbits.

The energy levels of the one-electron atom, deduced from the Schrödinger equation, are the same as those obtained from the Bohr model. However, it should be appreciated that the results of the Bohr model follow from ad-hoc, though interesting, assumptions, while those from the Schrödinger equation are based on fundamental physical principles.

The solutions of the nonrelativistic Schrödinger equation for the one-electron atoms provide a satisfactory basis for the understanding of the general features of the energy spectra of these atoms. However, these spectra have a fine structure, to understand which small relativistic corrections should be included and additional properties for the electron proposed.

4.2 ELECTRON SPIN

The l-degeneracy of the energy levels of the one-electron atom is removed by any additional interaction which is noncoulombic in form. Such an interaction may be provided by the relativistic corrections to the Schrödinger equation, or by the non-point structure of the nucleus. As a consequence, the energy levels for a given n split into several closely-spaced levels corresponding to different l values. This leads to multiplets of spectral lines which are described as the *fine structure* of the spectrum.

There are some fine-structure lines which cannot be described in terms of the l-multiplets. Examples of these are the alkali metal spectra which show doublets of closely spaced lines. Prominent among these is the sodium yellow line for the ($n = 3, l = 1$) transition to ($n = 3, l = 0$), which actually consists of two closely-spaced lines of wavelengths 5890 Å and 5896 Å. To explain these splitting, Goudsmit and Uhlenbeck (1925) proposed that the electron has an intrinsic angular momentum called the *spin* (the origin of spin is not the spatial motion of the electron), and an associated magnetic moment. If S describes the spin of the electron, the associated magnetic moment is

$$\mu = -b\,\mathbf{S} \tag{4.33}$$

where, since the electron has negative charge, it is assumed that b is positive constant of proportionality. It is the interaction of this magnetic moment with the magnetic field seen by the electron due to the motion of the nucleus around it, that contributes to the fine structure of the atomic energy levels.

A definitive support to the hypothesis of spin is provided by the experiment of Stern and Gerlach (1922). In this experiment, a beam of neutral silver atoms was passed through an inhomogeneous magnetic field in the z direction (Fig. 4.1). If the atom has a magnetic moment μ, it has a potential energy

$$U = -\boldsymbol{\mu} \cdot \mathbf{B} \tag{4.34}$$

in this field, and is subjected to a force in the z-direction,

$$F_z = \mu_x \frac{\partial B}{\partial z} \tag{4.35}$$

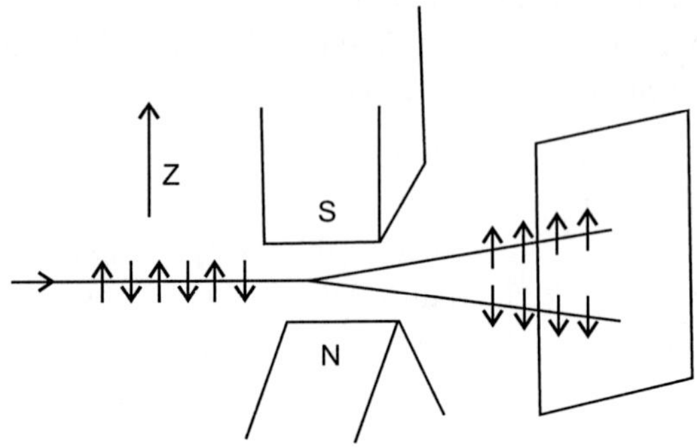

Fig. 4.1 Stern-Gerlach experiment to determine the spin components of the silver atom.

From the deflection of the beam on the screen, the value of μ_z can be obtained. Classically, the magnetic dipole moments would be randomly oriented which would give just a spreading of the beam. Stern and Gerlach observed, however, a splitting of the beam into two discrete components, indicating the existence of only two possible values of μ_z. Since the silver atom has one electron in the outermost shell, this suggests that the electron spin and its magnetic moment in Eq. (4.33) have only two possible values along a given direction. Furthermore, μ_z was found to have values

$$\mu_z = \pm \frac{e\hbar}{2m_e} \tag{4.36}$$

$e\hbar/2m_e$ is called the Bohr magneton for the electron, ($e\hbar/2m_e \approx 9.2731 \times 10^{-24}$ m^2 Cs^{-1}).

It was observed in Eq. (3.152) that the possible eigenvalues of $\mathbf{L}^2$ are $l\,(l + 1)\,\hbar^2$ and those of L_z are $m\hbar$ where $m\hbar = l\hbar, (l - 1)\hbar, ..., -l\hbar$. If this property is assumed to be valid for the spin angular momentum as well, it follows that since S_z has only two eigenvalues [see Eqs. (4.33), (4.36)], $\mathbf{S}_z$ and S^2 have eigenvalues

$$S_z = m_s\hbar, \qquad m_s = \pm \frac{1}{2} \tag{4.37}$$

$$\mathbf{S}^2 = s\,(s + 1)\,\hbar^2, \quad s = \frac{1}{2} \tag{4.38}$$

It is now easy to deduce from Eqs. (4.37), (4.36) and (4.33) that

$$\mu = -\frac{e}{m_e}\mathbf{S} \tag{4.39}$$

These relations introduce the new idea of *intrinsic spin angular momentum* whose quantum numbers take half-integral values in contrast to the integral values taken by the quantum numbers of angular momentum originating from the spatial motion of particles. The relation in Eq. (4.39) also differs (by a factor of 2) from the relation

$$\mu = -\frac{e}{2m_e}\mathbf{L} \tag{4.40}$$

expected for the magnetic moment of a negatively charged particle moving around in a circle with angular momentum **L**. All in all, the spin of an electron, with half-integral values for its quantum numbers, is a revolutionary idea with no classical analogue. It has found strong support not only in the wealth of experimental data it can explain, but also in the elegant formulation of the linearized relativistic equation of Dirac (1928) describing a spin 1/2 particle.

4.3 TOTAL ANGULAR MOMENTUM

The total angular momentum consists of two parts, the orbital angular momentum **L** and the spin angular momentum **S**. Designating the total angular momentum by *J*,

$$\mathbf{J} = \mathbf{L} + \mathbf{S} \tag{4.41}$$

As in the case of the orbital angular momentum, quantum numbers m_j and j can be associated with **J**, such that J_z and $\mathbf{J}^2$ have eigenvalues

$$J_z = m_j\hbar \tag{4.42}$$

$$\mathbf{J}^2 = j\,(j-1)\,\hbar^2 \tag{4.43}$$

It follows from Eq. (4.41), that

$$m_j = m_l + m_s \tag{4.44}$$

so that m_j has integral values if m_s has integral values, and half-integral values if m_s has half-integral values (m_l is used in place of m to have a more symmetric notation). For deducing the possible values for j, it is assumed that $l \geq s$. It is also noted that the magnitude of **J** is not affected by the choice of the direction of **L**, *i.e.* the choice of m_l. Taking the largest possible value m_l, $= l$, gives

$$m_j = 1 + m_s \tag{4.45}$$

Thus the largest and the smallest values of m_j are $l + s$ and $l - s$. This result, along with a similar analysis for $l \leq s$, implies that the allowed values of j and m_j, are

$$|\,l - s\,| \leq j \leq l + s \tag{4.46}$$

$$m_j = j, j-1, \ldots, -j \tag{4.47}$$

It therefore follows from Eqs. (4.47) and (4.44), that j takes on integral values if s takes integral values, and half-integral values if s takes half-integral values.

An electron n an atom is characterized by the quantum numbers n, l, s and j. In spectroscopic notation, the state of such an electron is designated by

$$n^{2s+1} L_j \tag{4.48}$$

The superscript $2s + 1$ gives the multiplicity of the state for $l \geq s$, as can be deduced from Eq. (4.46). The subscript in the notation describes the total angular momentum. In place of L, a letter which conventionally denotes a particular orbital angular momentum is used, *e.g.* s, p, d, f, g and h for $l = 0, 1, 2, 3, 4$, and 5, respectively. These small case latters are used to describe the states of individual electrons. For the states of the atom, capital letters S, P, D, F, G, H are used instead.

Table 4.1 Spectroscopic letters for different l values, small case letters for electron states and capital letters for atomic states

Values of l	→	0	1	2	3	4	5
Letter symbol	→	s, S	p, P	d, D	f, F	g, G	h, H

The ground state of sodium is described by an electron in the $n = 3$, $l = 0$, $j = 1/2$ state, and hence the atomic ground state may be described by $3^2S_{1/2}$. The excited electron in the $n = 3$, $l = 1$ state can have $j = 3/2$ or $j = 1/2$ and the two states would be designated by $3^2P_{3/2}$ and $3^2P_{1/2}$. If, as will be seen, these two states have different energies, one would observe a doublet of lines corresponding to transitions $3^2P_{3/2} \rightarrow 3^2S_{1/2}$ and $3^2P_{1/2} \rightarrow 3^2S_{1/2}$. This provides a basis for the explanation of the observed doublet of sodium lines with wavelengths 5890 Å and 5896 Å.

4.4 FINE STRUCTURE OF ONE-ELECTRON ATOMIC SPECTRA

In this section, the fine structure of the one-electron atomic spectra is considered. The fine structure arises from the small corrections due to essentially relativistic effects.

It was noted in Sec. 4.2 that an electron has an intrinsic magnetic moment $\mu = -(e/m_e)\,\mathbf{S}$. This interacts with the magnetic field seen by the electron in its rest frame, due to the motion of the nucleus around it. Since the nucleus moving in a circle of radius r produces a circular current $I = Zer/2\pi r$, the magnetic field seen by the electron is

$$\mathbf{B} = \left(\frac{Ze}{4\pi\varepsilon_0 m_e c^2}\right)\frac{m_c \mathbf{r}\times\mathbf{v}}{r^3} \tag{4.49}$$

Therefore, the energy of the spin magnetic moment of the electron interacting with this field is

$$V_1' = -\boldsymbol{\mu}\cdot\mathbf{B}$$

$$= \left(\frac{Ze^2}{4\pi\varepsilon_0 m_e^2 c^2}\right)\frac{\mathbf{S.L}}{r^3} \tag{4.50}$$

However, this is the energy seen in the frame of the electron which is being accelerated. It was shown by Thomas that the corresponding energy in the rest frame of the nucleus, is smaller by a factor of 1/2, so that the first correction to the nonrelativistic energy is (for details see Ref. 1)

$$V_1 = \left(\frac{Ze^2}{8\pi\varepsilon_0 m_e^2 c^2}\right)\frac{\mathbf{S.L}}{r^3} \tag{4.51}$$

It is easy to see that this term is smaller than the angular momentum term in Eq. (4.8) by an order of magnitude of $(Ze^2/4\pi\varepsilon_0 r)\,(1/m_c c^2)$, *i.e.* ratio of binding energy to rest energy. This illustrates that the spin-orbit interaction, given in Eq. (4.51), is a relativistic effect and gives corrections which are smaller than the nonrelativistic energies by an order of about 10^{-5}. This correction is there only for the states with $l \neq 0$.

The second correction is obtained from using the relativistic expression for the kinetic energy

$$T = (p^2c^2 + m_e^{\,2}c^4)^{1/2} - m_e c^2$$

$$\approx \frac{1}{2m_e}p^2 - \frac{1}{8m_e^3c^2}p^4 + \ldots \tag{4.52}$$

Thus, the leading correction gives rise to an extra term for the energy,

$$V_2 = -\frac{1}{8m_e^3c^2}p^4 \tag{4.53}$$

which is smaller than the kinetic energy by a factor of about $(p^2/2m_e)\,(1/2m_e c^2)$. Hence, this term also gives rise to corrections which are smaller by a factor of about 10^{-5} than the nonrelativistic energies. Being negative, it will lower the energy of all the states.

Finally, there is an additional correction which follows from the relativistic Dirac equation. It is called the *Darwin term* and has the form

$$V_3 = \frac{\pi Z_e^2 \hbar^2}{8\pi\varepsilon_0 m_e^2 c^2}\,\delta(\mathbf{r}) \tag{4.54}$$

Because of the presence of the Dirac delta function $\delta\,(\mathbf{r})$, this term contributes only to the $l = 0$ states ($\delta\,(\mathbf{r}) = 0$ for $r \neq 0$ but $\int\delta\,(\mathbf{r})\,d\tau = 1$), since the wave functions of states with $l \neq 0$ vanish at $r = 0$. This term is of the same order as the spin-orbit interaction, and therefore contribute corrections of the order of 10^{-5} compared to the leading terms.

Collecting all the corrections together, the additional energy is

$$V = V_1 + V_2 + V_3$$

$$= \left(\frac{Ze^2}{8\pi\varepsilon_0 m_e^2 c^2}\right)\frac{\mathbf{S.L}}{r^3} - \frac{1}{8m_e^3c^2}\,p^4 + \frac{\pi Z_e^2\hbar^2}{8\pi\varepsilon_0 m_e^2 c^2}\,\delta(\mathbf{r}) \tag{4.55}$$

where the contribution of the first term is only to the $l \neq 0$ terms. Since these terms are small, their contribution to the energy levels can be evaluated by using the first order perturbation theory described in Sec. 3.10 as

$$\Delta E_n \approx \int \phi_n^{\,*}\, V\phi_n\, d\tau \tag{4.56}$$

For the evaluation of the contribution of V_1 to ΔE_n, we note that

$$\mathbf{S.L} = \frac{1}{2}\,[(\mathbf{L} + \mathbf{S})^2 - \mathbf{L}^2 - \mathbf{S}^2] \tag{4.57}$$

This, together with the value of $\langle 1/r^3\rangle$ given in Eq. (4.32), leads to

$$\langle V_1\rangle = Z^2\alpha^2\,|E_n|\,\frac{j(j+1) - l(l+1) - 3/4}{nl\,(2l+1)\,(l+1)},\ l \neq 0 \tag{4.58}$$

$$= 0, \quad \text{for } l = 0$$

where α is the *fine structure constant* $\left(e^2/4\pi\varepsilon_0\hbar_c\right)$ and has an approximate value of (1/137). The contribution of V_2 is obtained from

$$\langle -p^4/8m_e^{\,3}c^2 \rangle = -\frac{1}{2m_e c^2}\langle\left(H_0 + \frac{Ze^2}{4\pi\varepsilon_0 r}\right)^2\rangle$$

$$= -\frac{1}{2m_e c^2}\left[E_n^2 + 2E_n\langle\frac{Ze^2}{4\pi\varepsilon_0 r}\rangle + \langle\left(\frac{Ze^2}{4\pi\varepsilon_0 r}\right)^2\rangle\right] \tag{4.59}$$

Using relations (4.30) and (4.31) gives

$$\langle V_2 \rangle = \frac{Z^2\alpha^2\,|E_n|}{4n^2}\left(3 - \frac{4n}{l+1/2}\right) \tag{4.60}$$

Finally, the contribution of the Darwin term depends on the wave function at the origin. Detailed analysis shows that $\psi(0) = \frac{1}{\pi^{1/2}}\left(\frac{Z}{a_1 n}\right)^{3/2} \delta_{l0}$ which then leads to

$$\langle V_3 \rangle = \frac{Z^2\alpha^2 |E_n|}{n} \qquad \text{for } l = 0$$

$$= 0 \qquad \text{for } l \neq 0 \qquad (4.61)$$

These relations allow us to obtain ΔE_n,

$$\Delta E_n = \frac{Z^2\alpha^2 |E_n|}{4n^2}\left(3 - \frac{4n}{j+1/2}\right) \qquad (4.62)$$

for the fine structure of the energy levels of the hydrogen atom.

The important properties of the fine structure given in Eq. (4.62) and demonstrated in Fig. 4.2 are:

1. As expected, the fine structure corrections are smaller than E_n by a factor of about $\alpha^2/4 \sim 10^{-5}$. The hydrogen atom energies E_n themselves may be written as $E_n = -\alpha^2 mc^2/2n^2$, which means that they are smaller than the rest energy by a factor of about α^2. All the shifts in the energy levels due to fine structure corrections are negative and the shift decreases as j increases. Furthermore, the corrections decrease rapidly as n increases, so that its effect is more easily noticeable for small-n states.

2. The fine structure corrections remove some of the degeneracy of the energy levels E_n. The states with different j values now have different energies. For a given n, the allowed j values range from 1/2 to $(n - 1/2)$ so that each n level is now split into n levels.

3. Some degeneracy still survives. For a given n, the level with $j = n - 1/2$ is nondegenerate but all the other levels have a degeneracy of order two corresponding to $l = j \pm 1/2$. For example, for $n = 2$, $2\,^2P_{3/2}$ is nondegenerate but $2^2P_{1/2}$ and $2^2S_{1/2}$ are degenerate. Actually there is a small separation between the $2^2P_{1/2}$ and $2^2S_{1/2}$ states also, known as the *Lamb shift*, which can be satisfactorily explained in terms of quantum electrodynamics.

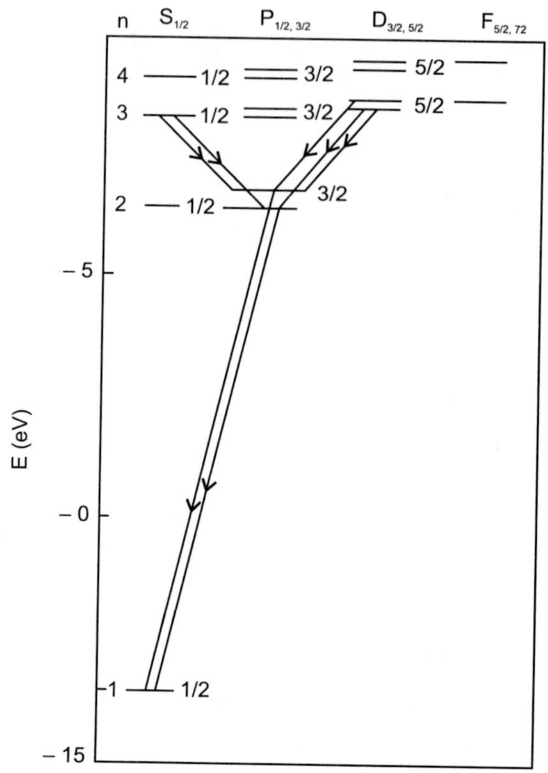

Fig. 4.2 Schematic diagram of the fine structure of the hydrogen levels (greatly exaggerated), showing some of the transitions allowed by the selection rules in Eq. (4.63).

4. Not all the transitions between the different levels are allowed. As will be seen later, there are selection rules for the allowed transitions. For the most prominent transitions, called the electric dipole transitions, the allowed transitions satisfy the selection rules

$$\begin{aligned} &\Delta l = \pm 1 \\ &\Delta j = \pm 1, 0, \qquad \text{but not } j = 0 \rightarrow j = 0 \\ &\Delta m_j = \pm 1, 0 \\ &\Delta n = \text{unrestricted} \end{aligned} \tag{4.63}$$

Thus, these transitions are allowed only between adjacent columns in Fig. 4.2. For example, there are two lines (a doublet) corresponding to transitions between $2P$ and $1S$ levels, two lines (a doublet) corresponding to transitions between $3S$ and $2P$ levels and three lines (a triplet) corresponding to transitions between $3D$ and $2P$ states.

4.5 HYPERFINE STRUCTURE

In the discussion so far, the nucleus of the atom was assumed to be a point particle without any structure. However, this assumption is insufficient to explain many experimental results, for example, the observation of hyperfine structure of atomic levels using high resolution spectrographs. To explain these, Pauli (1924) suggested that the nucleus also has an intrinsic angular momentum and an associated magnetic moment. These properties have now been firmly established by several experiments, and are essential elements in the description of atoms and nuclei.

Let **I** be the spin of the nuleus, with eigenvalues

$$I_z = m_i\hbar$$

$$\mathbf{I}^2 = I(I+1)\,\hbar^2 \tag{4.64}$$

Associated with I is a magnetic moment $\mu\mathbf{N}$,

$$\mu_N = g\frac{e}{m_p}\mathbf{I} \tag{4.65}$$

where m_p is the mass of the proton. Because the structure of the nucleus is more complicated than that of an electron, the value of g is generally different from 1, and is 2.79 for the proton. The nuclear magnetic moment is seen to be smaller than the electron magnetic moment by a factor of about $m_e/m_p \sim 1/1000$.

The atomic states are now designated by the total angular momentum **F**,

$$\mathbf{F} = \mathbf{J} + \mathbf{I} \tag{4.66}$$

with eigen values

$$F_z = \left(m_j + m_i\right)\hbar \tag{4.67}$$

$$\mathbf{F}^2 = F(F+1)\hbar^2, |\, j - I\,| \le F \le j + I$$

This means that each level with a given j has a multiplicity of $2I + 1$ if $j > I$ and a multiplicity of $2j + 1$ if $j \le I$. The allowed electric dipole transitions are found to satisfy the selection rules

$$\Delta l = \pm\, 0$$

$$\Delta F = \pm\, 1, 0 \quad \text{but not } F = 0 \rightarrow F = 0 \tag{4.68}$$

$$\Delta m\mathbf{F} = \pm\, 1, 0$$

The nuclear magnetic moment interacts with the magnetic field created at the nucleus by the electron. The magnetic field is due to (*i*) the orbital motion of the electron around the nucleus, and (*ii*) the intrinsic magnetic moment of the electron. The magnetic field at the nucleus, due to the orbital motion of the electron can be deduced from Eq. (4.49) as

$$\mathbf{B} = -\frac{e}{4\pi\varepsilon_0 m_e c^2}(\mathbf{L}/r^3) \tag{4.69}$$

Therefore, the interaction energy due to this field is

$$\langle V_{\text{orb}} \rangle = \frac{ge^2}{4\pi\varepsilon_0 m_e m_p c^2}\mathbf{I.L}\langle 1/r^3 \rangle \quad \text{for } l \neq 0 \tag{4.70}$$

$$= 0 \quad \text{for } l = 0$$

This is smaller than the fine structure terms by a factor of about m_e/m_p ~ 1/1000.

For calculating the field due to the intrinsic magnetic moment of the electron, we note that the vector potential due to a magnetic dipole moment μ is

$$\mathbf{A} = \frac{1}{4\pi\varepsilon_0 c^2}\mu\times\left(\frac{\mathbf{r}}{r^3}\right) \tag{4.71}$$

From this, the magnetic field comes out as

$$\mathbf{B} = \nabla\times\mathbf{A}$$

$$= \frac{1}{4\pi\varepsilon_0 c^2}\left[\mu\left(\nabla.\frac{\mathbf{r}}{r^3}\right) - (\mu.\nabla)\frac{\mathbf{r}}{r^3}\right] \tag{4.72}$$

Therefore, the energy of the nuclear magnetic moment μ_N interacting with this field is

$$V_{\text{spin}} = -\frac{1}{4\pi\varepsilon_0 c^2}\nabla.\left[(\mu_N.\mu)\frac{\mathbf{r}}{r^3} - \left(\mu_N.\frac{\mathbf{r}}{r^3}\right)\mu\right] \tag{4.73}$$

With this, the perturbative expression for the interaction energy comes out to be

$$\langle V_{\text{spin}} \rangle = \frac{1}{4\pi\varepsilon_0 c^2}\left\langle\left(\frac{\mu_N.\mu}{r^3} - 3\frac{\mu_N.\mathbf{r}\,\mu.\mathbf{r}}{r^5}\right)\right\rangle \quad \text{for } l \neq 0 \tag{4.74}$$

For $l = 0$, the angular integration in Eq. (4.74) gives zero for $r \neq 0$. For obtaining the correct value of the contribution from $r = 0$, Gauss theorem is used in the expectation value of the expression in Eq. (4.73), to get

$$\langle V_{\text{spin}} \rangle_{l=0} = -\frac{1}{4\pi\varepsilon_0 c^2}(\mu_N.\mu)\frac{8\pi}{3}|\psi(0)|^2$$

$$= \frac{2ge^2}{3\varepsilon_0 m_e m_p c^2}(\mathbf{I.S})|\psi(0)|^2 \quad (4.75)$$

where ψ (0) is the wave function at the origin. In particular, using the wave functions in Eq. (4.26), the hyperfine splitting between the $F = 1$ and $F = 0$ levels of the ground state of the hydrogen atom is

$$E_1 (F = 1) - E_1 (F = 0) = \frac{16m_e}{3m_p}(g\,\alpha^2)|E_1| \quad (4.76)$$

Transition between these states leads to a spectral line with a frequency

$$\nu = \frac{E_1 (F = 1) - E_1 (F = 0)}{h}$$

$$= 1.420 \times 10^9 \text{ s}^{-1} \quad (4.77)$$

which corresponds to a wavelength of about 21.1 cm. This is the famous 2*l* cm *line* observed by the radio astronomers in the spectrum of interstellar hydrogen.

The discussion of the hyperfine structure is concluded with the following comments:

1. Nuclear spin introduces an additional multiplicity of atomic levels. For the hydrogen atom, each energy level acquires and additional multiplicity of 2. The magnetic moment associated with the nuclear spin introduces a hyperfine splitting between the levels. These splittings are about 1000 times smaller than the fine structure splittings.

2. The hyperfine splitting may be observed in a high resolution spectrograph fitted with accessories like *Fabry-Perot etalons*, as a hyperfine structure. Transitions between hyperfine levels corresponding to microwaves may be observed in nuclear magnetic resonances (discussed in Chapter 6). They are also observed as stimulated emissions in a maser where the population has been inverted (discussed in Chapter 6).

3. The hyperfine splitting can be measured to a very high accuracy in the case of transition of the hydrogen atom in the ground state, between the $F = 1$ and $F = 0$ levels and of the ^{133}Cs atom in the $6S_{1/2}$ ground state, between the $F = 4$ and $F = 3$ levels. These correspond to frequencies $\nu = 1.4204057518 \times 10^9 \text{ s}^{-1}$ and $\nu = 9.192631770 \times 10^9 \text{ s}^{-1}$, respectively and are used as time standards of atomic clocks.

4. The $\nu = 1.420 \times 10^9\ s^{-1}$ frequency radiation (known as the 21 cm hydrogen line) corresponding to the hyperfine transition between the ground state levels with $F = 1$ and $F = 0$, is used to study the distribution and motion (in terms of Doppler shift) of the hydrogen in interstellar and intergalactic space.

4.6 EXAMPLES OF ONE-ELECTRON ATOMS

Some examples of one electron atoms are now considered and their special properties discussed.

Hydrogen

For hydrogen, $Z = 1$. The l-degeneracy is removed by the relativistic corrections producing fine structure (Fig. 4.2). The j-degeneracy is removed by quantum electrodynamic effects (Lamb shift). The spin of the proton doubles the number of states, while the interaction of the magnetic moment of the proton with the magnetic field produced by the electron, produces hyperfine structure. The microwave radiation from transition between the two hyperfine levels of the ground state, is especially important in producing maser action, in the investigation of interstellar and extragalactic hydrogen, and as a time standard in atomic clocks.

Heavier isotopes of hydrogen, such as deuterium and tritium, have spectra very similar to the hydrogen spectrum, except for small differences due to slightly different reduced masses (the energy levels are lower for larger reduced masses). However, the hyperfine structure will be quite different since the spin and the magnetic moment of their nuclei are quite different from those of the proton.

Atoms with Z > 1

The singly ionized He ($Z = 2$) atom and doubly ionized Li ($Z = 3$) atom, have energy levels which are larger by a factor of Z^2. Therefore, their energy levels with principal quantum numbers $n' = nZ$ will be similar to those of the hydrogen atom with principal quantum number n, except for small differences due to different reduced masses. The fine structures will be significantly larger, since they vary as Z^4.

Positronium

Positronium is a bound state of an electron and a positron which has the same mass as an electron and an equal but opposite charge as an electron (see Sec. 4.8). They are formed when a beam of positrons is stopped by a gas.

Since the reduced mass for the positronium is $m_e/2$, its energies will be nearly half those of the hydrogen atom. However, the magnetic moment of a positron is equal in magnitude to that of the electron, and hence much greater than that of the proton. Therefore, the hyperfine interaction is much larger for the positronium. In particular, the separation between the $F = 1$ and $F = 0$ levels corresponds to a frequency of $\nu = 2.034 \times 10^{11}$ s^{-1}. The electron and the positron in the positronium annihilate each other, emitting two photons in the $F = 0$ state and three photons in the $F = 1$ state. The lifetimes of the positronium in these two states are different: 1.25×10^{-10} s for the $F = 0$ state and 1.4×10^{-7} s for the $F = 1$ state.

Muonium

Muonium is a bound state of an electron and a μ^+ meson (μ meson or muon and μ^+ meson are similar to an electron and a positron respectively, except that they are heavier, their mass being about 206.84 m_e). They are produced when a beam of μ^+ is stopped by a gas. Their energy levels are similar to those of the hydrogen atom except for a small difference due to the difference in the reduced mass. The hyperfine energy levels of the muonium are of importance since they can be calculated precisely, and serve as a test of the theory.

Muonic Helium

Muonic helium is formed by replacing one of the electrons in a helium atom by a muon. Since the Bohr radius of the muon is smaller by a factor of about 207 than that of the electron, the electron essentially sees a nucleus of charge $2|e|$ with a muon moving around close to the nucleus. Therefore, the energy levels are similar to those of the hydrogen atom. However, the hyperfine splitting is due to the electron magnetic moment interacting with the muon magnetic moment. Using Eq. (4.76) as a first approximation but taking $g = 1$ and replacing m_N by m_μ, it is found that the hyperfine splitting for muonic helium corresponds to a frequency of $\nu = 4.515 \times 10^9$ s^{-1}, close to the experimental value of 4.465×10^9 s^{-1}.

Muonic atoms are very useful for probing the structure of nuclei since the Bohr radius of the muon is quite small, and therefore the probability of finding the muon inside the nucleus may be quite substantial.

Rydberg Atoms

When an electron in an atom is in a state with a sufficiently large principal quantum number n, it is influenced mainly by the net positive charge of the ionic core and not by its distribution. These excited states of atoms are similar to those of a hydrogen atom. They are termed *Rydberg states* and the atoms are called *Rydberg atoms*. It is the advent of tunable lasers (see Sec. 6.5) that has helped to excite and investigate the Rydberg states. They are of interest for the following reasons:

1. The departures of the energy spectrum of Rydberg atoms, from the hydrogenic spectrum, provide useful information about the structure of the ionic core and the interaction between the core and the valence electron.
2. Their behaviour in the presence of external fields can be understood by direct extensions of the analysis for the hydrogen atom. It may also be noted that because of the large size of Rydberg atoms, the effects of the external fields are greatly enhanced.

4.7 SCHRÖDINGER EQUATION FOR SPIN 1/2 PARTICLES

It is clear from the previous discussion that a function which only depends on spatial coordinates, cannot adequately describe the spin and magnetic moment of an electron. An additional variable that describes the spin state of an electron or more generally, any particle with spin 1/2, has to be introduced. Furthermore, one would like the introduction of the associated magnetic moment of $-\frac{e}{m}\mathbf{S}$ to be more appealing.

The spin eigenstates of S_z with eigenvalues $\pm\frac{1}{2}\hbar$ are designated by α and β, so that

$$S_z\alpha = \frac{1}{2}\hbar\alpha\,,\; S_z\beta = -\frac{1}{2}\hbar\beta \tag{4.78}$$

The spin operator $\mathbf{S}$ has the properties,

$$\mathbf{S}^2 = s\,(s+1)\,\hbar^2 = \frac{3}{4}\hbar^2 \tag{4.79}$$

$$S_x^{\,2} = S_y^{\,2} = S_z^{\,2} = \frac{1}{4}\hbar^2 \tag{4.80}$$

Furthermore, the raising and lowering operators $S_{\pm}$ defined as

$$S_{\pm} = S_x \pm iS_y \tag{4.81}$$

satisfy the properties [see Eq. (3.165)]

$$\begin{aligned} S_+\beta &= b_1\hbar\alpha\,,\; S_+\alpha = 0 \\ S_-\alpha &= b_2\hbar\beta\,,\; S_-\beta = 0 \end{aligned} \tag{4.82}$$

where b_1 and b_2 are constants which may be taken to be 1 (see Example 5). The operation of $\mathbf{S}$ is then determined by Eqs. (4.78) and (4.82). It follows from Eq. (4.82) that

$$S_+^{\,2} = S_-^{\,2} = 0 \tag{4.83}$$

and also $$S_xS_y + S_yS_x = -\frac{i}{2}(S_+^2 - S_-^2) = 0 \tag{4.84}$$

Together with the commutation relation $S_xS_y - S_yS_x = i\hbar S_z$ satisfied by all angular momentum operators [see Eq. (3.163)], this implies

$$S_xS_y = -S_yS_x = \frac{i}{2}\hbar S_z \tag{4.85}$$

By symmetry,

$$S_yS_z = -S_zS_y = \frac{i}{2}\hbar S_x \tag{4.86}$$

$$S_zS_x = -S_xS_z = \frac{i}{2}\hbar S_y$$

For writing down the Schrödinger equation for a free, spin 1/2 particle of mass m, it is proposed that the kinetic energy be written as

$$E = \frac{2}{m\hbar^2}(\mathrm{p}\cdot\mathrm{S})(\mathrm{p}\cdot\mathrm{S}) \tag{4.87}$$

This expression is equivalent to $\frac{1}{2m}\mathbf{p}^2$ for the free particle, as can be shown by using Eqs. (4.80), (4.85) and (4.86). On using the operator expressions for E and $\mathbf{p}$, it leads to the Schrödinger equation for a free particle,

$$i\hbar\frac{\partial\psi}{\partial t} = -\frac{2}{m}(\Delta\cdot\mathrm{S})(\Delta\cdot\mathrm{S})\psi \tag{4.88}$$

The interaction with the electrostatic potential ϕ, can be introduced by the prescription that

$$E \to E - q\phi \tag{4.89}$$

where q is the charge of the particle. However, since both $\left(\mathbf{p}, \frac{1}{c}E_{\text{tot}}\right)$ and $\left(\mathbf{A}, \frac{1}{c}\phi\right)$ transform as relativistic 4-vectors, requirements of relativistic covariance imply that the prescription in Eq. (4.89) should be accompanied by the replacement

$$\mathbf{p} \to \mathbf{p} - q\mathbf{A} \tag{4.90}$$

The fact that only the kinetic energy appears in Eq. (4.87) does not alter the essential results since the addition of a constant to the energy only redefines the zero of the energy. Equations (4.89) and (4.90) introduce what is known as the *minimal electromagnetic interaction*. With these prescriptions, the Schrödinger equation for a spin 1/2 particle in the presence of electromagnetic fields, comes out as

$$i\hbar\frac{\partial\psi}{\partial t} = \frac{2}{m\hbar^2}[(-i\hbar\nabla - q\mathbf{A})\cdot\mathbf{S}]^2\,\psi + q\phi\psi \tag{4.91}$$

where
$$\psi = \psi_1\,(\mathbf{r}, t)\,\alpha + \psi_2\,(\mathbf{r}, t)\beta \tag{4.92}$$

Using Eqs. (4.80), (4.85) and (4.86), this reduces to

$$i\hbar\frac{\partial\psi}{\partial t} = \frac{1}{2m}\left(-i\hbar\nabla - q\mathbf{A}\right)^2\psi\,\frac{q}{m}\mathbf{S}.(\nabla\times\mathbf{A} + \mathbf{A}\times\nabla)\psi + q\phi\psi \tag{4.93}$$

Finally, noting that ∇ in $\nabla\times\mathbf{A}$ operates on $\mathbf{A}$ as well as on ψ, and writing V for $q\phi$, gives

$$i\hbar\frac{\partial\psi}{\partial t} = \frac{1}{2m}\left(-i\hbar\nabla - q\mathbf{A}\right)^2\psi - \frac{q}{m}\mathbf{S}\cdot\mathbf{B}\psi + V\psi \tag{4.94}$$

This expression for the energy contains a term which corresponds to the interaction of a particle with magnetic moment

$$\mu = \frac{q}{m}\mathbf{S} \tag{4.95}$$

with the external magnetic field $\mathbf{B}$. Thus, the particle which satisfies Eq. (4.91), has an intrinsic spin $\mathbf{S}$ and an associated magnetic moment given by Eq. (4.95). For the electron $q = -|e|$. These results are in conformity with the experimental observations discussed in Sec. 4.2.

4.8 DIRAC EQUATION

The spin and the magnetic properties of an electron can be discussed in terms of the Schrödinger equation given in Eq. (4.91). However, it has been pointed out earlier that the fine structure is essentially of relativistic origin. Therefore, a relativistic description of the spin 1/2 particle has to be considered to provide a satisfactory explanation of the fine structure.

The relativistic equation for a spin 1/2 particle of mass m may be obtained from the relation

$$E^2 = \frac{4c^2}{\hbar^2}(\mathbf{p}\cdot\mathbf{S})\,(\mathbf{p}\cdot\mathbf{S}) + m^2c^4 \tag{4.96}$$

which is equivalent to $E^2 = p^2c^2 + m^2c^2$. On taking the momentum term to the left hand side and factorizing, it leads to the equation

$$\left(i\hbar\frac{\partial}{\partial t} - 2ic\nabla \cdot \mathbf{S}\right)\left(i\hbar\frac{\partial}{\partial t} + 2ic\nabla \cdot \mathbf{S}\right)\psi = m^2c^2\psi \tag{4.97}$$

where ψ has the form given in Eq. (4.92). It can be linearized by defining

$$\left(i\hbar\frac{\partial}{\partial t} - 2ic\nabla \cdot \mathbf{S}\right)\psi = mc^2 x \tag{4.98}$$

substitution of which in Eq. (4.97) leads to

$$\left(i\hbar\frac{\partial}{\partial t} - 2ic\nabla \cdot \mathbf{S}\right)x = mc^2\psi \tag{4.99}$$

Equations (4.98) and (4.99) together are equivalent to the Dirac equation (1928) for a spin 1/2 particle.

The free particle solutions can be written by noting that

$$\psi = (b_1\alpha + b_2\beta) \exp\left[- i\,(Et - \mathbf{k} \cdot \mathbf{r})\right] \hbar\,] \tag{4.100}$$

satisfies Eq. (4.97) provided

$$E^2 = k^2c^2 + m^2c^4 \tag{4.101}$$

The corresponding x is obtained from Eq. (4.98), as

$$x = \frac{1}{mc^2}\left(E - \frac{2c}{\hbar}\mathbf{k} \cdot \mathbf{S}\right)(b_1\alpha + b_2\beta) \exp\left[-i(Et - \mathbf{k} \cdot \mathbf{r})/\hbar\right] \tag{4.102}$$

The most striking property of these solutions is that negative energy solutions with $E = -(k^2c^2 + m^2c^4)^{1/2}$ are allowed in addition to the usual positive energy solutions with $E = (k^2c^2 + m^2c^4)^{1/2}$.

Further discussion of the Dirac equation is not within the scope of this book. We will be content with making a few remarks.

1. The problem of one-electron atoms can be considered by the replacement

$$i\hbar\frac{\partial}{\partial t} \rightarrow i\hbar\frac{\partial}{\partial t} + \frac{Ze^2}{4\pi\,\varepsilon_0 r} \tag{4.103}$$

in Eqs. (4.98) and (4.99). The various fine structure terms can then be deduced by carrying out suitable expansions.

2. The existence of negative energy states creates some complications. Since no negative energy particles are observed in nature, how are possible transitions to negative energy states explained ? Dirac overcame this difficulty by postulating that vacuum consists of a sea of electrons which fill all the negative energy levels. Hence, transitions to negative energy

states are forbidden by the Fermi-Dirac statistics (see Chapter 7) according to which no two identical particles with half integral spin can occupy the same state [Fig. 4.3 (*a*)].

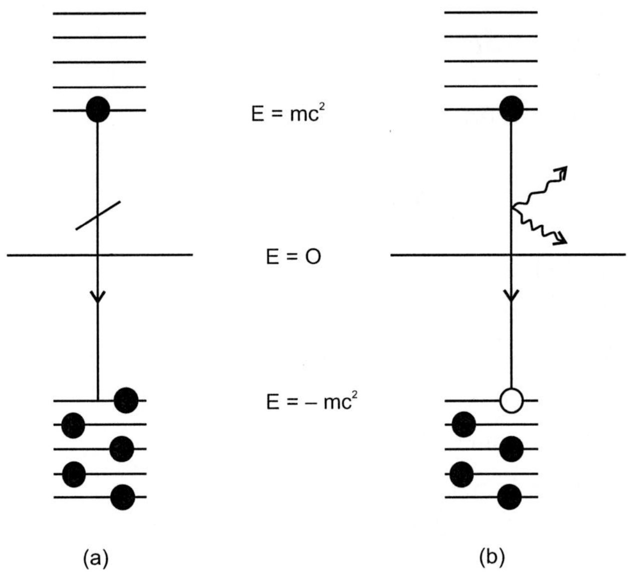

Fig. 4.3 (a) Forbidden transition to a filled negative energy state, and (b) allowed transition leading to annihilation of an electron and a hole with the emission of two photons. Filled circles are electrons and the open circle is a hole or a vacancy.

3. The Dirac hypothesis of the negative energy sea suggests that if enough energy is provided, a negative energy electron may become a positive energy electron. The energy of vacuum can be written as

$$E_v = \sum_i (-E_i) \tag{4.104}$$

where the summation is over all the negative energy states (ignoring the complication of the infinite sum). Suppose two photons with energies hv_1 and hv_2, come together and give all their energy to an electron with energy $-E_l$, which now has a positive energy E_n. Then, by energy conservation

$$hv_1 + hv_2 + \sum_i (-E_i) = E_n + \sum_{i \neq l} (-E_i) = E_n + E_l + \sum_i (-E_i) \tag{4.105}$$

The final state consists of an electron with energy E_n and another particle with energy E_l corresponding to the hole in the negative energy sea. A similar analysis for charge conservation shows that the final state consists of a negatively charged electron (with energy E_n) and a positively charged particle (corresponding to the hole in the negative-charge sea). The hole therefore has properties exactly opposite to those of the vacant negative energy electron state, i.e. it has positive energy and positive charge (also opposite momentum and spin). This hole state is called the *positron*, which is an example of what are known as *antiparticles*. The overall process is equivalent to two photons annihilating each other to produce an electron and a positron. The process is known as pair creation. Similarly, pair annihilation occurs when a positive energy electron drops into a vacancy in the negative energy sea, giving out radiation [Fig. 4.3 (*b*)]. Pair creation and annihilation are important processes in particle physics, though the associated particles may not always be two photons or electrons.

4. The presence of an external charge polarizes the sea of negative charges thus reducing the effective charge of the external particle. This is called *vacuum polarization*. As a result, an *s*-wave electron in the hydrogen atom, which is 'nearer' to the proton than a *p*-wave electron, sees a greater charge for the proton. Hence, the vacuum polarization lowers the *s*-wave levels compared to the *p*-wave levels. This contributes to the removal of *j*-degeneracy, in particular, the degeneracy between $2p_{1/2}$ and $2s_{1/2}$ states. However, there are additional contributions to the separation of these energy levels called the Lamb shift, for other effects such as self interaction, field fluctuations, etc. which can be treated within the framework of quantum electrodynamics. The predictions of the theory for the Lamb shift are in excellent agreement with the experimental observations.

4.9 EXAMPLES

A few examples to illustrate the properties of the one-electron atoms are discussed here.

Example 1

Though the solutions to the radial equation, Eq. (4.13), are in general complicated, the solutions for $l = n - 1$, are fairly simple.

Consider a solution of the form

$$R(r) = r^{n-1} \exp\left[-(-2m_r E/\hbar^2)^{1/2} r\right] \tag{4.106}$$

substitution of which in Eq. (4.13) gives the relation

$$\frac{\hbar^2}{2m_r}\left[2n\,(-2m_r E/\hbar^2)^{1/2}\left(\frac{1}{r}\right)-\frac{n(n-1)}{r^2}+\frac{l(l+1)}{r^2}\right]-\frac{Ze^2}{4\pi\varepsilon_0 r}=0 \tag{4.107}$$

This relation can be satisfied if

$$l = n - 1,$$

$$E = -\frac{m_r}{2\hbar^2}\left(\frac{Ze^2}{4\pi\varepsilon_0}\right)^2\left(\frac{1}{n^2}\right) \tag{4.108}$$

The corresponding solutions normalized according to Eq. (4.23), is

$$R_{n,n-1} = \left(\frac{1}{(2n)!}\right)^{1/2}\left(\frac{2Z}{a_1 n}\right)^{n+1/2} r^{n-1}\exp\,(-rZ/a_1 n) \tag{4.109}$$

where a_1 is the radius of the first Bohr orbit with $Z = 1$.

Example 2

The scaling properties of Eq. (4.13) provide a useful insight into the solutions. Consider a transformation

$$r \rightarrow \lambda r \tag{4.110}$$

which takes Eq. (4.13) to the form

$$\frac{\hbar^2}{2m_r}\left[-\frac{1}{r^2}\frac{d}{dr}r^2\frac{d}{dr}+\frac{l(l+1)}{r^2}\right]R(\lambda r)-\frac{(\lambda Z)e^2}{4\pi\varepsilon_0 r}R(\lambda r)$$

$$= \lambda^2 ER\,(\lambda r) \tag{4.111}$$

Taking $\lambda = 1/Z$, the equation for $Z = 1$ is obtained. Hence,

$$R(r, 1) = \frac{1}{Z^{3/2}}R(r/Z,Z)$$

$$E\,(Z = 1) = \frac{1}{Z^2}E(Z) \tag{4.112}$$

where Z is shown as an additional variable. The factor of $Z^{-3/2}$ in the first relation is due to normalization. Thus, we can obtain the solutions for Eq. (4.13) in terms of solutions for $Z = 1$. It also follows that

$$\langle r^n \rangle_Z = \frac{1}{Z^n}\langle r^n\rangle_{Z=1} \tag{4.113}$$

Example 3

A beam of sodium atoms with velocity 10^3 m/s, moves along the magnetic poles over a distance of 0.15 m, in the x-direction. We determine the separation between the two components of the beam, at a distance of 0.6 m from the magnet, given that the magnetic induction between the poles varies in the z-direction as

$$B = (1 - 100z) \text{ W/m}^2 \tag{4.114}$$

The force on the atoms is

$$\begin{aligned} F_z &= \pm \frac{e\hbar}{2m_e}(100) \\ &= \pm 9.28 \times 10^{-22} \text{ N} \end{aligned} \tag{4.115}$$

If t_1 is the time taken by the atoms to traverse the poles and t_2 is the time taken to move from the magnets to the plane of observation, the separation between the two components of the beam is

$$\Delta z = 2\left[\frac{1}{2}\left(\frac{|F_z|}{m_n}\right)t_1^2 + \left(\frac{|F_z|}{m_n}\right)t_1 t_2\right] \tag{4.116}$$

Since $t_1 \approx 1.5 \times 10^{-4}$ s and $t_2 \approx 6.0 \times 10^{-4}$ s, one gets

$$\Delta z = 4.9 \times 10^{-3} \text{ m} \tag{4.117}$$

Example 4

Some general properties of fine structure lines are now enumerated.

1. Transitions from the level with principal quantum number $n > 1$, to the ground state give rise to doublets corresponding to

$$np_{1/2} \to 1s_{1/2},\ np_{3/2} \to 1s_{1/2} \tag{4.118}$$

2. Transitions from the level with principal quantum number $n > 2$ to the level with $n = 2$, gives rise to seven lines corresponding to

$$\begin{aligned} &np_{1/2,\,3/2} \to 2s_{1/2} \\ &ns_{1/2} \to 2p_{1/2,\,3/2} \\ &nd_{3/2} \to 2p_{1/2,\,3/2} \\ &nd_{5/2} \to 2p_{3/2} \end{aligned} \tag{4.119}$$

3. If $n_0 \geq 2$, where n_0 is the principal quantum number of the final state, every increase in n_0 increases the number of fine structure line by 6. These additional lines correspond to

$$(n, l = n_0 + 1, j = n_0 + 3/2) \rightarrow (n_0 + 1, l = n_0, j = n_0 + 1/2)$$

$$(n, l = n_0 + 1, j = n_0 + 1/2) \rightarrow (n_0 + 1, l = n_0, j = n_0 \pm 1/2) \tag{4.120}$$

$$(n, l = n_0 - 1, j = n_0 - 1/2) \rightarrow (n_0 + 1, l = n_0, j = n_0 \pm 1/2)$$

$$(n, l = n_0 - 1, j = n_0 - 3/2) \rightarrow (n_0 + 1, l = n_0, j = n_0 \pm 1/2) \tag{4.121}$$

where $n_0 \geq 2$, and $n > n_0 + 1$. Hence, the number of fine-structure lines for transitions from $n \rightarrow n_0 + 1$ states is

$$N = 7 + 6\,(n_0 - 2),\; n \rightarrow n_0 + 1,\; n > n_0 + 1 \geq 3 \tag{4.122}$$

Thus, each line in the Paschen series consists of 13 lines, each line in the Brackett series consists of 19 lines, etc.

4. For a transition $n \rightarrow n'$, $n > n'$, the highest frequency corresponds to the transition

$$(n, j = 3/2) \rightarrow (n', j = 1/2) \tag{4.123}$$

and is given by

$$\nu = \nu_0 + \frac{Z^2\alpha^2\,|E_n|}{4n^2h}(3 - 2n) - \frac{Z^2\,\alpha^2\,|E_{n'}|}{4n'^2\,h}(3 - 4n') \tag{4.124}$$

where ν_0 is the frequency in the absence of fine structure correction given by

$$\nu_0 = \frac{E_n - E_{n'}}{h} \tag{4.125}$$

Example 5

The spin 1/2 space contains only two linearly independent states α and β as defined in Eq. (4.78). It is convenient to regard these states with $S_z = \pm \frac{1}{2}\hbar$ as two-component column vectors

$$\alpha = \begin{pmatrix} 1 \\ 0 \end{pmatrix} \tag{4.126}$$

$$\beta = \begin{pmatrix} 0 \\ 1 \end{pmatrix}$$

and describe the spin operators by 2×2 matrices. These matrices must satisfy the commutation relations in Eq. (3.163), and the properties stated in Eqs. (4.78), (4.80), (4.85) and (4.86). One set of such matrices is given by

$$S_x = \frac{1}{2}\hbar\begin{pmatrix} 0 & 1 \\ 1 & 0 \end{pmatrix}$$

$$S_y = \frac{1}{2}\hbar\begin{pmatrix} 0 & -i \\ i & 0 \end{pmatrix}$$

$$S_z = \frac{1}{2}\hbar\begin{pmatrix} 1 & 0 \\ 0 & -1 \end{pmatrix} \tag{4.127}$$

The operation of the spin operator is then given by the operation of these matrices on the column vectors given in Eq. (4.126). It is easy to see that these matrices give the results in Eq. (4.82) with the specific choice of $b_1 = b_2 = 1$.

PROBLEMS

1. Calculate the expectation value of $\langle -Ze^2/4\pi\varepsilon_0 r \rangle$ for the one-electron atoms in the ground state and hence deduce the expectation value $\langle p^2/2m_r \rangle$ of the kinetic energy.
2. For the ground state of the hydrogen atom, the wave function is of the form $\psi = b \exp(-r/a)$, where b is a constant and a is the Bohr radius. Determine the probability of finding the electron at a separation greater than $2a$. What is the corresponding classical probability?
3. Consider a wave function of the form $R(r) = (1 + br)\, e^{-gr}$ for a one-electron atom. What is the possible eigenvalue of this state?
4. What is the value of r at which $4\pi r^2\, |\phi\,(r)|^2$ has a maximum for the ground state? What is the probability density as a function of r, at this value?
5. The effect of the finite size of a nucleus may be taken into account by modifying the potential for $r < r_n$, such that the potential $V(r) = -Ze^2/4\pi\varepsilon_0 r_n$ for $r < r_n$, r_n being the radius of the nucleus. Treating the modification perturbatively, show that the correction to the ground state energy is approximately $\left(\dfrac{4r_n^2 Z^2}{3a_1^2}\right) |E_1|$. What is the order of magnitude of this correction?

6. Using scaling arguments, show that

$$E(m_r) = m_r E\,(1) \text{ and } R(r, m_r) = m_r^{3/2}\, R(rm_r, 1).$$

Hence show that $\langle r^n \rangle_{mr} = \dfrac{1}{m_r^n} \langle r^n \rangle_{m_r = 1}$.

7. A μ^- meson is similar to an electron except that its mass is 206.84 m_e. It can form a mesic. Atom with a proton. Compare the energy levels and the average radial distances $\langle r \rangle$ of such an atom with those of hydrogen.

8. Obtain the frequencies of the two Lyman lines corresponding to $n = 2 \rightarrow n = 1$ transition. Verify that these frequencies satisfy the relation given in Eq. (4.124).

9. What are the allowed electric dipole transitions between the fine structure states with $n = 3$ and $n = 2$?

10. Enumerate the $n = 2$ levels of the hydrogen atom in terms of the total angular momentum states, *i.e.* eigenstates of F^2. What are the allowed electric dipole transitions between these states?

11. Write Eqs. (4.98) and (4.99) in terms of $\psi + \chi$ and $\psi - \chi$. Show that for the free particle solutions with positive energy E and momentum **p**,

$$\psi - \chi = \frac{2c}{\hbar(E + mc^2)}\, \mathbf{p} \cdot \mathbf{S}\,(\psi + \chi)$$

which vanishes for $\mathbf{p} \rightarrow 0$. Therefore, corrections to nonrelativistic equations can be worked out conveniently in terms of $\psi \pm \chi$.

5

Atoms and Molecules

Structures of the Chapter

In Chapter 4, it was shown, that the quantum-mechanical framework provides a detailed and accurate description of the energy levels of the one-electron atom. In this chapter, the attempts to understand the properties of many-electron atoms and molecules will be discussed.

It is generally difficult to obtain accurate solutions to the problem of N interacting particles, where $N \geq 3$, whether in classical mechanics or in quantum mechanics (simple harmonic potential is an exception). However, there is a special symmetry property for spin 1/2 particle sin quantum mechanics, called the *Pauli exclusion principle*, which gives a very satisfactory qualitative and often quantitative understanding of many-electron atoms and molecules.

5.1 EXCHANGE SYMMETRY OF WAVE FUNCTIONS

In the treatment of two or more identical particles, such as electrons, within the framework of quantum mechanics, the uncertainty principle limits our ability to follow the motion of the particles without disturbing the system (in classical mechanics this disturbance can be made indefinitely small). Therefore, in general, it cannot be ascertained as to which of the identical particles has been found at a place.

Consider the Hamiltonian $H(1, 2, ...N)$ of N identical interacting particles, where the numbers represent all the variables of the particles, *i.e.* spatial coordinates and spin variables. Since the particles are identical, the following is true:

$$H(1, 2 ..., i ..., j, ... N) = H(1, 2, ..., j, ..., i, ...N) \tag{5.1}$$

Now, an operator P_{ij} which interchanges all the coordinates of articles i and j is defined as:

$$P_{ij}\,\psi(1, 2, ... i, ..., j,...N) = H(1, 2, ..., j, ..., i, ...N) \tag{5.2}$$

It is easy to show that P_{ij} is a hermitian operator (See Eq. 3.35). Furthermore, in view of Eq. (5.1), it is seen that

$$P_{ij}H(1, 2, ... i, ..., j, ... N) = H(1, 2, ..., i, ..., j, ...N)\,P_{ij} \tag{5.3}$$

This means that states which are simultaneous eigenstates of H and P_{ij} can be chosen (see Sec. 3.4).

The eigenvalues of P_{ij} can be deduced from Eq. (5.2) by noting that

$$P_{ij}^2 = 1 \tag{5.4}$$

which implies that the possible eigenvalues of P_{ij} are + 1 or –1. It is experimentally observed that the physical states indeed are (not just 'can be') eigenstates of P_{ij} and the eigenvalues are characteristic of the nature of the particles. This result is stated in terms of the following rules:

1. The wave function of a system is symmetric with respect to the interchange of the space and spin variables of any two identical particles i and j, with integral quantum number for their spin

$$\psi\,(1, 2, ..., i, ..., j, ...\, N) = \psi\,(1, 2, ..., j, ..., i, ...N) \tag{5.5}$$

2. The wave function of a system is antisymmetric with respect to the interchange of the space and spin variables of any two identical articles k and l, with half-integral quantum number for their spin

$$\psi\,(1, 2, ..., k, ..., l, ...N) = -\,\psi\,(1, 2, ..., l, ..., k, ...N) \tag{5.6}$$

The particles with integral quantum number for their spin are called *bosons* (because of Bose-Einstein statistics which governs these particles, Chapter 7) and the particles with half-integral quantum number for their spin are called *fermions* (because of Fermi-Dirac statistics which governs these particles, Chapter 7). Examples of bosons are the photon (spin 1), deuteron (spin 1), π-meson (spin 0), etc. while examples of fermions are the proton (spin 1/2), neutron (spin 1/2), electron (spin 1/2), Ω– (spin 3/2), muon (spin 1/2), etc. It may be noted that the symmetry requirement for fermions with spin 1/2 and for bosons can be deduced within the framework of quantum field theory with the assumptions of Lorentz invariance, etc. and is known as the *spin-statistics theorem*.

The symmetry properties stated in Eqs. (5.5) and (5.6) are of great importance and give rise to significant macroscopic phenomena such as superfluidity of liquid helium at low temperatures, blackbody radiation, magnetism, stimulated emission of radiation, shell structure in atoms and in nuclei, etc. Here, the implications of an antisymmetric wavefunction for the electrons, on the structure and energy levels of atoms and molecules are discussed.

A simple model is now considered to illustrate the ideas of symmetrization of states. In this model, the N identical particles do not interact with each other, but each of these particles has the same external interaction. The Hamiltonian for such a system is of the form

$$H = \sum_{i=1}^{N}\left[\frac{1}{2m}p_i^{\,2} + V(i)\right] \tag{5.7}$$

The separable eigenstates of this Hamiltonian with energy E are of the form

$$\psi\,(1, 2, ..., N) = \psi_{a_1}(1)\,\psi_{a_2}(2)...\psi_{a_N}(N) \tag{5.8}$$

where ψ_{a_i} are the normalized solutions of the equation

$$\left[-\frac{\hbar^2}{2m}\nabla i^2 + V(i)\right]\psi_{a_i}(i) = E_{a_i}\,\psi_{a_i}(i) \tag{5.9}$$

with the total energy being

$$E = \sum_{i=1}^{N} E_{a_i} \tag{5.10}$$

It is clear that any permutation of the particle indices in the solution in Eq. (5.8) will again give an eigenstate of the Hamiltonian, with energy E. Since all these states are degenerate, any linear combination of these states will also be an eigenstate with energy E. In particular, the symmetric and antisymmetric eigenstates are respectively

$$\psi_+(1, 2, ..., N) = \frac{1}{(N!)^{1/2}} \sum_{\text{perm}} \psi_{a_1}(1)\, \psi_{a_2}(2) \ldots \psi_{a_N}(N) \tag{5.11}$$

$$\psi_-(1, 2, ..., N) = \frac{1}{(N!)^{1/2}} \det. \begin{bmatrix} \psi_{a_1}(1)\, \psi_{a_1}(2) \ldots \psi_{a_1}(N) \\ \psi_{a_2}(1)\psi_{a_2}(2) \ldots \psi_{a_2}(N) \\ \vdots \qquad \ldots \qquad \ldots \\ \psi_{a_N}(1)\, \psi_{a_N}(2) \ldots \psi_{a_N}(N) \end{bmatrix} \tag{5.12}$$

where the summation is over all the permutations of the particle indices (the normalization is different if some of the a_i are the same). For the simple case of two identical particles, the symmetric and antisymmetric states are

$$\psi_\pm(1, 2) = \frac{1}{2^{1/2}} [\psi_{a_1}(1)\, \psi_{a_2}(2) \pm \psi_{a_1}(2)\, \psi_{a_2}(1)] \tag{5.13}$$

In these equations, *i.e.* Eqs. (5.11) to (5.13), the solutions ψ_+ with the plus sign are applicable to bosons while the solutions ψ_- with the minus sign apply to fermions. These solutions are of great importance as solutions for N identical particles, and serve as approximate starting solutions even for systems whose Hamiltonian cannot be written in the separable form in Eq. (5.7).

An extremely important point to note in Eqs. (5.12) and (5.13), is that the fermion wave functions vanish if any two of the a_i are equal. This means that no two identical, noninteracting fermions can be in states described by the same set of quantum numbers. This rule was first stated for electrons in an atom by Pauli (1925), *no two electrons in an atom can have the same set of quantum numbers* n, l, m_l and m_s, and is known as Pauli's exclusion principle. It is central to the understanding of the structure of atoms. We now begin the analysis of the structure and the energy levels of an atom, subject to the constraints of Pauli's exclusion principle.

5.2 SHELLS AND SUBSHELLS IN ATOMS

The Hamiltonian for the electrons in an atom with the nucleus placed at the origin, can be written as:

$$H = H_0 + H_1 + H_2 + H_3 \tag{5.14}$$

where

$$H_0 = \sum_i \left[\frac{1}{2m} p_i^2 - \frac{Ze^2}{4\pi\,\varepsilon_0 e_i} + V(r_i) \right] \tag{5.15}$$

$$H_1 = \frac{1}{2} \sum_{i \neq j} \frac{e^2}{4\pi\varepsilon_0 r_{ij}} - \sum_i V(r_i) \tag{5.16}$$

$$H_2 = \frac{1}{2} \sum_i \frac{1}{2m^2c^2 r_i} \left(\frac{dV}{dr_i} \right) \mathbf{l}_i.\mathbf{s}_i \tag{5.17}$$

and H_3 includes the corrections due to spin-spin interaction, relativisitic corrections, etc. The term H_0 contains the potential due to the nucleus and $V(r_i)$ which represents some averge potential due to the other electrons (the other electrons screen the nuclear charge). The term $V(r_i)$ is not important for small r_i but will become increasingly important as r_i increases. In a simple model, the screening contribution is assumed to be of the form

$$V(r) = \frac{(Z-1)e^2}{4\pi\varepsilon_0 r} (1 - e^{-r/b}) \tag{5.18}$$

which has the correct behaviour for $r \to 0$ and $r \to \infty$. Since b is expected to be large compared to the radius of the inner orbits, the potential can be expanded in powers of r to obtain an approximate expression for $V(r_i)$ as

$$V(r_i) \approx \frac{(Z-1)e^2}{4\pi\varepsilon_0} \left(\frac{1}{b} - \frac{1}{2b^2} r_i + \ldots \right) \tag{5.19}$$

The second term H_1 represents the deviations of the actual repulsive potential from the average potential $V(r_i)$. The term H_2 describes the spin-orbit interaction of the electrons. It is of the same form as Eq. (4.51) except that $Ze^2/4\pi\varepsilon_0\, r^2$ has been replaced by the more general expression dV/dr. Of these three terms, the general structure of the atoms is determined mainly by H_0. The terms H_1 and H_2, however, play an important role in the determination of the energy levels of the atom, in particular the fine structure of the levels.

For obtaining the structure of the atoms, one starts with only the kinetic energy of the electrons and the electrostatic interaction of the electrons with the nucleus, *i.e.* the first two terms in H_0 [Eq. (5.15)]. The energy levels of the electrons are then those of a one-electron atom, *i.e.*

$$E_n^{(0)} = -\frac{m}{2\hbar}\left(\frac{Ze^2}{4\pi\varepsilon_0}\right)^2 \frac{1}{n^2} \tag{5.20}$$

and the total energy is the sum of the energies of the N electrons,

$$E^{(0)} = \sum_{i=1}^{N} E_n^{(0)}(i) \tag{5.21}$$

However, the states that can be occupied by the electrons are constrained by Pauli's exclusion principle. The ground-state energy is therefore obtained by placing successive electrons in the lowest-energy, unocoupied states. It may be noted (see Eq. (4.27)) that for each value of the principal quantum number n, there are $2n^2$ states (including the factor of 2 due to the states) with the same energy. Thus, the first two electrons are to be placed in the $n = 1$ states, the next 8 electrons in the $n = 2$ states, the next 18 electrons in the $n = 3$ state, etc. Electrons with the same value of n form what are known as shells which are designated by the letters K for $n = 1$, L for $n = 2$, M for $n = 3$, etc.

It may be recollected (Sec. 4.1) that the degeneracy of the different l states (with $l \leq n - 1$) for a given value of the principal quantum number n, is a special property of the $1/r$ potential. The average potential $V(r_i)$, arising from the interaction with the other electrons will remove this degeneracy and states with different l value but the same n value, will have different energies, Since $V(r_i)$ is positive and becomes more important as r_i increases, it may be expected that the states with larger l values will be raised more than those with smaller l values. Explicit perturbative calculations can be made for the first two terms of the potential $V(r_i)$ given in Eq. (5.19). From Eq. (3.125),

$$E_{n,\,l} \approx E_n^{(0)} + \frac{(Z-1)e^2}{4\pi\varepsilon_0}\left[\frac{1}{b} - \frac{a_1}{4b^2 Z}(3n^2 - l\,(l+1))\right] \tag{5.22}$$

where Eq. (4.29) has been used for $\langle r \rangle$, a_1 being the radius of the first Bohr orbit with $Z = 1$. It is seen here that the screening effects due to other electrons remove the l-degeneracy, the energies now increasing as l increases. This implies that each shell is made up of subshells that have the same n value but different l values, the subshells with larger l values having higher energy. Indeed, it so happens that the energy of a subshell with sufficiently large l may be higher than that of another with larger n but a lower l. The relative positions of the various energy levels which follow from detailed calculations, and also from experimental observations, are shown in Fig. (5.1) and form the basis of the shell structure of the atoms.

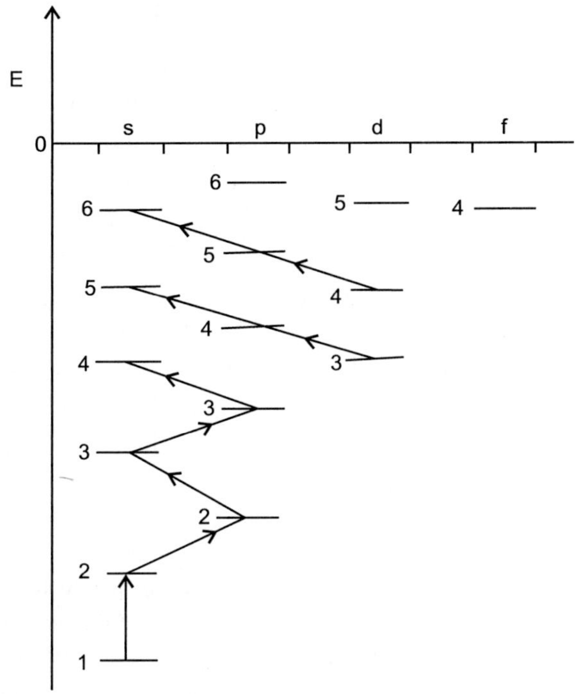

Fig 5.1 Schematic illustration of the energy levels of atomic subshells and the order in which the subshells are filled (given by the arows).

5.3 PERIODIC TABLE

For determining the manner in which the electrons in an atom are distributed among the states indicated in Fig. 5.1, two rules must be kept in mind: (*i*) In the ground state of the atom, the electrons occupy the lowest energy level available, and (*ii*) no two electrons can have the same quantum numbers n, l, m_l and m_s. The second rule implies that the maximum number of electrons each subshell can contain is equal to the degeneracy $2(2l + 1)$ of that level, *i.e.* 2 in an s-shell, 6 in a p-shell, etc. The order in which the shells are filled is according to the increasing values of their energies, and is shown in Fig. 5.1, the order being 1*s*, 2*s*, 2*p*, 3*s*, 3*p*, 4*s*, 3*d*, 4*p*, 5*s*, 4*d*, 5*p*, 6*s*, 4*f*, 5*d*, 6*p*, 7*s*. A few examples are given here to illustrate the results.

Examples

In its ground state H (hydrogen) has an electron with $n = 1$, $l = 0$, $m_l = 0$ and $m_s = 1/2$ or $-1/2$. This configuration is designated by 1*s*. The two electrons of the He (helium) are in states $n = 1$, $l = 0$, $m_l = 0$ and $m_s = \pm 1/2$, the configuration being $(1s)^2$. The configuration of Na is $(1s)^2\ (2s)^2\ (2p)^6\ 3s$, while that of Sr is $(1s)^2\ (2s)^2\ (2p)^6\ (3d)^2\ (3p)^6\ (3d)^{10}\ (4s)^2\ (4p)^6\ (5s)^2$ the total number of electrons being 38.

The electronic configurations are summarized in Table 5.1, showing the order in which the subshells are filled. The exceptions (shown in bold type) along with the configuration of the unfilled shells are Cr–$(4s)\ (3d)^5$, Cu – $4s\ (3d)^{10}$, Nb – $5s\ (4d)^4$, Mo–$5s\ (4d)^5$, Ru–$5s\ (4d)^7$, Rh –$5s\ (4d)^8$, Pd – $(4d)^{10}$ and no electrons in the $5s$ shell, Ag–$5s\ (4d)^{10}$, La – $(5d)$ and no electrons in the $4f$ shell, Gd–$5d\ (4f)^7$, Pt – $6s\ (5d)^9$ and Au–$6s\ (5d)^{10}$. Apart from these, of the heavy elements (Z = 89 to 102), Ac, Pa, U, Np, Pu, Am, Cm, Bk have the configuration $6d(5f)^{Z-89}$. The has the configuration $(6d)^2$ with no electrons in $5f$ shell, while Cf, E, Fm, Md, No have the configuration $(5f)^{Z-88}$ with no electrons in the $6d$ shell. Some important consequences of the electronic configurations of the atoms are now discussed.

Ionization Potential

Ionization potential of an atom is the least energy required to ionize the atom, and hence is also the *binding energy* of the least strongly bound electron. This electron would be expected to belong to the last shell (filled or unfilled). In Fig. 5.2 is shown the variation of the binding energy as Z changes, which can be understood from the following simple considerations.

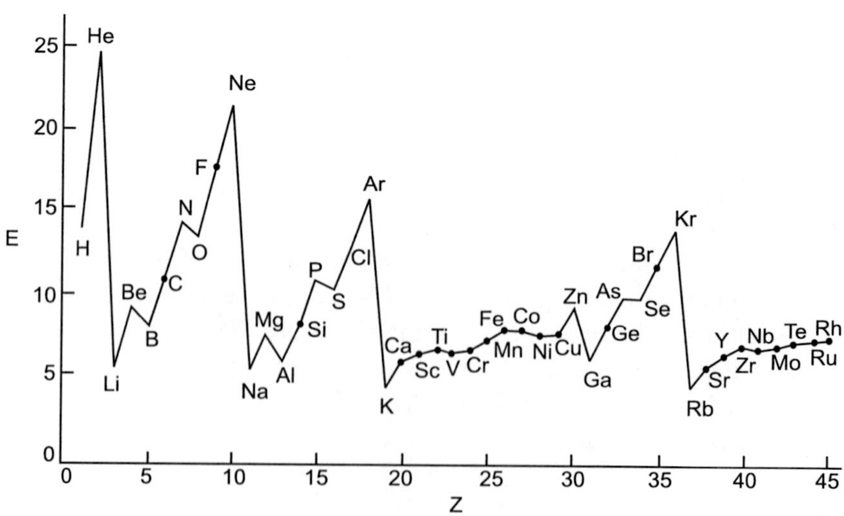

Fig. 5.2 Ionization potential in eV, as a function of Z.

Table 5.1 Electronic structure of elements

Subshell being filled	***Range of Z***	***Sequence of elements***
1*s*	1–2	H, He
2*s*	3–4	Li, Be
2*p*	5–10	B, C, N, O, F, Ne
3*s*	11–12	Na, Mg
3*p*	13–18	Al, Si, P, S, Cl, Ar
4*s*	19–20	K, Ca
3*d*	21–30	Sc, Ti, V, Cr, Mn, Fe, Co, Ni, Cu, Zn
4*p*	31–36	Ga, Ge, As, Se, Br, Kr
5*s*	37–38	Rb, Sr
4*d*	39–48	Y, Zr, Nb, Mo, Tc, R, Rh, Pd, Ag, Cd
5*p*	49–54	In, Sn, Sb, Te, I, Xe
6*s*	55–56	Cs, Ba
4*f*	57–70	La, Ce, Pr, Nd, Pm, Sm, Eu, Gd, Tb, Dy, Ho, Er, Tm, Yb
5*d*	71–80	Lu, Hf, Ta, W, Re, Os, Ir, Pt, Au, Hg
6*p*	81–86	Tl, Pb, Bi, Po, At, Rn
7*s*	87–88	Fr, Ra
5*f*–6*d*	89–102	Ac, Th, Pa, U, Np, Pu, Am, Cm, Bk, Cf, E, Fm, Md, No

The shells are filled in the order shown, the exceptions being shown in bold type. For the heaviest elements ($Z = 89$ to 102), the electrons are in both 5*f* and 6*d* subshells.

Consider an atom with Z electrons, i of while are in the last subshell. Then the potential seen by an electron in the last shell will be that due to the nucleus (with charge $Z|\ e\ |$) screened by Z-i electrons, *i.e.* essentially an attractive potential due to charge $i\ |\ e\ |$. Therefore, the ionization potential (or the binding energy of an electron in the last subshell) may be expected to increase as i increases. This trend is generally observed, with the ionization potential being a minimum for atoms with only one electron in the last shell, *e.g.* Li, Na, K, Ga, Rb, and a maximum for atoms with the last shell being complete, *e.g.* He, Ne, Ar, Kr, and Zn (less prominent). Of course, these arguments are very qualitative.

The situation becomes ambiguous if some of the subshells have approximately the same energy, of interpenetrate, and would require a more careful analysis.

Chemical Properties

The chemical properties of elements are related to the forces between the atoms. In this connection it is seen that (*i*) the atoms with slightly filled outer shells, are characterized by a small ionization energy. They can easily lose the electrons in the outer shell, and form positive ions. These elements are chemically active. (*ii*) Atoms with only a small number of vacancies in the outer shell (*i.e.* with almost-filled outer shell), though their net charge is zero, can easily accommodate electrons in the unfilled shell since the screening of the nucleus is mainly due to the inner shells. For example, Cl can bind an additional electron with a binding energy of 3.80 eV (even the hydrogen atom can bind an additional electron with a binding energy of 0.75 eV). With the acceptance of electrons, these elements form negative ions. They are chemically active. (*iii*) Since the chemical activity is related to the number of electrons in the outer shell, there exist groups of atoms with similar outer shells but with different inner shells, which have similar chemical properties. This is the origin of the periodic table (see Table 5.2) where the atoms are arranged in such a way that the atoms in vertical columns have similar outer shells and hence similar chemical properties. Examples of these are (*i*) Noble gases, helium, neon, argon, krypton, xenon and radon. They have closed outer shells. Their ionization potential is large, and they cannot easily form positive ions. Nor can they bind an additional electron which has to go into the next shell and hence would hardly experience any attraction. These atoms are therefore chemically quite inactive. (*ii*) Alkali metals, lithium, sodium, potassium, rubidium, caesium and francium. They have only one *s*-wave electron in the outer shell. Since the electron has a small binding energy, it can be easily lost. These atoms are positive monovalent, and are chemically active. (*iii*) Halogens, fluorine, chlorine, bromine, iodine and astatine with five electrons in the outer *p*-shell. They have an affinity for an additional electron which would complete the shell. The atoms are negative monvalent and are chemically active. (*iv*) Especially interesting are the groups of ten elements that correspond to progressive filling of an *nd* subshell but have a complete $(n + 1)s$ subshell.

Table 5.2 Periodic table of elements (deviations from the shown shell structure are indicated by encircling the elements).

	IA	IIA
1s	1 H	2 He
2s	3 Li	4 Be
3s	11 Na	12 Mg
4s	19 K	20 Ca
5s	37 Rb	38 Sr
6s	55 Cs	56 Ba
7s	87 Fr	88 Ra
	1	2

	IIIA	IVA	VA	VIA	VIIA	VIIIA			IB	IIB
3d	21 Sc	22 Ti	23 V	**24 Cr**	25 Mn	26 Fe	27 Co	28 Ni	**29 Cu**	30 Zn
4d	39 Y	40 Zr	**41 Nb**	**42 Mo**	43 Tc	**44 Ru**	**45 Rh**	**46 Pd**	**47 Ag**	**48 Cd**
5d	71 Lu	72 Hf	73 Ta	74 W	75 Re	76 Os	77 Ir	**78 Pt**	**79 Au**	80 Hg
6d	89 Ac	90 Th	**91 Pa**	**92 U**	**93 Np**	**94 Pu**				
	1	2	3	4	5	6	7	8	9	10

	IIIB	IVB	VB	VIB	VIIB	VIIIB
2p	5 B	6 C	7 N	8 O	9 F	10 Ne
3p	13 Al	14 Si	15 P	16 S	17 Cl	18 Ar
4p	31 Ga	32 Ge	33 As	34 Se	35 Br	36 Kr
5p	49 In	50 Sn	51 Sb	52 Te	53 I	54 Xe
6p	81 Tl	82 Pb	83 Bi	84 Po	85 At	86 Rn
	1	2	3	4	5	6

	1	2	3	4	5	6	7	8	9	10	11	12	13	14
4f	**57 La**	58 Ce	59 Pr	60 Nd	61 Pm	62 Sm	63 Eu	**64 Gd**	65 Tb	66 Dy	67 Ho	68 Er	69 Tm	70 Yb

Since they all have a similar outer shell, their chemical properties also are similar. The three series of elements are $21 \leq Z \leq 30$ for $n = 3$, $39 \leq Z \leq 48$ for $n = 4$, and $71 \leq Z \leq 80$ for $n = 5$. The partially filled inner $3d$ shell allows some of these elements to have large magnetic moments. In particular, Fe, Co and Ni are found to be ferromagnetic materials (see Sec. 8.6). (*v*) The rare-earths are a group of 14 elements corresponding to a progressive filling of the $4f$ subshell, though the $6s$ subshell is already complete. These elements have $57 \leq Z \leq 70$.

In summary, the electronic structure of atoms provides an insight into their properties.

5.4 ATOMIC SPECTRA

So far only the general features of the electronic structure of an atom have been considered. These follow from the model in which only the interaction of the electrons with the nucleus (which produces the single-electron levels with l-degeneracy) and a term which represents an average interaction between the electrons (which breaks the l-degeneracy) are included. In this model, the energy of the atom is the sum of the energies of the electrons. Since the model ignores the finer details of the interaction, the various energy levels are highly degenerate. For example, a level with one electron in the $4p$ shell and another in the $3d$ shell will have a degeneracy of 6×10 corresponding to a degeneracy of $2(2l + 1)$ for each electron. This degeneracy is partially removed if the interaction between the electrons and their spin-orbit interaction, *i.e.*, the terms H_1 and H_2 in Eq. (5.14) are included.

In this section, the effect of including the finer details of the mutual interaction between the electrons (the term H_1) and the spin-orbit interaction (the term H_2) will be considered. This will allow the characterization of different energy levels for a given shell configuration, in particular, the ground state. The purpose will be essentially to enumerate the various energy levels, and to some extent predict the ordering of the energy levels (the determination of the energies will require a much more elaborate calculation). These results will be useful not only for predicting the multiplicity of the spectral lines, but also for describing the behaviour of the atoms under different physical conditions, *e.g.* in the presence of an external magnetic field.

Once the mutual interaction and spin-orbit interaction terms are included, the total orbital angular momentum is no longer conserved, nor is the total spin angular momentum. However, since the interaction is a scalar, the total angular momentum $\mathbf{J} = \mathbf{L} + \mathbf{S}$ is still conserved (*i.e.* $[\mathbf{L}, H] \neq 0$, $[\mathbf{S}, \mathrm{H}] \neq 0$, but $[\mathbf{J}, H] = 0$), and the energy eigenstates can be designated by their total angular momentum. We will now use **l**, **s** and **j** to designate the individual electron angular momenta and **L**, **S** and **J** to designate the sums of the angular momenta. The problem of determining the multiplicity of energy levels corresponding to a given shell then reduces to the problem of determining the different possible total angular momentum states for a given shel configuration. In this context, it is noted that the total angular momentum of an assembly of electrons forming a complete shell is zero. This follows from the observation that the z components of the total orbital angular momentum and the total spin momentum for the assembly, are both zero, *i.e.*

$$\sum_{m_l=-l}^{l} m_l = 0$$

$$\sum_{m_s=-1/2}^{1/2} m_s = 0 \tag{5.23}$$

and therefore

$$\sum_{\text{shell}} m_j = \sum_{\text{shell}} (m_l + m_s) \tag{5.24}$$

$$= 0$$

Since the z-axis can be taken along any arbitrary direction, this implies that the total angular momentum is zero. Therefore, the total angular momentum of an atom is due to contributions from only the unfilled shells.

A detailed perturbative calculation of the contribution of H_1 and H_2 gives the important result that the atoms fall into two main categories:

1. For most atoms the residual mutual interaction between the electrons, *i.e.*, H_1, is more important than the spin-orbit interaction represented by H_2. This situation is treated as *LS* coupling of Russel-Saunders coupling.
2. For some atoms, mainly heavy atoms with large unclear charges, the spin orbit interaction, *i.e.* H_2, is more important than H_1. This is treated as *j-j* coupling.

Russel-Saunders or LS Coupling

In this scheme, the spin-orbit interaction is first neglected. Since H_1, being independent of **S**, commutes with **S** and therefore also **L**, the energy eigenstates can be designated by L and S quantum numbers. Having determined the different states and their qualitative ordering, the effect of spin-orbit interaction can then be introduced as a small perturbation to deduce the final multiplicity of the energy levels.

For determining the ordering of the energy eigenstates, it is noted that for the state with the largest total spin, the spins are essentially parallel to each other and therefore the spatial wave function will be antisymmetric under the exchange of spatial coordinates. There is a tendency for the electrons to avoid each oher, thus minimizing the mutual repulsion. As a result, the lowest energy state is generally the state with the largest total spin S. Extending these qualitative results, it is deduced that as the total spin, decreases, the energy of the states increases. Similar arguments can be used to determine the ordering of the L levels for a given S. Of the states with different L values, the state with the largest L will correspond to electrons moving in the same sense, *i.e.* they can avoid coming together. Therefore, *the ground state has the largest possible L value compatible with the largest possible S value*. This is known as *Hund's rule*. In addition, the energy of the states for a given S increases as L decreases. Finally, the degeneracy of the states with different J values but the same L and S values is removed by the spin-orbit interaction, *i.e.* H_2, whose effect can be represented by (this can be rigorously justified by using the theorem in Sec. 6.2),

$$H_2 = C_{LS}\,\mathbf{L}\cdot\mathbf{S} \tag{5.25}$$

where C_{LS} is constant for a given L and S. This term gives a perturbative contribution to the energy,

$$\Delta E_J = \frac{1}{2} C\,[J\,(J+1) - L\,(L+1) - S\,(S+1)] \tag{5.26}$$

where the subscripts of C have dropped. It is found that the constant C is positive for multiplets formed from a subshell that is half or less than half-filled, and negative for multiplets formed from a subshell which is more than half-filled. Therefore, for a subshell which is half-filled or less, the energy within the multiplet increases as J increases, and for a subshell which is more than half-filled, the energy within the multiplet decreases as J increases. This is known as the *multiplet rule*. In partiular, this means that the ground state of an atom has the smallest J value subject to Hund's rule if the subshell is half-filled or less, and the largest J value if the subshell is more than half-filled. The separation between the levels with values $J + 1$ and J (but the same L and S) is obtained as

$$E_{J+1} - E_J = C\,(J + 1) \tag{5.27}$$

This is known as the *Lande interval rule* which states that *the spacing between consecutive levels of a fine-structure multiplet is proportional to the larger of the two J values of the levels*. These ideas are made explicit by the following two examples.

Consider an atom with two valence electrons, one in the *ns* state and the second in the $n'\ l$ state where $l \neq 0$. This state has a total degeneracy of $4(2l + 1)$ corresponding to two states for the s electron and $2(2l + 1)$ states for the l electron. The total wave function is a product of the spatial part

$$u_{n,n',L}\,(\mathbf{r}_1, \mathbf{r}_2) = R_n\,(y_1)\,Y_0^{\,0}\,(\theta_1, \phi_1)\,R_n'\,(r_2)\,Y_l^{\,m}\,(\theta_2, \phi_2) \tag{5.28}$$

and the spin part

$$v = v^{m_s}(1)\, v^{m'_s}(2) \tag{5.29}$$

It is clear that the total orbital angular momentum of the state u corresponds to $L = l$. The state can be symmetrized or antisymmetrized with respect to the two electrons, so that there are two states with $L = l$ given by

$$u^{\pm}_{n,n',\,l} = \frac{1}{2^{1/2}}\left[u_{n,n',l}(\mathbf{r}_1, \mathbf{r}_2) \pm u_{n,n',l}(\mathbf{r}_2, \mathbf{r}_1)\right] \tag{5.30}$$

For the spin part, since both the electrons have $s = 1/2$, the allowed values of S are $S = 1, 0$. It can be shown that the three symmetric states

$$v_1^{+} = v^{m_s}(1)\, v^{m's}(2) + v^{m_s}(2)\, V^{m's}(1) \tag{5.31}$$

correspond, except for normalization, to the $S = 1$ states, while the antisymmetric state

$$v_0^{-} = v^{1/2}(1)\; v^{-1/2}(2) - v^{1/2}(2)\; v^{-1/2}(1) \tag{5.32}$$

corresponds (except for normalization) to the $S = 0$ state. The total allowed antisymmetric wave functions $\psi_{L,\,s}$ are,

$$\psi_{l,\,0} = u^{+}_{n,n',\,l}\; v_0^{-} \tag{5.33}$$

which are $2l + 1$ in number, and

$$\psi_{l,1} = u^{-}_{n,\,n',\,l}\; v_1^{+} \tag{5.34}$$

which are $3(2l + 1)$ in number. Our earlier discussion indicates that $\psi_{l,\,1}$ states (which have the largest allowed S value, $S = 1$) have lower energy. The $\psi_{l,\,1}$ states can have $J = l + 1, l, l - 1$. The degeneracy between these states is removed by the spin-orbit interaction [see Eq. (5.26)], such that the states with larger J values have greater energy. The resulting energy levels are shown in Fig. (5.3), and there are a total of $4(2l + 1)$ states. These states are characterized by the notation $^{(2S+1)}L_J$, *e.g.* if $l = 1$, the singlet state is $^1P_1$ while the triplet states are $^3P_{2,1,0}$. The case $l = 0$ needs special consideration. For $l = 0$ but $n \neq n'$, there are only the $S = 0, 1$ leels which also correspond to $J = 0$ and 1 respectively. They are denoted by $^1S_0$ and $^3S_1$ respectively. If $l = 0$ and $n = n'$, there is only one level with $S = J = 0$ ($S = 1$ is not allowed by the exclusion principle) and is represented by $^1S_0$.

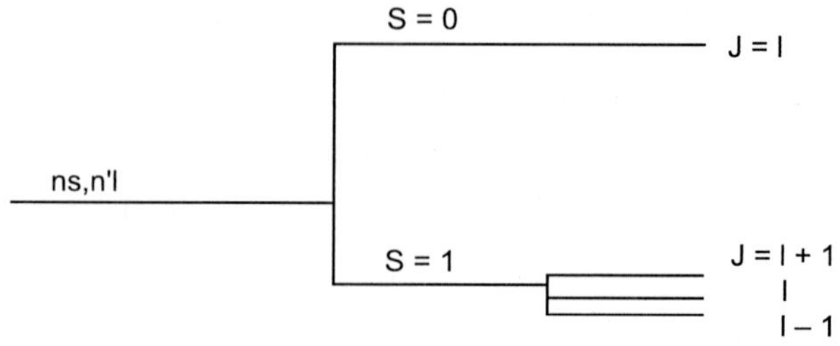

Fig. 5.3 Schematic illustration of the fine-structure splitting of a level with *ns* and *n'l* ($l \neq 0$) electrons. The states are split into $S = 1$ and $S = 0$ states by the electrostatic interaction between the electrons. The $S = 1$ state is further split into $J = l - 1, l, l + 1$ by the spin-orbit interaction.

The second example considered is an atom with one valence electron in the *np* state, and the second in the $n'l$ state, the total degeneracy being $12(2l + l)$. Since $l = 0$ case was considered in the first example, it is assumed here that $l \neq 0$, and also that $l \neq 1$. Then the spatial wave functions are still given by Eq. (5.30) except that the allowed values of the angular momentum quantum numbers now are $l + 1, l, l - 1$, *i.e.*

$$u^{\pm}_{n',\,n,\,L} = \frac{1}{2^{1/2}}\left[u_{n,n',L}(\mathbf{r}_1, \mathbf{r}_2) \pm u_{n,n',L}(\mathbf{r}_2, \mathbf{r}_1)\right],\; L = l+1, l, l-1 \tag{5.35}$$

The total wave functions are given by Eqs. (5.33) and (5.34) except that $u^{\pm}_{n,n'n'}$, l are replaced by $u^{\pm}_{n,\,n',\,L}$, with $L = l + 1, l, l - 1$. The spin-orbit interaction removes the J degeneracy, and the final energy levels are shown in Fig. 5.4. As before, they are described by the notation ${}^{(2S+1)}L_J$. If $l = 1$ but $n \neq n'$, there is only one level corresponding to $L = l - 1$, for each S, with $J = 0$ for $S = 0$, and $J = 1$ for $S = 1$. The other levels, *i.e.* $L = l, l + 1$ are singlets or triplets, as shown in Fig. (5.4). Finally, the case of $l = 1$ and $n = n'$ requires a special treatment. In this case, the allowed values of L are $L = 2, 0$ for $u^{+}_{n}, {}_{n,\,L}$ and $L = 1$ for $u^{-}_{n,\,n,\,L}$. Thus, the $S = 0$ state has $L = 2, 0$ states associated with it, while the $S = 1$ state has $L = 1$ associated with it. However, the $S = 1$ state splits into $J = 2, 1, 0$ states because of the spin-orbit interaction, for which the energy increases with J.

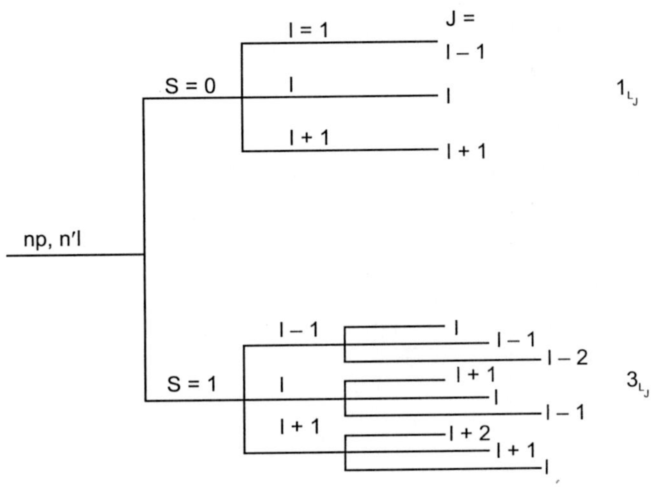

Fig. 5.4 Schematic illustration of the fine-structure splitting of a level with np, n′l (l > 1) electrons.

j–j Coupling

While the *LS* coupling scheme is applicable to the fine structure of most atoms, there are some atoms, mainly the heavy atoms with large nuclear charges, for which the spin-orbit interaction, i.e. H_2, is more important than the residual electrostatic interaction, *i.e.* H_1. For these atoms, the *j-j* coupling scheme is used.

In this case, the original single-particle level is split into two levels corresponding to $j_i = l_i \pm 1/2$ (except for $l_i = 0$ for which there will be only the $j_i = 1/2$ level) by the spin-orbit interaction H_2 such that

$$\Delta E_i = \frac{1}{2} C_{n_i l_i} [j_i (j_i + 1) - l_i (l_i + 1) - 3/4],$$
$$j_i = l_i \pm 1/2, \; l_i \neq 0 \tag{5.36}$$

where $C_{n_i l_i}$ is the expectation value of the coefficient of $\mathbf{l}_i.\mathbf{s}_i$ in H_2. The many-electron state with a given set of j_i contains states with different J values, which are degenerate. This degeneracy is removed by the residual electrostatic interaction between the electrons. The final levels are characterized by the quantum numbers j_i and the total J. A schematic illustration of the levels given in Fig. 5.5 for the fine-structure splitting of the levels with one electron in the *np* state and another in the $n'l$ state ($l \geq 2$). These levels are generally represented by $np\, n'l\, (j_1, j_2)_J$. For example, the ground state is $npn'l(1/2, l - 1/2)_{l-1}$. It is to be noted that if $l = 0$, the only allowed value of j_2 is is 1/2, so that $J = 1, 0$ for $j_1 = 1/2, j_2 = 1/2$ and $J = 2, 1$ for $j_1 = 3/2,\; j_2 = 1/2$. For $l = 1$ but $n' \neq n$, one has only $J = 2, 1$ for $j_1 = 3/2, j_2 = 1/2$. Finally, for $l = 1$ and $n' = n$, the electrons are in the same subshell. There is only one set of antisymmetric states corresponding to $j_1 = 1/2, j_2 = 3/2$ and $j_1 = 3/2, j_2 = 1/2$. Furthermore, the Pauli exclusion principle restricts the allowed states to $(3/2, 3/2)_{2,0}$, $\{(3/2, 1/2), (1/2, 3/2)\}_{2,1}$ and $(1/2, 1/2)_0$ for $(j_1, j_2)_J$, where the last state is the ground state. It should be observed that the number of final levels and the allowed J values are the same in both the LS coupling scheme and the j-j coupling scheme [compare Figs. (5.4) and (5.5)].

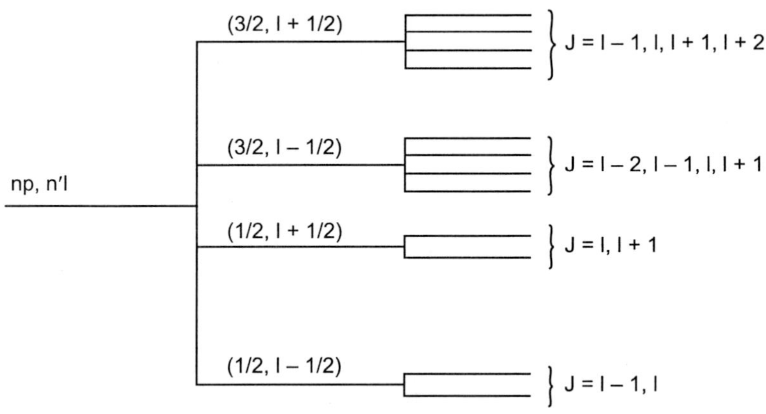

Fig. 5.5 Schematic representation of the fine-structure splitting of the np, n′l level in the *j-j* coupling scheme.

As far as the applicability of the LS or the j-j coupling schemes is concerned, it is noted that the LS coupling scheme is applicable for the lighter elements and the j-j coupling scheme is valid for the heavier elements, whereas the elements in-between have to be studied under the conditions of intermediate coupling (H_1 and H_2 of comparable strength). A good example is that with the two p electrons being in the ground state: carbon is described by the LS coupling scheme, lead by the j-j coupling scheme, whereas silicon, germanium and tin fall in the category of intermediate coupling.

Selection Rules

The most prominent transitions between the atomic levels just discussed are the *electric dipole transitions* (see Chapter 6). These transitions follow the selection rules:

1. Transitions occur only between states in which one electron changes its state. The l-value of this electron changes by one unit,

$$\Delta l = \pm 1 \tag{5.37}$$

2. The allowed changes in the quantum numbers of the whole state are

$$\left.\begin{array}{l}\Delta S = 0, \Delta L = 0, \pm 1 \\ \Delta J = 0, \pm 1, \text{but not } J = 0 \rightarrow J = 0 \\ \Delta m_J = 0, \pm 1\end{array}\right\} LS \text{ coupling} \tag{5.38}$$

and

$$\left.\begin{array}{l}\Delta J = 0, \pm 1 \text{ for the electron which} \\ \qquad \text{change its state} \\ \Delta J = 0, \pm 1 \text{ but not } J = 0 \rightarrow J = 0 \\ \Delta M_J = 0, \pm 1\end{array}\right\} j\text{–}j \text{ coupling} \tag{5.39}$$

A nice illustration of the energy levels and the allowed transitions is provided by the mercury atom (Fig. 5.6). This atom has two valence electrons both of which are in the 6*s* shell for the ground state. In the excited state, one of the electrons will go into $n'l$ state. The energy levels for each configuration are essentially those given in Fig. 5.3 except for $l = 0$ and $n' \neq n$ in which case there are only two states $S = J = 0$ and $S = J = 1$, and for $l = 0$ and $n' = n$ in which case only the $S = J = 0$ state exists. The energy levels according to the *LS* coupling scheme and the observed transitions are shown in Fig. 5.6 where the energies are so normalised that the energy of the singly-ionized state is zero. It should be noted that the transition $(6s)\,(6p)^3P_1 \rightarrow (6s)\,(6s)\,^1S_0$ violates the selection rule $\Delta S = 0$ for *LS* coupling. Its observation is due to the fact that all the atoms with $S = 1$ will go down to the $6\,^3P_1$ level being the lowest-energy triplet state, which therefore can have a high population density, and that the *LS* coupling scheme is only an approximate scheme, *i.e.* the levels contain mixtures of $S = 0$ and $S = 1$ terms. The discussion for mercury can be directly extended to the helium atom which has a ground-state configuration of $(1s)^2$.

Some Regularities in Atomic Spectra

It is clear that the structures of atomic energy levels and their spectra, are in general quite complicated. There are, however, some observed regularities which can be understood in terms of the electronic structure of the atoms.

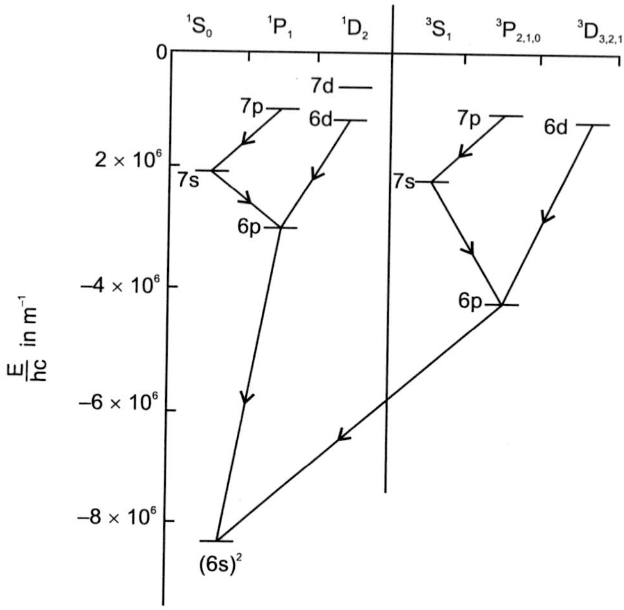

Fig. 5.6 Some energy levels and allowed transitions for the mercury atom (the fine structure due to the spin-orbit interaction removing the J degeneracy is not shown).

1. *Rydberg series:* As mentioned in Sec. 4.6, an atomic electron in an highly excited state is influenced mainly by the net charge of the ion. Consequently, its energy levels a approach those of the hydrogen atom. For example, the energy of the Na valence electron is –0.5461 eV in the $n = 5$, $l = 2$ state and –0.5432 eV in the $n = 5$, $l = 3$ state, whereas that of hydrogenic electron in the $n = 5$ state is –0.5430 eV. This means that the spectral lines corresponding to transitions between states with large quantum numbers approach those of the hydrogen atom. This result is implicit in the observation of Rydberg that many of the spectral series can be expressed in the form

$$\nu = R\left[\frac{1}{(n'-\delta')^2} - \frac{1}{(n-\delta)^2}\right], n > n' \tag{5.40}$$

where δ and δ' are small constants.

2. *Spectra of a chemical group*: Atoms belonging to the same chemical group, for example, alkali atoms Na, K, Rb, Cs, have a similar valence structure though their cores are different. It is, therefore, observed that

their spectra, particularly their fine structure, corresponding to transitions between states with same quantum numbers, are qualitatively similar.

3. *Isoelectronic sequences*: Atoms and ions with the same electronic configuration form what is known as an *isoelectronic sequence, e.g.* A, K^+, Ca^{++}. The members of the sequence differ only in the charge of the nucleus. As such, their spectra will be similar, except that their frequencies increase systematically with the increase in the charge of the nucleus. This was noted in Sec. 4.6 for H, He^+, Li^{++}.

5.5 X-RAY SPECTRA

In Sec. 5.4, atomic spectra governed by the outer or valence electrons were discussed. At the other extreme are the *x*-ray spectra which are produced by the electrons in the inner shells of heavy atoms. For these electorns, the interaction with the nucleus dominates over other interactions, which allows a relatively simple description of the *x*-ray spectra.

X-rays are produced when solid targets are bombarded with fast electrons. An *x*-ray tube consists of an evacuated bulb in which a heated cathode serves as a source of electrons (*thermionic emission*). The electrons are accelerated through a potential difference of the order of 50 kV, and made to strike the anode made of heavy metals (Cu, W, Pt, etc.). While most of the electron energy is liberated at the anode in the form of heat (the anode therefore, has to be cooled), about 1 to 3% of it is converted into high frequency radiation– the *x*-ray radiation which has a wavelength of the order of 10^{-10} m.

The spectrum of *x*-rays from an *x*-ray tube, is a superposition [see Fig. (5.7)] of a continuous spectrum called *white radiation* and a line spectrum called *characteristic spectrum* (it characterizes the material of the anode).

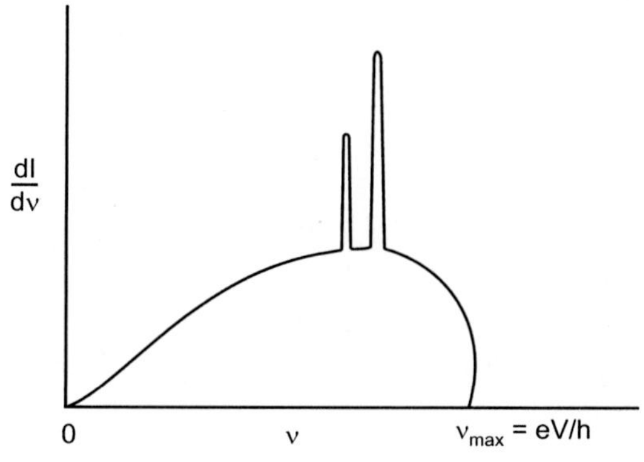

Fig. 5.7 The spectral distribution of intensity per unit frequency showing two characteristic lines superposed over a continuous spectrum.

When the high-velocity electrons reach the anode, they are subjected to large accelerations in the vectorial sense by the strong electrostatic interaction with the nuclei of the anode, which causes them to emit electromagnetic radiation. This radiation due to deceleration, called **bremsstrahlung**, forms the continuous radiation of the x-rays from an x-rays tube. As might be expressed from this mechanism, the maximum energy of the radiation an electron with energy e times V (V is the voltage difference in the tube) can emit is eV so that the highest frequency in the continuous spectrum is

$$\nu_{max} = \frac{e}{h} V \tag{5.40a}$$

This relation povides a means of obtaining an accurate measurement of the ratio e/h. As V increases, the intensity increases at all frequencies, and ν_{max} increases in proportion to V. It is also found that as the nuclear charge Z increases (V being the same), ν_{max} remains unaltered but the intensity increases (since the decelerating forces increase with Z).

In contrast to the continuous spectrum, the line spectrum is independent of the accelerating voltage V, but depends only on the material of which the anode is made. When the fast-moving electrons strike the atoms of the anode, they will occasionally knock out an electron in one of the inner shells, creating a vacancy there. Subsequently, an electron from an outer shell will undergo a transition to the vacant level by emitting a photon of energy equal to the difference in the energies of the two levels. This gives rise to the observed characteristic line spectrum whose frequencies depend only on the energy levels of the atoms in the anode.

Emission Spectrum

The line spectrum of x-rays is due to transitions between states in the inner shells of heavy metals ($Z > 30$), for which the nuclear interaction is dominant. Therefore, for these states, it is reasonable to use an approximate Hamiltonian

$$H_i = \frac{1}{2m} p_i^2 - \frac{Ze^2}{4\pi\varepsilon_0 r_i} + H' \tag{5.41}$$

$$H' = \frac{(Z-1)e^2}{4\pi\varepsilon_0 r_i}[1 - \exp(-r_i/b)] + \frac{Ze^2}{8\pi\varepsilon_0 m^2 c^2}\left(\frac{s_i \cdot l_i}{r_i^3}\right) \tag{5.42}$$

where the first term in H' is the screening interaction given in Eq. (5.18), and the second term is the spin-orbit interaction. Treating H' perturbatively, [see Eq. (3.125)], the energy levels are

$$E_{n,l} = E_n^{(0)} + \frac{(Z-1)e^2}{4\pi\varepsilon_0}\left[\frac{Z}{a_1 n^2} - \left\langle \frac{e^{-r/b}}{r} \right\rangle\right]$$

$$+ \frac{Z^2\alpha^2 |E_n^{(0)}|}{4n^2}\left(3 - \frac{4n}{j+1/2}\right) \quad (5.43)$$

where the last term includes the relativistic corrections discussed in sec. 4.4 and

$$\left\langle \frac{e^{-r/b}}{r} \right\rangle = \left(\frac{Z}{a_1}\right)\frac{(n+l)!(2b_0)^{2l+2}}{n^{2l+4}(2l+1)!N!}(1 + 2b_0/n)^{-2n}$$

$$F(-N, -N, 2l+2, 4b_0^2/n^2) \quad (5.44)$$

where $$F(\alpha, \alpha, \beta, x) \equiv 1 + \frac{\alpha^2}{\beta}\frac{x}{1!} + \frac{[\alpha(\alpha+1)]^2}{\beta(\beta+1)}\frac{x^2}{2!} + \ldots \quad (5.45)$$

with $b_0 = bZ/a_1$ (a_1 is the radius of the first Bohr orbit with $Z = 1$), and $N = n - l - 1$. This expression is quite simple for $n = 1$ and 2:

$$\left\langle \frac{e^{-r/b}}{r} \right\rangle = \frac{Z}{a_1}\left(\frac{2b_0}{2b_0+1}\right)^2 \quad \text{for } n = 1$$

$$= \frac{Z}{4a_1}\frac{b_0^2(2+b_0^2)}{(1+b_0)^4} \quad \text{for } n = 2, l = 0 \quad (5.46)$$

$$= \frac{Z}{4a_1}\left(\frac{b_0}{1+b_0}\right)^4 \quad \text{for } n = 2, l = 1$$

A more detailed analysis (based on what is called as the *Fermi-Thomas model*) indicates that the screening parameter b has the form

$$b = c\, a_1\, Z^{-1/3} \quad (5.47)$$

and the experimental first give the result $b_0 \approx 0.80\, Z^{2/3}$.

In x-ray spectroscopy, the shells are designated by the capital letters K, L, M, etc. corresponding to the principal quantum number n = 1, 2, 3, etc. respectively, and the subshells by the sub indices I, II, III, etc. in the order of increasing energy. For example, in the K shell (n = 1) there is only one level, whereas in the L shell (n = 2) there are three subshells, $L_{\rm I}(n = 2, l = 0)$, $L_{\rm II}(n = 2, l = 1, j = 1/2)$ and $L_{\rm III}(n = 2, l = 1, j = 3/2)$. The energy level of the K and L shells of some elements, obtained from the expression in Eq. (5.43) are given in Table (5.3). The agreement between the predicted values and the experimental values is generally good (except in the case $E_{L_{\rm II}} - E_{L_{\rm I}}$ of heavy elements), especially considering the fact that the predictions use only first order perturbative calculations. It is important to observe that the spin-relativity separation (*e.g.* $E_{L_{\rm III}} - E_{L_{\rm II}}$)

Table 5.3 The energy levels (in keV) of K and L shells for some atoms, obtained from Eq. (5.43), along with the experimental values in brackets

	$-E_K$	$-E_{L_I}$	$E_{L_{II}} - E_{L_I}$	$E_{L_{III}} - E_{L_{II}}$
Cu	8.908	0.853	0.122	0.032
	(8.996)	(1.104)	(0.149)	(0.020)
Mo	20.06	2.53	0.173	0.141
	(20.04)	(2.87)	(0.239)	(0.103)
Ag	25.56	3.46	0.192	0.221
	(25.56)	(3.81)	(0.282)	(0.173)
W	70.01	11.79	0.285	1.358
	(69.64)	(12.12)	(0.556)	(1.340)
Pb	88.24	15.51	0.310	2.048
	(88.16)	(15.89)	(0.661)	(2.170)
U	114.74	21.11	0.341	3.245
	(115.80)	(21.80)	(0.821)	(3.781)

increases rapidly as Z increases, whereas the screening separation (*e.g.* $E_{L_{II}} - E_{L_I}$) increases only gradually. This supports out earlier statement that the spin-orbit interaction and hence the *j-j* coupling becomes important for large-Z elements.

When the bombarding electrons have sufficient energy to knock out inner-shell electrons, vacancies are created in the inner-shells. The electrons from outer shells undergo tansitions to these vacant states emitting photons whose energy is equal to the difference in the energies of the two levels:

$$h\nu = E_i - E_f \tag{5.48}$$

E_i and E_f being the initial and the final energies of the states. This gives rise to the observed characteristic spectrum of the x-rays. The allowed transitions satisfy the usual selection rules for electric dipole trasitions

$$\begin{aligned} \Delta l &= \pm 1 \\ \Delta j &= \pm 1, 0, \text{ but not } j = 0 \rightarrow j = 0 \\ \Delta n &= \text{unrestricted} \end{aligned} \tag{5.49}$$

X-ray spectra are grouped into several series, the K series for transitions to the K shell (*i.e.* $n = 1$), the L series for transitions to the L shell (*i.e.* $n = 2$), etc. Within each series, the lines are characterized by the indices α, β, γ, etc. according to decreasing intensity, *e.g.* K_α for the transition from the L shell to the K shell,

K_β for the transition from the M shell to the K shell, etc. The ultiplets are given an additional number index, *e.g.* $K_{\alpha 1}$ for $L_{III} \rightarrow K$, $K_{\alpha 2}$ for $L_{II} \rightarrow K$, etc. The first few allowed transitions are shown in Fig. 5.8.

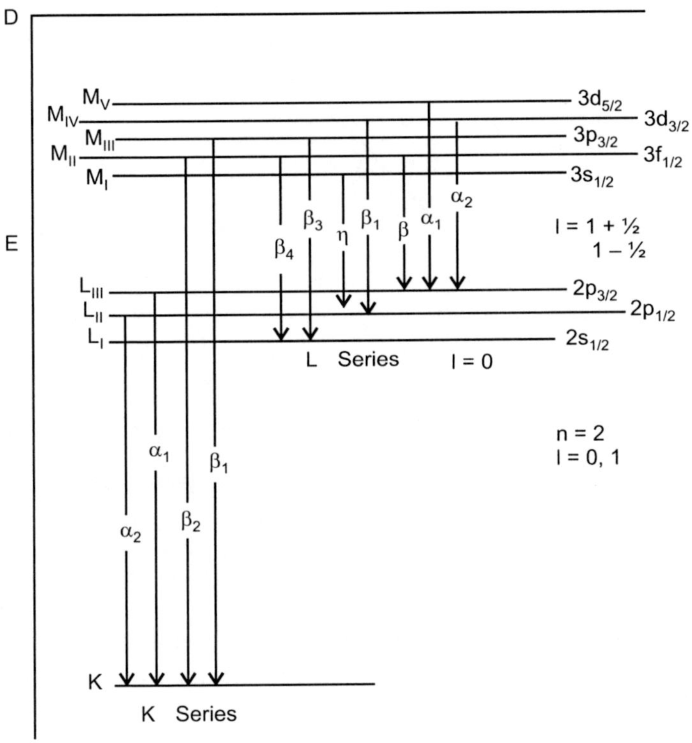

Fig. 5.8 Schematic illustration of the transitions for some *K* and *L* lines in the *x*-ray spectrum.

Moseley's Law

The relation between the x-ray fequencies and the Z values of the atoms was investigated by Moseley (1913-1914). It was found that the square root of the frequency is essentially a linear function of Z. This is seen in the plot of $(h\nu/E_0)^{1/2}$ against Z (Fig. 5.9), where it is also observed that the intercept on the Z-axis is of the order of unity for the K lines. Thus,

$$(h\nu/E_0)^{1/2} = c(Z - \sigma) \tag{5.50}$$

where σ is called the *shielding factor*. This relation is known as Moseley's law. This law can be discussed in terms of the one-electron energy levels including the screening effect. If $(Z - \sigma_i)\, e$ and $(Z - \sigma_f)\, e$ are the effective screened charges for the initial and the final states, the frequency of the radiation for transition between these states is given by

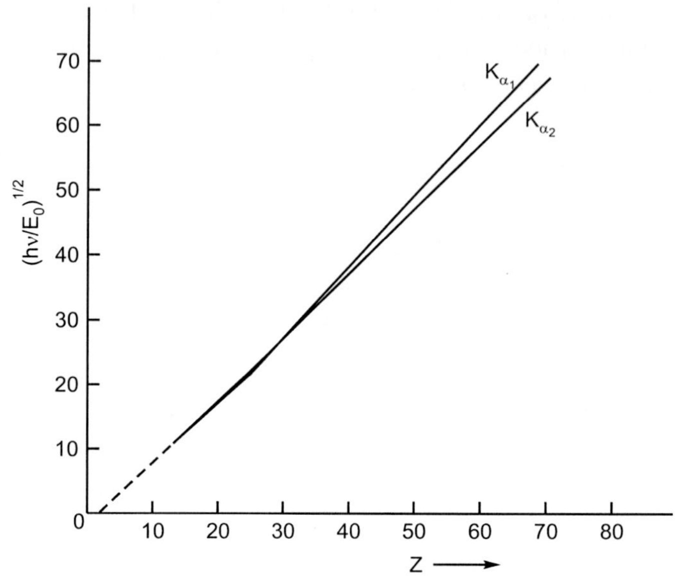

Fig. 5.9 Moseley diagram for the plot of $(h\nu/E_0)^{1/2}$ against Z for the K_a lines.

$$h\nu \approx |E_0| \left[\frac{(Z-\sigma_f)^2}{nf^2} - \frac{(Z-\sigma_l)^2}{n_i^2} \right] \quad \text{...(5.51)}$$

where E_0 is the energy of the ground state of the hydrogen atom. This expression may be written in the approximate form in conformity with the experimentally

$$h\nu \approx |E_0| \left(\frac{1}{nf^2} - \frac{1}{n_i^2} \right) (Z-\sigma)^2 \quad (5.52)$$

observed relation in Eq. (5.50). It is found that $\sigma_n \approx 1$ for the ground state $n = 1$ and $\sigma_n \approx 7.5$ for $n = 2$. It may also be noted that the separation between $K_{\sigma 1}$ and $K_{\sigma 2}$ lines, being related to the spin-relativity separation, increases rapidly as Z increases.

X-ray Absorption Spectrum

X-rays can pass through matter. The intensity is reduced in the process, the reduction depending upon the nature of the material (which forms the basis of many practical applications), and on the frequency. High frequency *x*-rays are generally absorbed less than low frequency *x*-rays.

The amount of absorption of *x*-rays by a given material is studied in terms of the mass absorption coefficient which is defined by the relation

$$dI = -\mu \rho I dx \quad (5.53)$$

where dI is the reduction in intensity while passing through a thickness dx of a material of density ρ, I is the intensity of x-rays of frequecy v, and μ is the frequency-dependent *mass absorption coefficient*. The relation can be integrated to give

$$I(x) = I_0 e^{-\mu\rho x} \tag{5.54}$$

where I_0 is the intensity of the incident radiation. The absorption characteristics can be discussed in terms of the frequency dependence of μ (Fig. 5.10).

As the frequency is lowered from a high value, the absorption coefficient μ increases (low frequency x-rays are less penetrating) gradually, till a frequency v_K is reached when it suddenly drops to a low value. As the frequency is further lowered, μ increases, but drops again at frequency v_{LI}. This process continues but becomes obscured at very low frequencies. The edges in the plot of μ at the different frequecies (Fig. 5.10) are known as the K-absorption edge, L_I-absorption edge, etc.

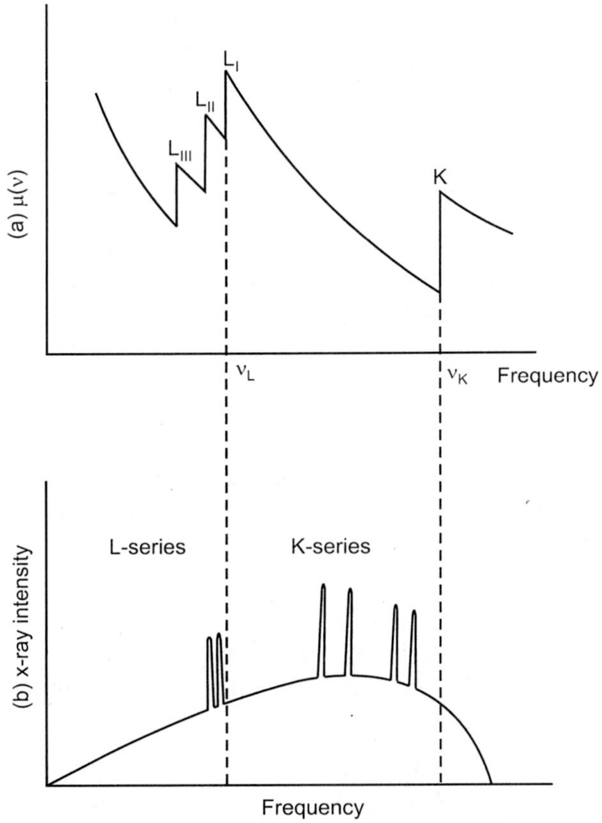

Fig. 5.10 A comparison of absorption and emission characteristics of the x-rays (only a few illustrative lines are shown).

The main contribution to the attenuation of x-rays is the photoelectric absorption of the photons with the emission of electrons. As the frequency is lowered beyond ν_K, the photon energy is not sufficient to eject a K-shell electron. The closure of this channel lowers the absorption coefficient μ and gives rise to the K-absorption edge. When the energy is lower than ν_{LI}, the electrons in the L_I shell also cannot be ejected, giving rise to the L_I absorption edge. Thus, there are three absorption edges corresponding to the L_I, L_{II} and L_{III} subshells, five absorption edges corresponding to the M shell (see Fig. 5.8), etc. The frequencies of these absorption edges are given by

$$h\nu = |E| \tag{5.55}$$

where E is the energy of the K shell, the L_I shell, etc. Comparison of this expression with that in Eq. (5.48) for characteristic emission frequencies shows that the frequency of the absorption edge is always greater than the corresponding characteristic emission frequencies (this may be observed in Fig. 5.10). The reason for this is that emission lines correspond to transition between two shells (or subshells). On the other hand, the frequency of the absorption edge, corresponds to a transition between the lower shell or subshell and the continum states. In comparison with Eq. (5.51), the frequency of the absorption edges is given by

$$h\nu \approx |E_0| \frac{(Z-\sigma_f)^2}{n_f^{\,2}} \tag{5.56}$$

which is greater than the corresponding characteristic emission frequency in Eq. (5.51), and is equal to the limit of the characteristic emission frequency for $n_i \to \infty$.

Auger Effect

So far, it has been assumed that the vacancy in an inner shell, say the K shell, is filled by an electron from an outer shell, along with the emission of a photon. It is however found experimentally, that the fluorescence yield w defined as:

$$w = \frac{n_p}{n_e} \tag{5.57}$$

n_p being the number of K photons and n_e being number of K electrons knocked out (*i.e.* number of K-shell vacancies), is smaller than 1, ranging from a value of 0.1 for light elements to about 0.95 for uranium.

An explanation of the above observation was found by Auger (1925) who noted that the ejection of the K-shell electron is often accompanied by the ejection of another electron. This is due to the fact that the electron which undergoes a transition to the vacant K shell may knock out another electron usually from the same shell, *i.e.* its initial shell. This is known as an *Auger transition* and the emitted electron is known as an *Auger electron*. It should be emphasized that the Auger electron is not knocked out by the photo-electric absorption of a photon emitted by the electron which undergoes a transition to the K shell, but emerges directly as a part of the process of readjustment of the atom. For example, the vacancy in the K shell may be filled by an electron in the L_I shell and the electron in the L_{II} shell may be knocked out, with the result that there will be two vacancies in the L shell. Thus, the de-excitation of the atom may be accompanied either by the emission of a photon (characteristic radiation) or an electron (Auger electron). The two processes together essentially account for the number of vacancies in the K shell. Finally, it is noted that the basic process in the Auger effect is also known as *auto-ionization* or *internal conversion* (in nuclear transitions).

5.6 MOLECULAR BONDING

When atoms approach one another, attractive forces come into play which generally, though not always, bind them into molecules. The mechanism generating these attractive forces differs from molecule to molecule, but is usually a combination (*i*) van der Waals forces, (*ii*) ionic (or heteropolar) bonds, and (*iii*) covalent (or homopolar) bonds. In most cases, however, the ionic bonds or covalent bonds are dominant. Here, a brief discussion of these mechanisms is given for diatomic melocules. It is to be noted that the molecular forces arise primarily from the interaction of the outer electrons.

Van der Waals Forces

When two atoms approach each other, though neutral, they induce fluctuating but correlated electric dipole moments in each other. This gives rise to a dipole-dipole attractive potential which is of the form $- a/r^6$. When the atoms are so close that the electronic orbits overlap, the Pauli exclusion principle forces them into higher orbits which essentially brings in strong repulsive forces. The effective potential may be written in the form:

$$V(r) = \frac{b}{r^n} - \frac{a}{r^6} \tag{5.58}$$

where a and b are positive numbers and n is a large number (~ 10). This represents what is known as van der Waals interaction. Of course, when the nuclei are very close to each other, the dominant force is the repulsion between the nuclei. However, the forces in this region are not important for molecular bonding.

Ionic Bonds

These bonds are formed when it is energetically favourable for electrons to be transferred from one atom to another, with the resulting ions held together by electrostatic attraction between them. Such bonds are known as *ionic* or *heteropolar* bonds and are exemplified by the bonds in NaCl, KBr, HCl, etc.

As a specific case, we consider the KCl molecule. The K atom has its valence electron in the 4s shell with a binding energy of only 4.34 eV. Now, a Cl atom which has five valence electrons in the $3p$ shell, can attract another electron (because of its incomplete shell) and bind it with a binding energy of 3.80 eV. However, if an electron is transferred from a K atom to the Cl atom, resulting in K^+ and Cl^- ions, there will be an additional electrostatic attraction between the ions. Including the van der Waals repulsion (the $-1/r^6$ attraction may be neglected as compared to the electrostatic attraction), the energy of the system is

$$E = -3.80 - \frac{14.4}{r} + \frac{b}{r^n} \tag{5.59}$$

where E is expressed in eV and r is in Å (the small kinetic energy of the atoms has not been included). If this energy is less than –4.34 eV (the binding energy of the electron in the K atom), then it is favourable for the electron from the K atom to be transferred to the Cl atom, with the resulting ions held together by the electrostatic attraction between them. This gives rise to ionic bondng. The details are shown in Fig. 5.11, the system together having a minimum energy of –8.76 eV at a separation 2.79 Å. It is observed that the dissociation energy, *i.e.* the energy required to separate the KCl molecule into K and Cl atoms is (8.76 –4.34) eV, *i.e.* 4.42 eV.

Covalent Bonds

In some cases, the valence electrons of the atoms have no particular preference for either of the two atoms, and are shared by both the atoms. This is especially true in the case of identical atoms forming molecules, *e.g.*

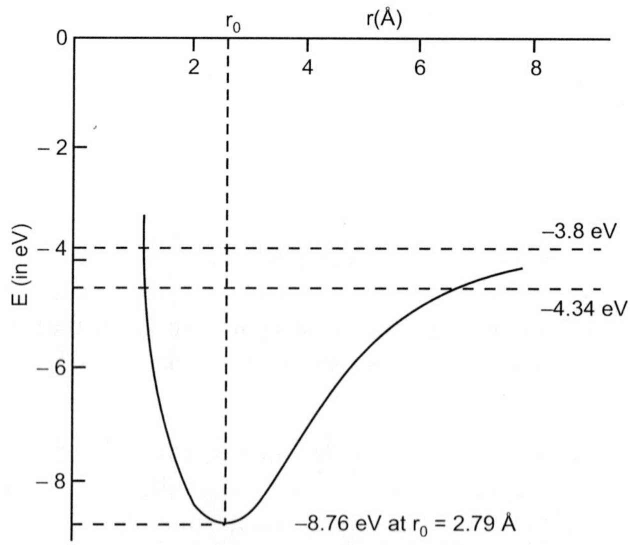

Fig. 5.11 The plot of E in eV against the distance of separation r is Å, between $K^+$and Cl^- ions, for the KCl molecule.

H_2, O_2, N_2, etc. The bonds resulting from the sharing of the valence electrons are known as *covalent* or *homopolar bonds.*

Consider a particle moving in the presence of two similar, one-dimensional, attractive potential V_1 and V_2 which are centred at positions x_1 and x_2. If the positions x_1 and x_2 are separated by a large distance d, the ground state energy E_0 will be essentially degenerate, the degenerate eigenstates being ψ_1 and ψ_2 which are eigenstates with only V_1 or V_2 being present, respectively. As the separation distance d decreases, one may consider as possible eigenstates,

$$\psi_{\pm} = \frac{1}{\sqrt{2}} (\psi_1 \pm \psi_2) \tag{5.60}$$

where the overlap integral is ignored in the normalization. If it is also assumed that ψ_1 is small at x_2 and ψ_2 is small at x_1, the expectation value of the energy is

$$E_{\pm} \approx E_0 \pm \int \psi_1^* (x)\ V_2 \psi_2 (x)\ dx \tag{5.61}$$

Since the potential is attractive ($V_{1,2}$ may be taken to be negative), the integral is expected to be negative. Therefore, the ground state splits into two states, the symmetric state ψ_+ with energy E_+ and the antisymmetric state ψ^- with energy E –. For the symmetric state, which has a lower energy, it is seen that the wavefunction is larger in between the centres of potentials than it is on the outside.

Take the above model to be reasonable, it is seen that when two hydrogen atoms are brought together, the two valence electrons prefer to be in between the two atoms, and have a lower energy than when they are attached to the atoms in isolation. The equilibrium situation is reached at some finite distance of separation (the repulsion term in Eq. (5.58) becomes important at short distances), in which the electrons are shared by the two atoms and the electrons prefer to be in between the atoms. This gives rise to what are known as *covalent* or *homopolar bonds*. The equilibrium distance in the case of the H_2 molecule is about 0.74 Å and the corresponding binding energy is 4.75 eV. It is important to note that in covalent bonds, since both the electrons prefer to be in between the atoms, the spatial wave function is symmetric under the exchange of the two electrons. It is, therefore, required by Pauli's exclusion principle that the spins must point in opposite directions and the total spin of the two electrons must be zero. The state with parallel spins is antisymmetric in the spatial part of the wave function, and will in general have a higher energy (see Fig. 5.12). The antiparallel spin state is called the *bonding state* and the parallel spin state is called the *antibonding* state. The situation is similar for other diatomic molecules.

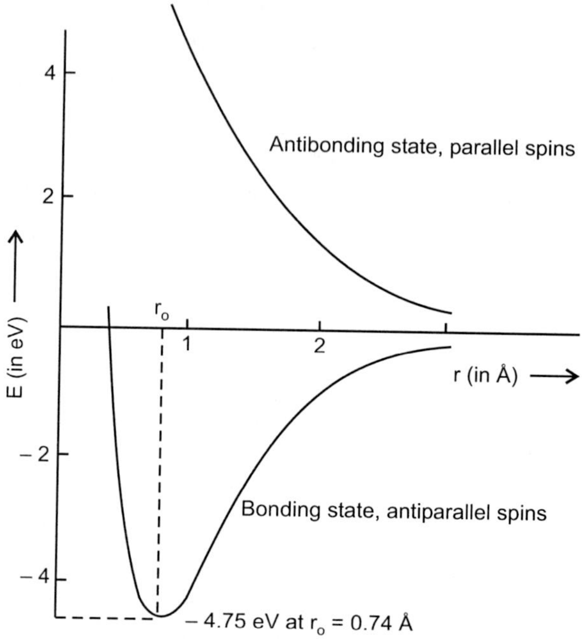

Fig. 5.12 The energy of the H_2 molecule for the bonding and anti-bonding states.

5.7 MOLECULAR SPECTRA

Ikn sec. 5.6, the interatomic forces which lead to molecular bonding were discussed. Apart from the ground state, a molecule can be in a higher energy state, and transitions between the various energy levels give rise to the observed molecular spectra.

The energy of a diatomic molecule arises from different modes: (*i*) The electronic configuration of the electrons in the molecule, (*ii*) the vibration of the atoms about the equilibrium position, and (*iii*) the rotation of the molecule about its centre of mass. For an approximate consideration of molecular spectra, the total energy may be written as a sum of three components, and their excitations treated independently:

$$E = E_e + E_v + E_r \tag{5.62}$$

Electronic excitations involve the largest changes in energy, but are also the most difficult to deduce from theory. Here only the vibrational and rotational energies of the molecule will be considered.

If the molecule is treated as consisting of two point masses, the nuclei, the energy in Figs. 5.11 and 5.12 for the bonding state, may be regarded as the potential energy of these point masses. Expanding this potential energy about the minimum,

$$V(r) = V_0 + \frac{1}{2}\frac{d^2V}{dr^2}\bigg|_{r=r_0} (r - r_0)^2 + \ldots \tag{5.63}$$

where the constant term only defines the zero of the energy of the system. Neglecting the higher order terms in the expansion, the Hamiltonian for vibrational and rotational motion is

$$H_{vr} \approx \frac{1}{2M}\mathbf{p}_r^2 + \frac{1}{2I}\mathbf{J}^2 + \frac{1}{2}\ k\ (r - r_0)^2 \tag{5.64}$$

where the first term is the kinetic energy of the vibrational motion (M is the reduced mass) and the second term is that of the rotational motion $\mathbf{J}$ being the rotational angular momentum (I is the moment of inertia about the centre of mass), and k is d^2V/dr^2. The energy eigenvalues of this Hamiltonian are easily obtained from Eqs. (3.120) and (3.156), leading to the total energy E,

$$E = E_e + \left(n + \frac{1}{2}\right)\hbar\left(\frac{k}{M}\right)^{1/2} + \frac{\hbar^2}{2I}\ J(J+1),$$

$$n = 0, 1, 2, \ldots,\ J = 0, 1, 2, \ldots \tag{5.65}$$

When the molecule undergoes a transition, there is a change in the energy of the state. In an emission process, the frequency of the photon is given by

$$h\nu + E_e - E_e' + (n - n')\ \hbar\left(\frac{k}{M}\right)^{1/2} + \frac{\hbar^2}{2I} J\ (J+1) - \frac{\hbar^2}{2I}\ J'\ (J' + 1) \tag{5.66}$$

It is found, both from theory and experiments, that the separation between electronic energy levels is of the order of 5 eV while that between vibrational energy levels is about 1 eV and that between rotational levels is about 10^{-5}–10^{-3} eV. Therefore, for weak excitations, only changes in rotational states are observed whereas changes in vibrational and electronic states require stronger excitations to be observed. Here, we will confine out discussion to changes in the rotational and vibrational states (symmetric molecules such as $\mathbf{H}_2$ requires a special treatment).

Selection Rules

The electric dipole transitions (which are the most prominent transitions) for vibrational and rotational states, satisfy the selection rules

$$\Delta n = 0, \pm 1$$

$$\Delta J = \pm 1 \tag{5.67}$$

For transitions with $\Delta n = 0$, the emission frequency is given by

$$h\nu = \frac{\hbar^2}{2I} J(J+1) - \frac{\hbar^2}{2I}\ (J - 1)\ J$$

$$= \frac{\hbar^2}{I} J, J\ 1,2,... \text{ for } \Delta n = 0 \tag{5.68}$$

This gives a band of spectral lines (Fig. 5.13), known as the rotational band, with equally spaced frequencies

$$\nu = \left(\frac{h}{4\pi^2 I}\right) J \tag{5.69}$$

The spacing is of the order of 10^{12} s^{-1}, which falls in the very far infrared region. The spacing allows us to evaluate I and hence the equilibrium distance $[r_0 = (I/M)^{1/2}]$. For HCl, the spacing is $\Delta\nu \approx 6.2 \times 10^{11}$ s^{-1} which gives the values of $I \approx 2.7 \times 10^{-47}$ kg. m^2 and therefore $r_0 \approx 1.29$ Å.

For transitions with $\Delta n = 1$,

$$h\nu = \hbar\,(k/M)^{1/2} \pm \frac{h^2}{I} J, J = 1, 2,... \text{ for } \Delta n = 1 \tag{5.70}$$

where the plus sign is for $J' = J - 1$ and the minus sign is for $J' = J + 1$. This again gives us a band of spectral lines which have the same spacing as the lines in the rotational band, but with the centre at $\nu_0 = \frac{1}{2\pi}(k/M)^{1/2}$ (which has a value of above 8.67×10^{13} s^{-1} for HCl) and with the central frequency missing. This is known as the *vibrational-rotational band* (Fig. 5.13).

Symmetric Molecules

Symmetrical molecules do not have an electric dipole moment and the associated dipole transitions, and hence do not exhibit the pure rotational or vibrational-rotational bands just described. The changes in their states are due to higher-order effects, so that the radiation emitted is much weaker. These higher-order transitions obey the selection rules:

$$\Delta J = 0, \pm 1, \pm 2 \tag{5.71}$$

Since the nuclei of a symmetrical molecule are identical, the total nuclear wave function must satisfy the requirements of exchange symmetry, *i.e.* the total wave function must be symmetric for an integral nuclear spin I, and antisymmetric for an half-integral nuclear spin I, under the interchange of the nuclei. The exchange symmetry of the spatial part of the wave function is determined by the rotational states (*i.e.* the Y_l^m (θ, ϕ) functions) which for the exchange of the nuclei (*i.e.* $\theta \to \pi - \theta$, and $\phi \to \pi + \phi$) are even for even l and odd for odd l. Here l plays the role of J. Of the spin states of nuclei with spin I, there are $(I + 1)(2I + 1)$ states which are even and $I(2I + 1)$ states which are odd, under the exchange of spin. States with even spin state are called *ortho-modifications* while those with odd spin state are called *para-modifications.*

For nuclei with an integral I, the *ortho* states are associated with even l values and the *para* states with odd l values, while for nuclei with an half-integral I, the *ortho* states are associated with odd/values and the *para* states with even l values. Since nuclear spin has only a weak interaction, it does not change in a normal transition.

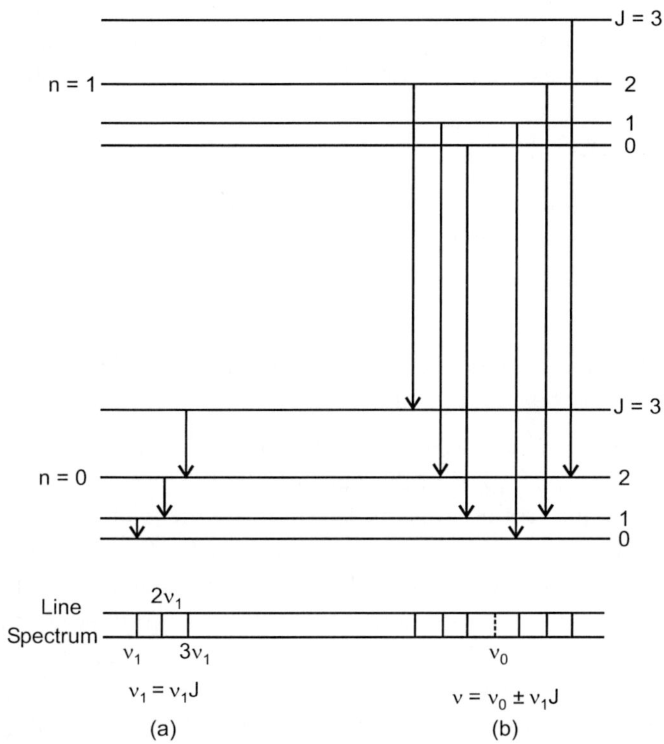

Fig. 5.13 Rotational (*a*) and (*b*) vibrational–rotational spectral lines (spacing of rotational levels is greatly enlarged).

The above discussion implies that the transitions for symmetric molecules take place only with ΔJ even, which in view of Eq. (5.71) implies $\Delta J = 0, \pm 2$. Since the transition frequencies increase with J, and the transitions alternate between *ortho*-transitions and *para*-transitions as J increases, the spectral bands contain lines which alternately arise from *ortho*-modifications and *para*-modifications. As a result, the intensities of the lines also alternate (for $I = 0$, one has only the *ortho*-modification and the alternate lines will be missing *e.g.* 0_2). The intensities at ordinary temperatures are proportional to the number of states in the two modifications and hence the ratio of the two intensities is

$$\frac{N_{para}}{N_{ortho}} = \frac{I}{I+1} \tag{5.72}$$

This relation can be used to determine the nuclear spin.

For the hydrogen atom, $I = 1/2$. The *ortho*-modification has $I = 1$ (with 3 states) and the *para*-modification has $I = 0$ (with 1 state). At room temperature the two modifications will occur in the ratio of 3 : 1. However, at low temperatures most of the hydrogen molecules will go into the para state which has a slightly lower energy, so that its spectral band will have alternate lines missing. When *para*-hydrogen is heated, it retains its modification for a long time (several weeks). As the modification changes, the spectral bands, as well as some physical characteristics such as the heat capacity, show interesting variations.

5.8 EXAMPLES

Here, a few examples will be discussed to illustrate and extend some of the ideas introduced in this chapter.

Example 1

Hund's rule can be used to deduce the ground state of the elements. In particular, consider the period from Na to Ar.

Sodium has one 3s electron in the valence shell, and hence its ground state is $^2S_{1/2}$. Magnesium has $(3s)^2$ in the valence shell. Since this corresponds to a closed subshell, its ground state is $^1S_0$. For Al, there is one electron in the 3p subshell so that its ground state is $^2p_{1/2}$ (the smallest J value allowed is 1/2). For Si, the valence shell has $(3p)^2$ so that the ground state has $S = 1$. The largest allowed orbital angular momentum has $L = 1$ (since the space part is antisymmetric, the largest M_L corresponds to the two electrons having $m_l = 1$ and $m_l = 0$, so that the largest value of L is 1). Thus, the ground state is $^3P_0$ (smallest value of J is zero). For P, the valence shell is $(3p)^3$ so that the ground state has $S = 3/2$. The only allowed value of L is $L = 0$ (the antisymmetric spatial wave function corresponds to electrons having $m_l = 1$, $m_l = 0$, and $m_l = -1$). Therefore, the ground state is $^4S_{3/2}$.

For sulphur the shell is more than half-filled. Since a closed shell has $J = L = S = 0$, it is easier to consider the unfilled shell as hole states (two holes for sulphur). As in the case of holes in the Dirac sea, these holes may be regarded as having positive charge. The spin of the two-hole state for sulphur is $S = 1$, the orbital angular momentum is $L = 1$ (as for the two electron state), and $J = 2$ (the holes have positive charge so that the constant C is Eq. (5.26) is negative and the ground state has the largest allowed J value). Therefore, the ground state is denoted by $^3P_2$. for Cl, there is only one hole which gives for its ground state, $^2P_{3/2}$ (largest J value). Finally or Ar, the subshell is closed, giving its ground state as $^1S_0$.

As an example of two unfilled subshells, consider molybdenum whose unfilled shells are $(4d)^2$ (5s). The largest spin has $S = 3$, and the only allowed value of L

is 0 (m_l = 2, 1, 0, –1, –2 for the five *d*-shell electrons) so that the ground state is $^7S_3$.

Example 2

In Sec. 5.4, it was shown that the number of J states for a two-electron system is the same in LS and in j-j schemes, if one of the electrons has $l = 0$ or 1. This result is now extended to $l > 1$.

Let $1 < l_2 < l_1$ for electrons 1 and 2. The allowed values of L are $L = l_1 + l_2, ..., l_1 - l_2$, while the allowed values of S are $S = 0, 1$. The corresponding J values in the LS coupling scheme are:

$$\begin{aligned} S = 0 &: J = l_1 + l_2, ..., l_1 - l_2 \\ S = 1 &: J = l_1 + l_2 + 1, ..., l_1 - l_2 + 1 \\ &\quad J = l_1 + l_2, ..., l_1 - l_2 \\ &\quad J = l_1 + l_2 - 1, ..., l_1 - l_2 - 1 \end{aligned} \tag{5.73}$$

In the j-j coupling scheme, the allowed values of j_1 and j_2 are $l_1 \pm 1/2$ and $l_2 \pm 1/2$ respectively. Therefore, the allowed J values are:

$$\begin{aligned} j_1 = l_l + 1/2,\ j_2 = l_2 + 1/2 &: J = l_1 + l_2 + 1, ..., l_1 - l_2 \\ j_1 = l_1 + 1/2, j_2 = l_2 - 1/2 &: J = l_1 + l_2, ..., l_1 - l_2 + 1 \\ j_1 = l_l + 1/2,\ j_2 = l_2 + 1/2 &: J = l_1 + l_2, ...,l_1 - l_2 - 1 \\ j_1 = l_l + 1/2,\ j_2 = l_2 - 1/2 &: J = l_1 + l_2 + 1, ..., l_1 - l_2 \end{aligned} \tag{5.74}$$

It is observed that each J is repeated the same number of times in Eq. (5.73), *i.e.* the LS coupling scheme, and in Eq. (5.74) *i.e.* the j-j coupling scheme. For two inequivalent electrons but with $l_1 = l_2 > 1$, the allowed J values are those given in Eqs. (5.73) and (5.74) except that the state with $J = l_1 - l_2 - 1$ is not allowed.

For equivalent electrons, $l_1 = l_2 > 1$, the allowed J values in the LS coupling scheme are:

$$\begin{aligned} S = 0 &: J = l_1 + l_2, l_1 + l_2 - 2, ..., 0 \\ S = 1 &: J = l_1 + l_2,\ l_1 + l_2 - 2, ...,2 \\ &\quad J = l_1 + l_2 - 1, l_1 + l_2 - 3, ..., 1 \\ &\quad J = l_1 + l_2 - 2, l_1 + l_2 - 4, ..., 0 \end{aligned} \tag{5.75}$$

In the j-j coupling scheme, the allowed j_1 and j_2 values are $j_1 \pm 1/2$ and $j_2 \pm 1/2$ respectively. Therefore, the J values are

$$j_1 = l_1 + 1/2, j_2 = l_2 + 1/2 : J = l_1 + l_2, l_1 + l_2 - 2, ..., 0$$

$$\left.\begin{matrix} j_1 = l_1 + 1/2,\ j_2 = l_2 - 1/2 \\ j_1 = l_1 - 1/2,\ j_2 = l_2 + 1/2 \end{matrix}\right\} : J = l_1 + l_2, l_1 + l_2 - 1, ..., 1$$

$$j_1 = l_1 - 1/2, j_2 = l_2 - 1/2 : J = l_1 + l_2 - 2, l_1 + l_2 - 4, ..., 0 \quad (5.76)$$

It is again observed that each J is repeated the same number of times in the LS coupling scheme and n the j-j coupling scheme.

Example 3

The spin-orbit interaction splits the levels of the LS couplng scheme into multiplets. The multiplet structure of the first few observed lines in mercury is as follows:

The triplet levels are split into (6s) (np) $^3P_{2,1,0,}$ (6s) (nd) $^3D_3$,3,2,1, etc. wereas (6s) (ns) $^3S_1$ has only one level. The allowed transitions are:

$$\begin{aligned}
&(6s)\,(6p)\,{}^1P_1 \rightarrow (6s)\,(6s)\,{}^1S_0,\ \lambda = 1849.6\ \text{Å} \\
&(6s)\,(6p)\,{}^3P_1 \rightarrow (6s)\,(6s)\,{}^1S_0,\ \lambda = 2536.5\ \text{Å} \\
&(6s)\,(7s)\,{}^1S_0 \rightarrow (6s)\,(6p)\,{}^1P_1,\ \lambda = 10{,}139.7\ \text{Å} \qquad (5.77) \\
&(6s)\,(7s)\,{}^3S_1 \rightarrow (6s)\,(6p)\,{}^3P_0,\ \lambda = 4046.6\ \text{Å} \\
&(6s)\,(7s)\,{}^3S_1 \rightarrow (6s)\,(6p)\,{}^3P_1,\ \lambda = 4358.4\ \text{Å} \\
&(6s)\,(7s)\,{}^3S_1 \rightarrow (6s)\,(6p)\,{}^3P_2,\ \lambda = 5460.7\ \text{Å}
\end{aligned}$$

the other transitions between these multiplets being forbidden by the selection rules, *e.g.* (6s) (6p) $^3P_0 \rightarrow$ (6s) (6s) $^1S_0$ is not allowed.

Example 4

The spin-orbit interaction breaks the degeneracy of a given LS level into levels with different J values. It may be observed that the average of the **L.S** interaction, summed over all the states of a given LS level, is zero, *i.e.*

$$\sum_{M_S} \sum_{M_L} \langle \mathbf{L} \cdot \mathbf{S} \rangle = 0 \quad (5.78)$$

(this follows from the fact that with a given orientation of **S**, for every term with a given **L**, there is another term with –**L**). Now, the summation over the states can equally well be carried over M_J and J, which implies that

$$\sum_{J} \sum_{M_J} \langle \mathbf{L.S} \rangle = 0 \quad (5.79)$$

For a given J, the expectation value is the same for all M_J values, so that this relation is equivalent to

$$\sum_J (2J+1)\langle \mathbf{L}\cdot\mathbf{S}\rangle = 0 \tag{5.80}$$

Since spin-orbit interaction in proportional to **L.S**, Eq. (5.80) gives

$$\sum_J (2J+1)\,(\Delta E)_{\text{spin-orbit}} = 0 \tag{5.81}$$

Therefore, the weighted average E_{LS} can be used

$$E_{LS} = \sum_J (2J+1)\,E_J \Big/ \sum_J (2J+1) \tag{5.82}$$

to study the atomic energy levels without spin-orbit interaction.

Example 5

The calculation of the energy levels of a many-electron atom is in general quite difficult. For the helium atom, a perturbative estimation of the ground state energy can be made.

The Hamiltonian for the helium atom is

$$H = \frac{1}{2m_e}(p_1^2 + p_2^2) - \frac{Z'e^2}{4\pi\varepsilon_0}\left(\frac{1}{r_1}+\frac{1}{r_2}\right) + \frac{e^2}{4\pi\varepsilon_0}\frac{1}{|\mathbf{r}_1 - \mathbf{r}_2|} \tag{5.83}$$

Taking the unperturbed Hamiltonian as

$$H_0 = \frac{1}{2m_e}(p_1^2 + p_2^2) - \frac{Z'e^2}{4\pi\epsilon_0}\left(\frac{1}{r_1}+\frac{1}{r_2}\right) \tag{5.84}$$

with Z' representing the screened charge of the nucleus, and the perturbation as

$$1V = -(Z-Z')\frac{e^2}{5\pi\varepsilon_0}\left(\frac{1}{r_1}+\frac{1}{r_2}\right) + \frac{e^2}{4\pi\varepsilon_0}\frac{1}{|\mathbf{r}_1 - \mathbf{r}_2|} \quad ...(5.85)$$

A good perturbative estimaton of the energy can be obtained of the perturbation λV is small. It is plausible to 'optimize' the smallness of the perturbation by requiring that the expectation value of λV is zero,

$$\langle \lambda V\rangle = 0 \tag{5.86}$$

which will determine Z'.

The unperturbed ground-state wave function is

$$\psi_0(r_1, r_2) = \frac{Z'^3}{\pi a_1^3}\exp\left[-\frac{Z'}{a_1}(r_1+r_2)\right] \tag{5.87}$$

and the ground state energy is

$$E_0 = -\frac{e^2 Z'^2}{4\pi\varepsilon_0 a_1} \tag{5.88}$$

The calculation of the expectation value of λV is straight forward, and gives

$$\langle \lambda V \rangle = -\frac{(Z - Z')\,Z'\,e^2}{2\pi\varepsilon_0 a_1} + \frac{5e^2 Z'}{32\pi\varepsilon_0 a_1} \tag{5.89}$$

From the condition in Eq. (5.86,)

$$Z' = Z - \frac{5}{16} \tag{5.90}$$

which gives an estimation of the screening of the nuclear charge. With this value for Z', the ground state energy is

$$E = -\frac{e^2}{4\pi\varepsilon_0 a_1}\left(Z - \frac{5}{16}\right)^2 \tag{5.91}$$

On substracting this expression from the energy of the ionized state, the ionization energy is

$$E = -\frac{e^2}{4\pi\varepsilon_0 a_1}\left[\left(Z - \frac{5}{16}\right)^2 - \frac{1}{2}Z^2\right] \tag{5.92}$$

For the helium atom $Z = 2$, so that

$$I \approx 23.1 \text{ eV} \tag{5.93}$$

which is in very good agreement with the experimental value of 24.6 eV. For the singly ionized lithium, Li^+, the ionization energy from Eq. (5.92) with $Z = 3$, comes out to be 74.1 eV which may be compared with the experimental value of 75.6 eV.

Example 6

The highest-energy characteristic x-ray lines are obtained from U^{92}. The energies of K-lines are estimated from

$$E \approx 13.6\, Z^2 \text{ eV} \tag{5.94}$$

$$\approx 115 \text{ keV}$$

with a corresponding wavelength of about 0.11 Å.

Example 7

A knowledge of the molecular dissociation energy enables the estimation of its repulsive energy.

For a KCl molecule, the dissociation energy is 4.42 eV ($E = -8.76$ eV), so that from Eq. (5.59),

$$-8.76 = -3.80 - \frac{14.4}{r} + \frac{b}{r^n} \tag{5.95}$$

The equilibrium condition implies that at the equilibrium separation r_0

$$\frac{14.4}{r_0^2} + \frac{bn}{r_0^{n+1}} \tag{5.96}$$

Substituting this in Eq. (5.95)

$$-4.96 = -\frac{14.4}{r_0}\left(1 - \frac{1}{n}\right) \tag{5.97}$$

From the information that $r_0 \approx 2.79$ Å, $n \approx 25$. This is an overestimation and suggests that the terms that have been neglected (such as the van der Waals attraction) are important in the determination of n. The equilibrium value of r_0 gives the result that the net repulsion is about 0.2 eV at $r = 2.79$ Å.

PROBLEMS

1. Show that the expectation value $\langle(\mathbf{r}_1 - \mathbf{r}_2)^2\rangle$ is greater for ψ_- than for ψ_+ we here $\psi_\pm = \frac{1}{2^{1/2}}[\psi_i(r_1)\,\psi_j(r_2) \pm \psi_i(r_2)\,\psi_j(r_1)]$, ψ_i and ψ_j being orthogonal to each other. This would suggest that the two particles are closer together in the symmetric states.

2. From the relation $\sum_m |Y_l^m(\theta, \phi)|^2 = \frac{2l+1}{4\pi}$, show that the charge density of a closed shell is isotropic.

3. Show that the sum of the degeneracies for the $(ns)\,(n'l)$ system is $4\,(2l + 1)$ and for the $(np)\,(n'\,l)$ sytstem it is $12\,(2l + 1)$ (assume that the electrons are inequivalent). What are the sums of the degeneracies for equivalent electrons?

4. Show the energy levels of C in a diagram similar to Fig. 5.6, and indicate the first few allowed transitions.

5. Discuss the energy levels in the j-j coupling scheme for two valence electrons $(nd)\,(n'\,d)$. What happens if $n = n'$?

6. What are the ground-state terms for elements from K to Zn?

7. The wavelengths corresponding to transitions $(6s)\,(6d)\,{}^3D_2 \rightarrow (6s)\,(6p)\,({}^3P_1, {}^3P_2)$ in mercury are 3125.66 Å and 3654.83 Å respectively. What is the value of C_{LS} in Eq. (5.25) for the $L = 1$ and $S = 1$ state?

8. Discuss the energy level diagram of the valence electron in the sodium atom. If the $^2P_{1/2}$ to $^2S_{1/2}$ transition corresponds to a wavelength of 5895.923 Å, what is the minimum energy of the bombarding electrons required to excite this Na line? (Assume that the Na atoms are in the ground state.)

9. Using experimental information (Table 5.3) about the energy levels of Ag, determine the minimum potential required across the *x*-ray tube, to excite the *K* lines and the *L* lines. What are the wavelengths of the K_α lines? What are the frequencies of *K* and *L* absorption edges?

10. If the $K_{\alpha 1}$ radiation from silver is incident on a material, what is the largest *Z* value of the material for which the *K* electrons can be ejected (use Moseley's law)? What is the kinetic energy of the ejected electron for Cu?

11. Given that the *K*-absorption edges for lead is 0.140 Å, and the minimum voltage required for producing *K* lines in lead is 88.6 keV, determine the ratio of *h/e*.

12. For *Cu*, determine the kinetic energy of the Auger electron for the transition in which two vacancies are created in the L_I shell in filling up a *K*-shell vacancy (some simplifying assumptions may be required.)

13. Assuming that Na^+ and Cl^- behave like hard balls or radii 1.0 Å and 1.8 Å respectively (as far as repulsive forces are concerned), estimate the dissociation energy for a NaCl molecule. The ionization potential of Na is 5.1 eV and the electron affinity for Cl is 3.8 eV.

14. For the HCl molecule, lines are found as *v/c* equal to 2944, 2926, 2908, 2866, 2844, 2821 cm^{-1}. Determine the force constant for the vibrational motion, and the distance of separation for the ions.

15. The rule of equal spacing is not strictly valid for the vibrational-rotational band. Calculate the change in the energy if the moment of inertia in the two vibrational states is different, say I_0 and I_1. Estimate $(I_1 - I_0)/I_0$ for HCl for the states corresponding to the spectrum observed in Problem 14.

6

Interaction with External Fields

Structures of the Chapter

In Chapter 5, the structure and the energy levels of atoms and molecules were discussed. Here, their interaction with external electromagnetic fields will be considered. While the time independent fields allow the investigation and modification of the energy levels to suit our **convarience**, it is the time dependent fields which lead to transitions between the states. These effects are of great importance not only in deducing atomic and molecular properties, but also in devising useful practical applications such as the *lasers* and *masers.*

6.1 THE HAMILTONIAN

The Hamiltonian for the *i*th electron in the presence of external magnetic and electric fields is given by

$$H^{(i)} = \frac{1}{2m}[-i\hbar\,\nabla_i - q\mathbf{A}(\mathbf{r}_i, t)]^2 - \frac{q}{m}\mathbf{s}_i\,.\,\mathbf{B} + V(\mathbf{r}_i, t) \quad (6.1)$$

where V/q and **A** are the electrostatic and electromagnetic potentials. For the special case of the external magnetic and electric fields being constant,

$$\mathbf{A}\,(\mathbf{r}_i, t) = -\frac{1}{2}\mathbf{r}_i \times \mathbf{B}$$

$$V\,(\mathbf{r}_i, t) = -q\,\mathbf{r}_i\,.\,\mathbf{E} + V_{\text{int}} \quad (6.2)$$

where **B** and **E** are the constant magnetic and electric fields, respectively, and V_{int} is the potential due to the nucleus and the other electrons. Substitution of these expressions in Eq. (6.1), after some simplification, leads to

$$H^{(i)} = -\frac{\hbar^2}{2m}\nabla^2 i - \frac{q}{2m}(\mathbf{l}_i + 2\mathbf{s}_i)\,.\,\mathbf{B} + \frac{q^2}{8m}(\mathbf{r}\times\mathbf{B})^2$$

$$-\,q\,\mathbf{r}_i\,.\,\mathbf{E} + V_{\text{int}} \quad (6.3)$$

where, except for the extremely strong magnetic fields (such as $B \sim 10^9$ G), the quadratic term in **B** can be neglected. Therefore, the Hamiltonian for the atom is given by

$$H = H_0 + H_1 + H_2 + H_3 + H' \quad (6.4)$$

where H_0, H_1 and H_3, defined in Eq. (5.14), describe the atom in the absence of the external fields, and with $q = -e$, $e > 0$,

$$H' = \frac{e}{2m}\sum_i (\mathbf{l}_i + 2\mathbf{s}_i)\,.\,B + e\,\Sigma_i\,\mathbf{r}_i\,.\,\mathbf{E} \quad (6.5)$$

We first consider the case of $\mathbf{E} = 0$, which is known as the *Zeeman effect* for weak **B** field and as the *Paschen-Back* effect for strong **B** field. The interaction with a constant external electric field leads to what is known as the *Stark effect* and is discussed as an example in Sec. 6.8. The interaction with radiation is important in transitions and is discussed in Sec. 6.3.

6.2 ATOMS IN A MAGNETIC FIELD

For atoms in a constant magnetic field, the atoms energies are perturbed by the interaction

$$H' = \frac{e}{2m}(\mathbf{L} + 2\mathbf{S}) \, . \, \mathbf{B} \tag{6.6}$$

where **L** and **S** are the total orbital and spin angular momenta respectively. Without loss of generality, the magnetic field can be taken to be in the z-direction. It is convenient to consider the effect of H' given in Eq. (6.6) separately for a weak magnetic field (Zeeman effect) and a strong magnetic field (Paschen-Back effect). Most of our considerations will be for atoms with *LS* coupling in their unperturbed states, through these considerations can easily be extended to *j-j* coupling as well.

In obtaining the expression for the energy levels in the presence of an external magnetic field, the following important result will be required:

Theorem. If **J** is a sum of two angular momentum operators **L** and **S** which operate in different spaces,

$$\mathbf{J} = \mathbf{L} + \mathbf{S}, \; [\mathbf{L}, \mathbf{S}] = 0 \tag{6.7}$$

then

$$\int \psi^*_{\mathbf{J} M_J} \, \mathbf{S} \psi_{J, M'_J} \, d\tau = a \int \psi^*_{\mathbf{J} M'_J} \, \mathbf{J} \psi_{J, M'_J} \, d\tau \tag{6.8}$$

where a is a constant, independent of M_J and M_J'.

This theorem (see Example 1 in Sec. 6.8 for the proof) essentially implies that **S** is proportional to **J** within the sub space of states with a given value of J. A similar result is also valid for **L**.

Zeeman Effect

For the weak magnetic field, H' is regarded as a small perturbation. In the *LS* coupling scheme, the states are characterized by the quantum numbers L, S, J and M_J, so that the perturbation in energy is obtained by using Eq. (6.8) as

$$\Delta E = \frac{e}{2m} g \, \mathbf{B} \, . \langle L, S, J, M_J \, | \, \mathbf{J} \, | \, L, S, J, M_J \rangle \tag{6.9}$$

where the notation indicates taking an average with respect to the states with given L, S, J, and M_J; and g is defined by the relation

$$\mathbf{L} + 2\mathbf{S} = g \, \mathbf{J} \tag{6.10}$$

in the subspace of states with a given J. The constant of proportionality g, is determined by taking the scalar product of Eq. (6.10) with **J**, and calculating the expectation value between the states with given L, S, J and M_J:

$$\begin{aligned} &\langle L, S, J, M_J \, | (\mathbf{L} + 2\mathbf{S}) \, . \, \mathbf{J} \, | \, L, S, J, M_J \rangle \\ &\qquad = g \left\langle L, S, J, M_J \, | \, J^2 \, | \, L, S, J, M_J \right\rangle \end{aligned} \tag{6.11}$$

where gives (using $2\mathbf{L} \, . \, \mathbf{J} = \mathbf{L}^2 + \mathbf{J}^2 - \mathbf{S}^2$ and $2\mathbf{S} \, . \, \mathbf{J} = \mathbf{S}^2 + \mathbf{J}^2 - \mathbf{L}^2$)

$$g = \frac{J(J+1) + L(L+1) - S(S+1)}{2J(J+1)}$$
$$+ \frac{J(J+1) + S(S+1) - L(L+1)}{J(J+1)} \tag{6.12}$$

This on simplification leads to

$$g = \frac{3}{2} + \frac{S(S+1) - L(L+1)}{2J(J+1)} \tag{6.13}$$

The quantity g is called the *Lande g-factor*. The shift in the energy, due to the magnetic field taken along the z-direction, is given by

$$\Delta E = \frac{eh}{2m} gB\, M_J \tag{6.14}$$

which implies that the degenerate states with a given J split into $2J + 1$ equidistant levels. This is known as Zeeman effect, and is illustrated in Figs. (6.1) and (6.2).

The Lande g-factor is often derived from what is called the *vector model*. In this model, the $\mathbf{L} + 2\mathbf{S}$ vector is supposed to precess rapidly around $\mathbf{J}$ so that for the purpose of taking averages, only the component of $\mathbf{L} + 2\mathbf{S}$ along $\mathbf{J}$ is considered,

$$\langle \mathbf{L} + 2\mathbf{S} \rangle = \left\langle \frac{(\mathbf{L} + 2\mathbf{S}) \,.\, \mathbf{J}}{\mathbf{J}^2} \mathbf{J} \right\rangle \tag{6.15}$$

which again leads to the expression in Eq. (6.11) with g given by Eq. (6.13).

The weak-field approximation is valid if the energy shift in Eq. (6.14) is small compared to the fine-structure splitting. The energy shift for ordinary fields, is rather small, *e.g.* for $B \approx 10^4$ G (*i.e.* 1Wb/m^2), the energy splitting is of the order of 0.58×10^{-4} eV for $g = 1$.

The selection rules for transitions between the M_J multiplets of two levels, are:

$$\Delta M_J = 0, \pm 1 \tag{6.16}$$

and the shift in the frequency of the radiation emitted is given by

$$\Delta\omega = \frac{eB}{2m}(gM_J - g'M'_J) \tag{6.17}$$

$$M_J' - M_J = 0, \pm 1$$

It is observed that the frequency shifts of the spectral lines have a simple relation if the levels do not have fine structure, *i.e.* $S = 0$ (singlet states). In this case, since $\Delta S = 0$ for electric dipole transitions, one has $J = L$ for which

$$g = g' = 1, (S = S' = 0) \tag{6.18}$$

The shifts of the spectral lines in this case are

$$\Delta\omega = 0, \pm \frac{eB}{2m} \tag{6.19}$$

This is known as *normal Zeeman effect*, and results in each line splitting into three lines symmetrically placed about the unshifted line, one of which is the unshifted line (see Fig. 6.1). It may be noted that the shifts for ordinary magnetic fields are quite small, $\Delta\omega \sim 8 \times 10^{10}$ rad/s for $B \sim 10^4$ G, compared to $\omega \sim 3 \times 10^{15}$ rad/s for visible light.

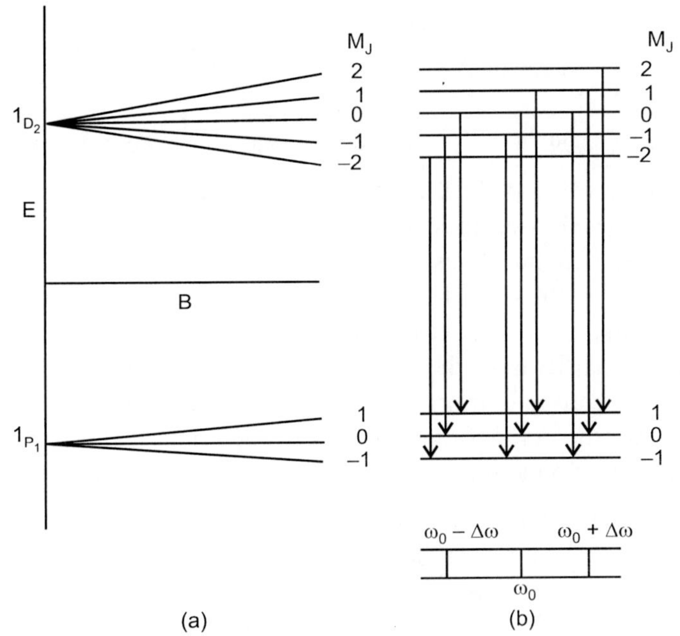

Fig. 6.1 (*a*) The energy levels of $^1P_1$ and $^1D_2$ states as a function of the magnetic field, and (*b*) the splitting of energy levels into three components illustrating normal Zeeman effect.

If the states have fine structure arising from spin-orbit interaction, the spectral lines break into more than three components, and the frequency shifts are given by rational fractions of the normal Zeeman shift,

$$\Delta\omega = \frac{p}{q}\Delta\omega_0$$

$$\Delta\omega_0 = \frac{eB}{2m} \tag{6.20}$$

where p and q are integers. This case is known as *anomalous Zeeman effect*. As a specific example, consider the splitting of the alkali doublet lines which correspond to the transitions $^2P_{1/2} \rightarrow {}^2S_{1/2}$ and $^2P_{3/2} \rightarrow {}^2S_{1/2}$. The Landé g-factors for these states are obtained from Eq. (6.13) as

$$
\begin{aligned}
{}^2S_{1/2} &: g = 2 \\
{}^2P_{1/2} &: g = 2/3 \\
{}^2P_{3/2} &: g = 4/3
\end{aligned} \tag{6.21}
$$

Substituting the values in Eq. (6.17) gives

$$\Delta\omega = (\pm\, 2/3, \pm\, 4/3)\, \Delta\omega_0 \qquad \text{for } {}^2P_{1/2} \rightarrow {}^2S_{1/2} \tag{6.22}$$

and

$$\Delta\omega = (\pm\, 1/3, \pm\, 1, \pm\, 5/3)\, \Delta\omega_0 \quad \text{for } {}^2P_{3/2} \rightarrow {}^2S_{1/2} \tag{6.23}$$

The energy levels as a function of the field B, the allowed transitions, and the splitting of the spectral lines, are shown in Fig. 6.2. It is worth noticing that some of the fine structure lines cross each other for $(e\hbar\, B/m) \sim \Delta E$. At these values of B, there is a mixing of states which, under some circumstances, causes a sharp change in the intensity of radiation emitted. This phenomenon is known as **Hanle effect** (particularly, the case of crossover at $B = 0$), and has been used to determine the constants involved in the fine structure multiples. Of course, for $e\hbar\, B/m \sim \Delta E$, the perturbative analysis is not strictly valid ($H' \sim H_2$), and a more complicated, nonperturbative analysis has to be carried out.

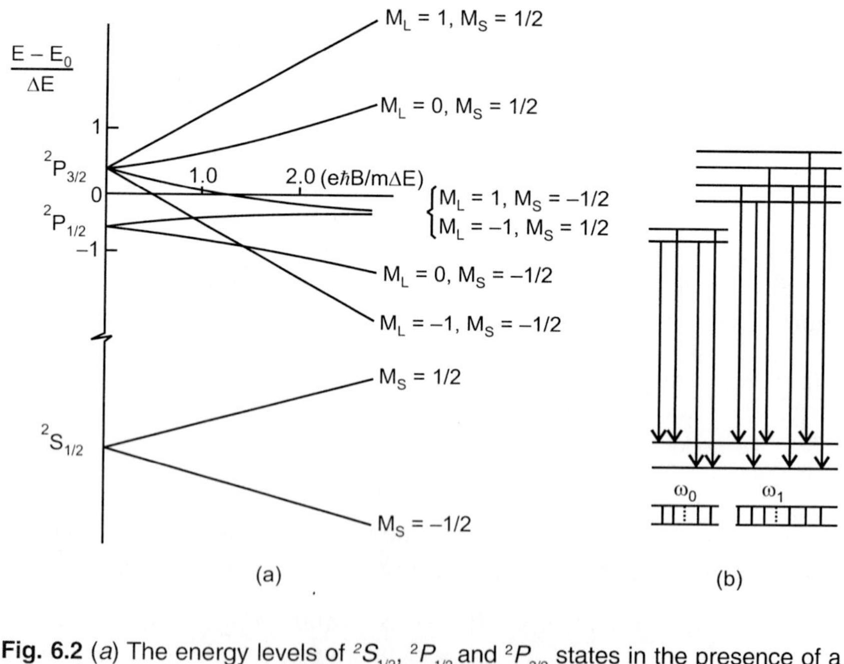

Fig. 6.2 (*a*) The energy levels of ${}^2S_{1/2}$, ${}^2P_{1/2}$ and ${}^2P_{3/2}$ states in the presence of a magnetic field, in units of the fine structure splitting ΔE. (*b*) Splitting of the spectral lines for transitions from ${}^2P_{1/2}$ and ${}^2P_{3/2}$ states to ${}^2S_{1/2}$ states at $e\hbar\, B/m = 0.5\ (\Delta E)$.

Paschen-Back Effect

If the magnetic field is so strong that the splitting of the energy levels due to the magnetic field is larger than the fine structure separation, is known as Paschen-Back effect. In this case, H' in Eq. (6.6) is treated as the main term and H_2 representing the spin-orbit interaction as a small perturbation, which makes the calculations relatively simple.

In the absence of spin-orbit interaction, the unperturbed states may be specified by the quantum numbers L, S, M_L and M_S. Taking the magnetic field along the z-direction, the energy shift due to the magnetic field is given by Eq. (6.6) as:

$$\Delta E' = \frac{e\hbar}{2m}(M_L + 2M_S)B \tag{6.24}$$

and the line-splitting is an integral multiple of $\Delta\omega_0$. The selection rules for electric dipole transitions in this case are:

$$\Delta M_L = 0, \pm 1, \Delta M_S = 0 \tag{6.25}$$

so that we essentially get back the normal Zeeman shifts

$$\Delta\omega = 0, \pm\frac{eB}{2m} \tag{6.26}$$

This is expected since the only role played by spin here is to change the energy levels by an amount $e\hbar M_b B/m$ which, in view of the selection rules in Eq. (6.25), does not affect the frequency associated with the transitions.

The effect of spin-orbit interaction can be included perturbatively by using the theorem stated earlier but applied to $\mathbf{l}_i$ whose sum if $\mathbf{L}$, and also to $\mathbf{s}_i$ shows sum is $\mathbf{S}$. The expectation value of the spin orbit interaction can then by written as

$$\langle L, S, M_L, M_S \,|\, H_2 \,|\, L, S, M_L, M_S\rangle$$
$$= a\,\langle L, S, M_L, M_S, \,|\, \mathbf{L}.\mathbf{S} \,|\, L, S, M_L, M_S\rangle \tag{6.27}$$

where a is independent of M_L and M_S. Since $L_xS_x + L_yS_y$ [which can be written as $\frac{1}{2}$ $(L_+S_- + L_-S_+)$] changes the values of M_L and M_S, the only term which contributes in Eq. (6.27) is the L_zS_z term. Therefore, the total energy shift is given by

$$\Delta E = \frac{e\hbar}{2m}(M_L + 2M_S)\,B + a\hbar^2 M_L M_S \tag{6.28}$$

Consider the effect of this term on the ^2P levels shown in Fig. 6.2(a). The allowed values for M_L are 1, 0, – 1 and for $2M_S$, 1, – 1, so that $M_L + 2M_S$ can take on the values 2, 1, 0, – 1, – 2. Thus, the ^2P levels are split into five equidistant levels by the first term in Eq. (6.28) [the lines in Fig. 6.2(a) for $B \to \infty$]. This equidistance is removed by the second term which, for these levels, has a value of $a\hbar^2$, 0, $-a\hbar^2/2$, 0, $-a\hbar^2$. The shifts in the transition frequencies are now given by

$$\Delta\omega = (\Delta\omega_0 + a\hbar M_S)\,\Delta M_L \tag{6.29}$$

For the $^2P \rightarrow {}^2S$ transitions, the frequency shifts for $M_S = 1/2$ are slightly larger in magnitude (a is generally positive) than those for $M_S = -1/2$. Thus, we get five lines, two doublets *i.e.*, for $\Delta M_L = \pm 1$, $M_S = \pm 1/2$, and a singlet which is the unshifted line corresponding to $\Delta M_L = 0$.

j-j Coupling

In the analysis so far, it has been assumed that LS coupling is valid. The theory can easily be modified to apply to heavy atoms where *j-j* coupling is dominant. The effect of the magnetic field on two electrons with *j-j* coupling is briefly discussed here.

In *j-j* coupling, the states are characterized by j_1, j_2, J and M_J. The energy shift due to the interaction of the two electrons with a weak external magnetic field is given by

$$\Delta E = \frac{e}{2m}\langle j_1, j_2, J, M_J\,|\mathbf{l}_1 + 2\mathbf{s}_1 + \mathbf{l}_2 + 2\mathbf{s}_2\,|\, j_1, j_2, J, M_J\rangle \cdot \mathbf{B} \tag{6.30}$$

Using the results of the theorem in Eq. (6.8), we can write

$$\langle j_1\,|\,\mathbf{l}_1\,|j_1|j_1\rangle = a_1\,\langle j_1\,|\,\mathbf{j}_1\,|\,j_1\rangle \tag{6.31}$$

$$\langle j_1\,|\mathbf{s}_1\,|\,j_1\rangle = b_1\,\langle j_1\,|\,\mathbf{j}_1\,|\,j_1\rangle$$

and similar relations for $\mathbf{l}_2$ and $\mathbf{s}_2$. The constants a_i and b_i are determined by the following steps similar to those leading to Eq. (6.13) giving

$$a_i = \frac{1}{2} + \frac{l_i\,(l_i+1) - s_i\,(s_i+1)}{2j_i\,(j_i+1)} \tag{6.32}$$

$$b_i = \frac{1}{2} = \frac{s_i\,(s_i+1) - l_i\,(l_i+1)}{2j_i\,(j_i+1)}$$

where $i = 1, 2$. Then one gets

$$\Delta E = \frac{e}{2m}\langle j_1, j_2, J, M_J\,|(a_i + 2b_1)\,\mathbf{j}_i$$

$$+ (a_2 + 2b_2)\; j_2\,|\, j_1, j_2, J, M_J\rangle \cdot \mathbf{B} \tag{6.33}$$

Again, applying the theorem in Eq. (6.8) to $\mathbf{j}_i$ whose sum is $\mathbf{J}$, gives

$$\langle J\,|\,\mathbf{j}_{1,2}\,|\,J\rangle = A_{1,2}\,\langle J\,|\,\mathbf{J}\,|\,J\rangle \tag{6.34}$$

with

$$A_1 = \frac{1}{2} + \frac{j_1(j_1+1) - j_2(j_2+1)}{2J(J+1)}$$

$$A_2 = \frac{1}{2} + \frac{j_2(j_2+1) - j_1(j_1+1)}{2J(J+1)} \tag{6.35}$$

Finally, one gets

$$\Delta E = \frac{e\hbar B}{2m} M_J [A_1 (a_1 + 2b_1) + A_2 (a_2 + 2b_2)] \tag{6.36}$$

which again results in $(2J + 1)$ equidistant energy levels.

In the strong field case, the unperturbed states are characterized by the quantum numbers j_1, m_{j_1} and j_2, m_{j_2} so that the energy shifts can be obtained from the relations in Eq. (6.31) as

$$\Delta E = \frac{e\hbar B}{2m} [(a_1 + 2b_1) m_{j_1} + (a_2 + 2b_2) m_{j_2}] \tag{6.37}$$

To this, the contribution of the spin-orbit interaction can be added, which also can be estimated by using arguments similar to those used in the discussion for the *LS* coupling. It gives a contribution proportional to $m_{j_1} m_{j_2}$.

In all discussions so far, only the effect of the linear term in B in Eq. (6.3) has been considered. The quadratic term becomes important for atoms for which the magnetic dipole moment is zero, *e.g.*, He, Ne, etc. which have $L = S = 0$. The magnetic properties of these materials such as magnetic susceptibility, are determined by the quadratic term. The quadratic term is also important in astrophysics where enormously large magnetic fields are encountered (in pulsars and neutron stars) and is some solid state problems. The energy shift due to the quadratic term is known as the quadratic Zeeman effect.

6.3 INTERACTION WITH RADIATION

The properties of atoms and molecules and their interactions are observed when there are changes in their states. Their interaction with radiation is the most important mechanism through which these changes take place. Quantum mechanics provides a very satisfactory description of the interaction of matter with radiation. Here, an elementary treatment of emission and absorption of radiation is presented.

Consider radiation of angular frequency ω, incident on a particle with charge q. The wavelength of radiation normally encountered is of the order of 10^3 Å which is quate large compared with atomic sizes (~ 1 Å). Therefore, the electric field can be regarded as being constant over atomic distances (this is called the *electric dipole* approximation). Such an electric field is provided by the scalar potential $-Ez\cos\omega t$, where it is assumed that the electric field is in the z-direction and has an amplitude E. The interaction of the charged particle with this scalar potential leads to the potential energy term

$$V = -q\,Ez\cos\omega t \tag{6.38}$$

The effect of this interaction on a bound state can be treated perturbatively.

Let the particle be in a bound eigenstate ϕ_0 of the Hamiltonian H_0, before the radiation is incident on it. After the radiation is introduced at $t = 0$, the particle can undergo a transition to any of the other eigenstates ϕ_n where

$$H_0\,\phi_n = E_n\phi_n \tag{6.39}$$

which satisfy the orthonormality conditions

$$\int \phi_m{}^*\phi_n d\tau = 0 \text{ for } m \neq n$$

$$= 1 \text{ for } m = n \tag{6.40}$$

The state of the particle can then be represented by

$$\phi = \sum_n a_n(t)\exp(-iE_n t/\hbar)\phi_n \tag{6.41}$$

where $|\,a_n(t)\,|^2$ represents the probability of finding the particle in state ϕ_n at time t, with the boundary conditions

$$a_0(0) = 1,\ a_n(0) = 0 \text{ for } n \neq 0 \tag{6.42}$$

Substituting the expression in the Schrödinger equation,

$$(H_0 - qEz\cos\omega t)\,\phi = i\hbar\frac{\partial\phi}{\partial t} \tag{6.43}$$

gives

$$i\hbar\sum_n \exp(-i\omega_n t)\frac{\partial a_n(t)}{\partial t}\phi_n = -qEz(\cos\omega t)\,\phi,\ \omega_n = E_n/\hbar \tag{6.44}$$

Multiplying both sides by ϕ_m^* and integrating, and using the orthonormality conditions gives

$$i\hbar\frac{\partial a_m(t)}{\partial t} = -qE\exp(i\omega_m t)(\cos\omega t)\int \phi_m{}^* z\phi d\tau \tag{6.45}$$

If the incident radiation is not very strong, we can approximate ϕ in Eq. (6.45) by its unperturbed expression, *i.e.*, $\phi \approx \exp(-i\,\omega_0 t)\,\phi_0$, and get as a first order approximation

$$a_m(t) = \frac{iqE}{\hbar} z_{m0} \int_0^t \exp[\omega_m - \omega_0)t]$$
$$\cos(\omega t)\, dt,\ m \neq 0, \tag{6.46}$$

where we have used the boundary conditions in Eq. (6.42), and

$$z_{m0} = \int \phi_m * z\phi_0 d\tau \tag{6.47}$$

Carrying out the integration over t leads to

$$a_m(t) = \frac{qE}{2\hbar} z_{m0} \left[\frac{\exp[i(\omega_m - \omega_0 + \omega)t] - 1}{\omega_m - \omega_0 - \omega} + \frac{\exp[i(\omega_m - \omega_0 - \omega)t] - 1}{\omega_m - \omega_0 + \omega} \right] \tag{6.48}$$

For sufficiently large t [*i.e.*, $t \gg 1/\omega_m - \omega_0$)], the magnitude of the coefficient is appreciable only for $\omega \approx \pm (\omega_0 - \omega_m)$. The first case, namely

$$\hbar\omega \approx E_0 - E_m \tag{6.49}$$

corresponds to emission of radiation of energy $\hbar\omega$, while the second case,

$$\hbar\omega \approx E_m - E_0 \tag{6.50}$$

corresponds to absorption of radiation of energy $\hbar\omega$. In both the cases, the probability of finding the particle in state ϕ_m at t, is

$$|a_m(t)|^2 \approx \frac{q^2E^2}{\hbar^2} |z_{m0}|^2 \frac{\sin^2[(\omega_{m0} - \omega)t/2]}{(\omega_{m0} - \omega)^2} \tag{6.51}$$

where $\omega_{m0} = |\omega_m - \omega_0|$. Since $|z_{m0}| = |z_{0m}|$, it also follows that if the particle is originally in state m, the probability of finding the particle in state ϕ_0 at t, $|a_0(t)|^2$, is given by the same expression as in Eq. (6.51). Thus, the external field E induces or stimulates transitions $0 \to m$ and transitions $m \to 0$ with the same probability. This important conclusion may be stated as an equality

$$P_{n \to m}(E) = P_{m \to n}(E) \tag{6.52}$$

for induced transition probabilities. Another important fact to be noted is that transitions are allowed between states which do not conserve energy, *i.e.*, $|E_m - E_0| \neq \hbar\omega$. However, for $t \gg 1/\omega_{m0}$, the probability $|a_m(t)|^2$ is significant only for $\omega_{m0} - \omega \sim 1/t$, (for $t \to \infty$ it is proportional to t^2), and rapidly goes to small values as $|\omega_{m0} - \omega|$ becomes larger than $1/t$. Thus energy conservation is applicable to the extent of the uncertainty

$$(\Delta E)(t) \sim \hbar \tag{6.53}$$

which is a manifestation of the uncertainty relation discussed in Eq. (3.64). For $t \to \infty$, the uncertainty $(\omega_{m0} - \omega)$ tends to zero which essentially restores the energy conservation relation $|E_m - E_0| = \hbar\omega$.

In the analysis so far, it has been assumed that the incident radiation has only one frequency. In reality, it has a frequency distribution. The changes required by this situation can be introduced by replacing the energy flux $\frac{1}{2}\varepsilon_0 c E^2(\omega)$ by

$$\frac{1}{2}\varepsilon_0 c\, E^2(\omega) \rightarrow \int \frac{dI}{d\omega} d\omega \tag{6.54}$$

where $dI/d\omega$ is the flux density per unit frequency. This gives

$$|\, a_m (t)\,|^2 = \frac{2q^2}{\varepsilon_0 c\hbar^2} \int |\, z_{m0}\,|^2 \frac{\sin^2 [(\omega_{m0} - \omega)t/2]}{(\omega_{m0} - \omega)^2} \frac{dI}{d\omega} d\omega \tag{6.55}$$

For large t, the integrand being sharply peaked at $\omega = \omega_{m0}$, $dI/d\omega$ can be evaluated at $\omega = \omega_{m0}$, and the remaining integration can be carried out to give

$$|\, a_m(t)\,|^2 = \frac{q^2\, \pi t}{\varepsilon_0 c\hbar^2} |\, z_{m0}\,|^2 \left.\frac{dI}{d\omega}\right|_{\omega = \omega_{m0}},$$

$$\left(\int \frac{\sin^2 ax}{x^2} dx = \pi a \right) \tag{6.56}$$

The transition probability per unit time is $|\, a_m (t)\,|^2/t$:

$$W_{0 \rightarrow m} = \frac{q^2 \pi}{\varepsilon_0 \hbar^2} |\,(n\,.\,r)_{m0}\,|^2\, u\,(\omega_{m0}) \tag{6.57}$$

where $cu\,(\omega) = dI/d\omega$ and z has been replaced by the more general quantity $n\,.\,r$ (n is a unit vector in the direction of the field). This expression is valid for absorption and emission induced or stimulated by the external field, and describes what are known as *electric dipole transitions.*

The important point to be noted in the transition probability, is that the transition takes lace only for the frequency $w = |\,E_m - E_0\,|/\hbar$; $E_m > E_0$ for absorption and $E_m < E_0$ for induced emission. The transition probability increases as the flux density $dI/d\omega$ increases. Finally, it should be mentioned that the approximation of the electromagnetic field being constant in space is not valid if $(n\,.\,r)_{n0}$ is zero (known as *forbidden transitions*) In this case, a more general approach, which takes into account the space-dependence of the electric field, and also the associated magnetic field, does allow the transitions but with reduced rates. They are known as *electric quadruple transitions, magnetic dipole transitions,* etc.

6.4 SPONTANEOUS TRANSITIONS

If atomic transitions were induced only by external fields [according to Eq. (6.57)], an excited, isolated atom would remain in the excited state for an indefinitely long period. However, excited atoms are found to undergo transition to a lower-lying state even in the absence of external fields. Einstein (1915) showed on the basis of thermodynamic arguments, that in addition to the induced or stimulated emissions, spontaneous emission of radiation by an excited state must also be present and deduced the relations between the different transitions.

Let P_{mn} be the probability for stimulated transition from state n to state m. It is reasonable to assume (see Sec. 6.3) that stimulated transitions are proportional to the density of radiation $u\,(\omega_{nm})$. The probabilities for stimulated transitions between states m and n, can then be written as

$$P_{mn} = B_{mn}\, u\,(\omega_{nm}) \tag{6.58}$$

$$P_{nm} = B_{nm}\, u\,(\mathrm{w}_{nm}) \tag{6.59}$$

In addition, let A_{nm} be the probability for spontaneous transitions from the higher state m to the lower state n ($E_m > E_n$). The number of transitions N_{ji} from state i to state j is given by the product of the number of atoms in state i and the total probability for the transition. Therefore N_{nm} and N_{mn} are given by

$$N_{nm} = N_m\,[B_{nm} u\,(\omega_{nm}) + A_{nm}]$$

$$N_{mn} = N_n\, B_{mn}\, u\,(\omega_{nm}) \tag{6.60}$$

The constants A and B are known as the *Einstein coefficients.* When the system is in thermal equilibrium

$$N_{nm} = N_{mn} \tag{6.61}$$

and the Boltzmann distribution ratio is

$$\frac{N_n}{N_m} = \exp\,[-\,(E_n - E_m)/kT] = \exp\,(\hbar\omega_{nm}/kT) \tag{6.62}$$

Using these conditions in Eq. (6.60),

$$B_{nm}\, u\,(\omega_{nm}) + A_{nm} = \exp\,(\hbar\,\omega_{nm}/kT)\, B_{mn}\, u\,(\omega_{nm}) \tag{6.63}$$

or

$$u\,(\omega_{nm}) = \frac{A_{nm}}{B_{mn}\exp\,(\hbar\,\omega_{nm}/kT) - B_{nm}} \tag{6.64}$$

This expression can be compared with Planck's law for blackbody radiation [see Eq. (2.13)]

$$u\,(\omega_{nm}) = \frac{\hbar\,\omega^3/\pi^2 c^3}{\exp\,(\hbar\,\omega/kT) - 1} \tag{6.65}$$

One therefore deduces that

$$B_{mn} = B_{nm} \tag{6.66}$$

in conformity with the earlier conclusion in Eq. (6.52), and

$$\mathrm{A}_{nm} = \frac{\hbar\omega^3}{\pi^2\,c^3} B_{nm} \tag{6.67}$$

The quantity B_{nm} can be obtained by comparing Eqs. (6.58) and (6.59) with Eq. (6.57) ($\mathrm{W}_{n \to m} = P_{mn}$). It is to be noted that the $u(\omega_{nm})$ deduced corresponds to *isotropic* blackbody radiations that the expression in Eq. (6.57) should be averaged over different directions which essentially leads to the replacement:

$$| (\mathbf{n} \cdot \mathbf{r})_{nm} |^2 \to \frac{1}{3} | (\mathbf{r})_{nm} |^2 \tag{6.68}$$

Equating $W_{m \to n}$ (with the above replacement) and P_{nm} then yields

$$B_{nm} = \frac{q^2 \pi}{3\varepsilon_0 \hbar^2} | (\mathbf{r})_{nm} |^2 \tag{6.69}$$

so that

$$A_{nm} = \frac{2q^2 \omega^3}{3\varepsilon_0 hc^3} | (\mathbf{r})_{nm} |^2 \tag{6.70}$$

This is the expression for the probability of spontaneous electric dipole transitions.

Selection Rules

It is observed that the induced and the spontaneous transition probabilities [Eqs. (6.57) and (6.70)] depend on the same matrix element, $(\mathbf{r})_{nm}$. Therefore these electric diole transitions are allowed only if this matrix element is nonzero. This imposes certain conditions on the allowed transitions. In particular, it is to be noted that since $\mathbf{r}$ is odd under parity transformation (*i.e.*, $\mathbf{r} \to -\mathbf{r}$) the product of ψ^*_n and ψ_m also should be odd. If these are single-particle, angular momentum states [see Eq. (3.153)] with orbital angular momentum quantum numbers l_n and l_m, the parity of the product of these states is $(-1)^{l_n + l_m}$ so that $(l_m + l_n)$ and therefore $(l_n - l_m)$ are odd. In addition, the angular dependence of $\mathbf{r}$ is of the form $Y_l^m(\theta, \phi)$ from which it can be shown that $| l_n - l_m | = 1$ and $| j_m - j_n | = 1, 0, j_m$ and j_n being the total angular momentum quantum numbers. Thus the allowed electric dipole transitions satisfy selection rules:

$$\Delta l = \pm 1, \quad \Delta j = \pm 1, 0, \Delta s = 0 \tag{6.71}$$

where the $\Delta s = 0$ result follows from the fact that spin is unchanged in the transitions. More detailed arguments also shown that

$$\Delta m_j = \pm 1, 0, \quad j = 0 \nrightarrow j = 0 \tag{6.72}$$

Lifetimes and Linewidths

The existence of a finite probability for the spontaneous transition from an excited state to a lower state means that the excited state has a finite lifetime. The finite lifetime gives rise to an uncertainty in the energy of the state,

$$\Delta E \sim \hbar/\tau \tag{6.73}$$

This uncertainty is reflected in the emitted radiation having a spread in the distribution of its frequency and leads to the observed width of the spectral lines called the *natural linewidth*,

$$\Delta\omega = 1/\tau \tag{6.74}$$

The average lifetime τ and therefore the linewidth are related to the transition probability.

If A is the transition probability, the number of particles $(-dN)$ which undergo transition in time dt is [see Eq. (1.80)]

$$dN = -AN(t)\,dt \tag{6.75}$$

which on integration gives

$$N(t) = N(0)\,e^{-At} \tag{6.76}$$

Now, the average lifetime of the particles is

$$\begin{aligned}\tau &= \sum_i t_i \frac{N(t_i) - N(t_i + \Delta t)}{N(0)} \\ &= A\int_0^\infty te^{-At}\,dt \\ &= \frac{1}{A}\end{aligned} \tag{6.77}$$

so that the linewidth in Eq. (6.74) is equal to the transition probability A. For most atomic systems which admit to electric dipole transitions, $\tau \sim 10^{-8}$ s. This gives rise to a spread of $\Delta\omega \sim 10^{8}\ \mathrm{s}^{-1}$. For $\lambda \sim 5000$ Å, the corresponding spread in the wavelength is

$$\Delta\lambda \sim 10^{-4}\ \text{Å} \tag{6.78}$$

For some excited states which are stable against electroid dipole transitions, *e.g.*, the 2 $^2S_{1/2}$ state in the hydrogen atom, known as *metastable* states, the lifetime is usually about 10^5 times larger, *i.e.*, $\tau \sim 10^{-3}$ s. Metastable states play a very important role in lasers and masers.

There are other effects which also contribute to the observed linewidth. One of them is due to Doppler effect. Since the atoms are moving around (thermal motion), the observed radiation is Doppler shifted from the frequency ω_0 expected for atoms at rest. If the velocity of the particle makes an angle of α with the line of observation, the observed frequency is

$$\omega \sim \omega_0\left(1 - \frac{v}{c}\cos\alpha\right) \tag{6.79}$$

For $v \sim 6000$ m/s (corresponding to atomic hydrogen at about 1400 K), and $\lambda \sim 5000$ Å, the Doppler shift is

$$\begin{aligned}\Delta\omega &\sim 7.5 \times 10^{10}\ \text{rad/s.} \\ \Delta\lambda &\sim 0.1\ \text{Å}\end{aligned} \tag{6.80}$$

Another phenomenon that contributes to the linewidth is atomic collisions which effectively change the lifetime of the excited states. The observed linewidth $\Delta\omega$ is the sum of the linewidths arising from the different effects.

6.5 LASERS AND MASERS

It was noted in Secs. 6.3 and 6.4 that, for an atom (or a molecule) in the presence of radiation, the stimulated probabilities for absorbing a photon of resonant energy ($\hbar\omega = E_m - E_n$) or emitting a photon of the same energy are equal. This is in addition to the probability for spontaneous emission of radiation. Neglecting the spontaneous transitions, the net rate of absorption per unit time is

$$\frac{dE}{dt} = (N_n - N_m)\, B_{nm}\, u\,(\omega_{nm}),\ E_m > E_n \tag{6.81}$$

At thermal equilibrium one has

$$\frac{N_n}{N_m} = \exp\,(\hbar\,\omega_{nm}/kT) \tag{6.82}$$

so that $N_n > N_m$, and radiation will be absorbed. However, if the initial condition is such that $N_n < N_m$, known as *population inversion*, then $\frac{dE}{dt} < 0$ and there is a net emission of radiation of frequency $\omega = (E_m - E_n)/\hbar$. This leads to an amplification of the incident wave and forms the basis of lasers and masers (those words stand for light/microwave amplification by stimulated emission of radiation). It is of importance to note that the emitted wave has the same frequency, phase, and polarization as the incident wave, so that the amplification is in the wave amplitude, not just the intensity.

Though population inversion is not an equilibrium condition, it is sometimes convenient to think of it as corresponding to a negative temperature in Eq. (6.82), for which $N_n < N_m$.

Similarly, since the radiation is increasing in intensity, the situation can be described by the usual formula

$$I(x) = I(0)\, e^{-\alpha x} \tag{6.83}$$

for the propagation through an absorbing medium but with a negative co-efficient of absorption.

The mechanism of population inversion characterizes the different lasers and masers. A few of them are discussed here.

The ammonia maser (*Townes,* 1954): This is based on the separation of the components of an ammonia beam. The NH_3 molecule has a pyramidal structure with N at the apex. However, the N atom can be on either side of the base and can tunnel from one side to the other through the base. As a result, the two degenerate ground states split into two closely-spaced energy levels, which are described essentially by the even and odd linear combinations of the two states and have an energy separation given by

$$\Delta E/h \approx 2.387 \times 10^{10}\ \text{s}^{-1} \tag{6.84}$$

which is in the microwave range ($\lambda \approx 1.25$ cm). At room temperature, these two levels are almost equally populated. However, the two states have different electric dipole moments. Therefore, an ammonia beam can be separated into two beams by subjecting it to a suitable electric field. The beam with higher energy is taken into a resonating chamber (whose role will be discussed shortly) where the spontaneously emitted photons will stimulate emissions of similar photons by other atoms, and the chain process rapidly builds up the amplitude.

Ruby laser (*Meiman, 1960*): A ruby consists of crystalline aluminium oxide (Al_2O_3) in which some of the aluminium atoms are replaced by the chrominium atoms. The energy levels of Cr^{3+} ions are shown schematically in Fig. 6.3 (*a*). The Cr ions in a ruby rod which is about 1 cm in diameter and 5 cm in length, are excited from level E_1 to a group of levels E_3 by the absorption of light from a xenon flash tube adjacent to the ruby rod (duration of flash is less than 10^{-3} s). Since there are many levels near E_3, most of the Cr ions will go into the excited state. The process of imparting energy to the working substance of a laser is known as *pumping*—in the present case it is *optical* pumping since the input energy is in the optical range.

The excited ions quickly undergo nonradiative transitions with a transfer of energy to the lattice thermal motion, to the level E_2. Now, the E_2 level is a metastable state with a lifetime of about 3×10^{-3} s (usual atomic lifetimes are of the order of 10^{-8} s), so that the population of the E_2 level becomes greater than that of the E_1 level, and population inversion is obtained.

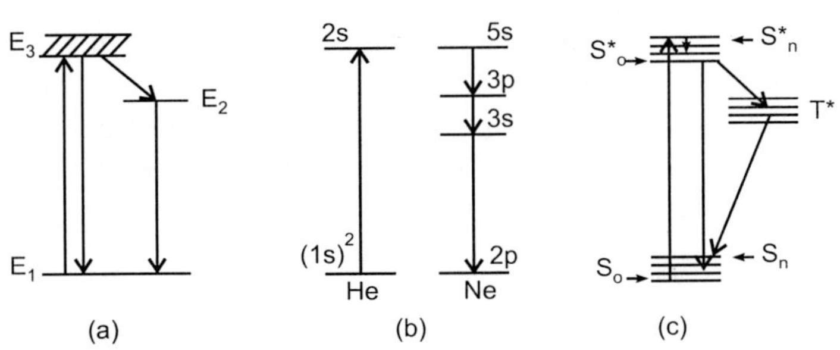

Fig. 6.3 A schematic representation of transitions in (*a*) a ruby laser where most ions from E_3, go to E_2 though a few go to E_1, (*b*) helium-neon laser, and (*c*) tunable dye laser.

Some photons are produced by spontaneous transition from E_2 and E_1, and have a wavelength of 6943 Å (ruby rod). The ends of the ruby rod are thoroughly polished and coated with layers of silver so as to act as reflecting mirrors, one end reflecting nearly 100% and the other between 90% to 100% of the incident radiation. Therefore, photons that are not moving parallel to the ruby rod escape from the side, but those moving parallel to it are reflected back and forth. These stimulate the emission of similar other photons and the chain reaction quickly develops a beam of photons all moving parallel to the rod, which is monochromatic (well-defined frequency) and is coherent (well-defined phase and polarization). When the beam develops sufficient intensity, it emerges through the partially silvered end. The ruby laser is a *solid-state* laser and operates in pulses (several pulses per minute). The larger amount of heat released in the crystal is cooled by liquid air.

Helium-neon laser: An example of a continuously operating laser is the helium-neon laser (Fie. 6.4). In this laser (Javan, 1960), an electric discharge is created by a dc current in a tube containing a mixture of helium and neon in the ratio of 5 : 1. The discharge raises some of the helium atoms into the 2s level [see Fig. 6.3 (*b*)] which is a metastable state (*i.e.,* it has a long lifetime). The energy of this level (20.61 eV) is almost the same as the energy of the 5s level (20.66 eV) in neon. Hence, the energy of the helium atoms is easily transferred to the neon atoms when they collide. This preferential transfer of the neon atoms to the 5s state results in a population inversion between the 5s and the 3p states. The spontaneous transitions from the 5s state to the 3p state, produce photons of wavelength 6328 Å, which then trigger stimulated transitions. Photons travelling parallel to the tube are reflected back and forth between the mirrors placed at the ends, the rapidly build up into an intense beam which escapes through the end with the lower reflectivity. The energy taken out by the laser beam is continuously replaced by the dc supply, so that it is a continuously operating laser. The usual efficiency of conversion of energy into the laser beam energy is quite small, about 10^{-3}%.

The hydrogen maser: Since the nucleus of the hydrogen atom has $I = 1/2$, the ground state of the atoms splits into two levels with total angular momentum quantum numbers $F = 0$ and $F = 1$, which have a small energy difference. Of the two levels, the one with $F = 0$ has a slightly lower energy. The hydrogen maser is based on stimulated transition from the $F = 1$ state to the $F = 0$ state.

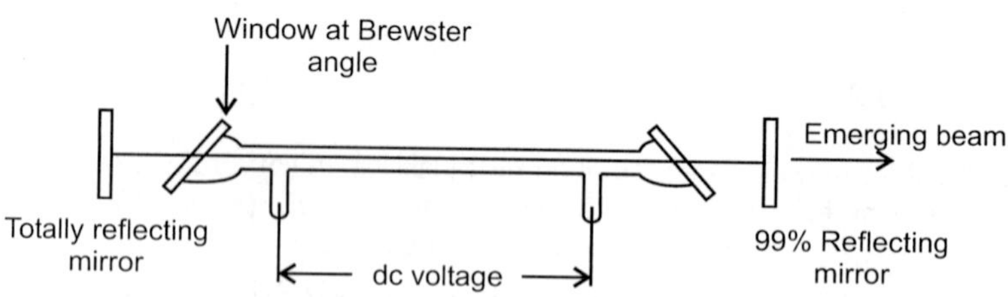

Fig. 6.4 Helium-neon laser. The windows are at Brewster angle to avoid losses by partial reflection.

For obtaining a population inversion in this maser, a beam of hydrogen atoms is passed through a region of suitable magnetic field. This allows the selection of a beam that is richer in the $F = 1$ atoms. The beam is taken into a resonance cavity where the $F = 1$ atoms undergo stimulated transitions to the $F = 0$ state. This maser is remarkable for the stability of the frequency of its radiation, $\nu_H = 1.420\,405\,7518 \times 10^9\ s^{-1}$ and can be used as a time standard with nearly the same accuracy as the caesium clock.

Dye lasers: A major constraint in the lasers discussed so far is that the frequency of the outcoming radiation is essentially fixed, though a small variation can be achieved by varying temperature, etc. This constraint was removed by the development of dye lasers which use solutions of organic dyes as the active medium for stimulated emission, *e.g.*, rhodamine 6 G in methanol solution. The output wavelength of these lasers is continuously tunable over a large range of frequencies, which makes them very versatile. The main features of a dye laser are described here in terms of a schematic representation of the energy levels in Fig. 6.3(*c*).

The large degrees of freedom in an organic dye molecule give rise to relatively broad energy bands with closely spaced vibrational and rotational levels. At ordinary temperatures, most of the molecules occupy energy levels close to the ground state level S_0 in the lowest energy band S characterized by the property that the molecule is in a singlet state, *i.e.*, total spin $S = 0$. If a solution of the dye is exposed to an intense radiation from a laser, usually a nitrogen laser, or a flash lamp, the molecules undergo transition to one of the excited singlet states S^*. Nonradiative transitions quickly bring them down (in about 10^{-11} to 10^{-12} s) to the bottom of the S^* band, *i.e.*, S_0^*. Since there are very few molecules in the upper part of S, population inversion is obtained between the lower part of the energy band S^* and the upper part of the S band. This gives rise to stimulated transitions to almost any part of the S band. These transitions can be tuned to any frequency within this range by using a suitable diffraction grating and a partially reflecting mirror, placed on the opposite sides of the active medium.

Alternatively, in some cases the molecules may undergo nonradiative transitions from S^* to the metastable triplet states T^* with total spin $S = 1$. Then stimulated transition can occur between the lowest triplet state T_0^*, and S states. Since T_0^* is metastable, usually the transitions from T_0^* states to the S states take place after a time delay. For nitrogen-laser-pumped dye lasers, it is the transitions from the S_0^* to the S band that produce the dominant laser action.

Over the years, a large number of materials that can produce laser action have been developed. A special mention should be made of semiconductor lasers which are sturdy, compact and inexpensive, and therefore suitable for practical applications. It is reasonable to expect that many more laser materials will be developed in the coming years.

Resonance cavity: The laser material is usually kept between mirrors (plane or concave) so that the photons are reflected back and forth many times to build up an intense photon beam (the effective path length is increased by a factor equal to the number of reflections). One of the mirrors is partially transparent, between 90% to 100% reflecting, which allows the beam to be taken out. If the tube windows (Fig. 6.4) are at the Brewster angle, the emerging beam will be plane polarized.

Apart from increasing the intensity of the beam, the mirrors help in producing a monochromatic beam in that they serve as walls of a resonance cavity which sustains only those wavelengths λ which satisfy the relation

$$p\lambda = 2t, \qquad p = 1, 2, \ldots \tag{6.85}$$

where t is the distance between the mirrors. The separation between two successive modes is

$$\begin{aligned} \lambda_p - \lambda_{p+1} &\approx \lambda^2/2t, \\ &\approx 0.002 \text{ Å for } \lambda = 6328 \text{ Å}, t = 1 \text{ m} \end{aligned} \tag{6.86}$$

which is quite a bit smaller than the spread in wavelength due to Doppler effect [Eq. (6.79)]. Thus, there are several frequencies, each with a very narrow width, supported by the resonance cavity, within the Doppler width of the central frequency. Some special technique can be used to select one of these frequencies, such as reducing t which will increase $\lambda_p - \lambda_{p+1}$, or lowering the temperature which will decrease the Doppler width.

6.6 APPLICATIONS OF LASERS

Since the laser beam is made up of stimulated emission, it is monochromatic, has a high temporal coherence, *i.e.,* the phases at different times are related, and a high spatial coherence, *i.e.,* the phases at different positions are related. It is parallel and has a very high intensity since the beam has a small cross section. The temporal coherence is determined by the frequency width of the beam,

$$\Delta t \sim 1/\Delta\omega \tag{6.87}$$

and can be as large as 10^{-6} or larger in some lasers. Spatial coherence implies that the beam is essentially described by a single plane wave with a width equal to the cross-section of the beam. This gives rise to a directionality constrained only by the width. If a beam with a cross-sectional area $(\Delta x)^2$ is travelling in the z-direction, uncertainty relation implies

$$\Delta p_x \approx \hbar/\Delta x \tag{6.88}$$

so that the angular spread is

$$\begin{aligned} \alpha = \frac{\Delta p_x}{p_z} &\approx \left(\frac{\hbar}{\Delta x}\right)\left(\frac{\lambda}{h}\right) \\ &\approx 10^{-4} \text{ rad for } \lambda \approx 5000 \text{ Å}, \Delta x \approx 10^{-3} \text{ m} \end{aligned} \tag{6.89}$$

It is the properties of narrow frequency range, coherence, directionality and high intensity which make the laser beam extremely useful. Some of the uses and applications of lasers are discussed below.

1. Tunable dye lasers allow the excitation and analysis of atomic and molecular energy levels with a high accuracy. Indeed, they have revolutionized the field of optical spectroscopy. In particular, Lamb shift has been observed optically, two-photon transitions have been observed in atomic and molecular systems, Rydberg stateshave been analysed, and significant tests of unified theories of electromagnetic and weak interactions have been made.
2. The coherence and the high intensity of laser beams enable us to measure small changes in the frequencies of radiation resulting from Raman scattering by measuring beats produced by the interference between the scattered and the original beams.
3. The narrow frequency width and coherence of lasers makes them very useful in precision measurements. Interferometers with laser beams allow measurement of distances to a very high accuracy, as also surface variations, refractive index, etc. For measuring the velocity of fluids, a laser beam is scattered by the fluid. This Doppler-shifted beam then interferes with the original beam producing beats. The beat frequency enables us to measure the velocity of the moving medium.
4. The well-defined directionality of a laser beam makes it valuable in communications, surveying and tracking systems.
5. Some lasers are capable of producing narrow beams of extremely high intensity. Such beams find use in precision cutting and boring, soldering and welding. They are used in tumor destruction and in eye surgery for 'welding' detached retina. There is also the possibility that they can induce controlled thermonuclear fusion.
6. Lasers have important applications in nonlinear optics and in holography. These are discussed in some detail.

Nonlinear Optics

For ordinary light sources, the electric field is so small that the induced polarization P is approximately proportional to the electric field E, and the various properties of the medium such as polarizability a, refractive index, etc. are independent of the field intensity (here, for simplicity the vector nature of P and E is neglected). Thus, we have what is known as *linear optics* for which the superposition principle holds *i.e.*, $P_1 = \alpha E_1$, $P_2 = \alpha E_2$ implies $P_1 + P_2 = \alpha (E_1 + E_2)$. However, with laser fields of high intensity, there is no longer a linear relation between P and E, and the description is in terms of *nonlinear optics*.

For high E values, a series expansion for P can be written, giving

$$\begin{aligned} P &= \alpha_1 E(t) + \alpha_2 E(t)^2 + \ldots \\ &= \alpha_1 E_0 \cos \omega t + \alpha_2 E_0^2 \cos^2 \omega t + \ldots \end{aligned} \quad (6.90)$$

Since $\cos^2 \omega t = [1 + \cos(2\omega t)]/2$, the second term builds a field component with a frequency of 2ω. This is called *frequency doubling*. For example, when a dielectric medium is irradiated with a powerful ruby laser beam with $\lambda = 6943$ Å, an ultraviolet component with $\lambda \approx 3472$ Å is observed to emerge from the medium. In general, higher harmonics with frequencies 3ω, 4ω, etc. also may be present.

If two beams with different frequencies, ω_1 and ω_2, at least one of them being a laser beam, are incident on the medium, the non-linear term will have terms with frequencies $2\omega_1$, $2\omega_2$, $\omega_1 + \omega_2$ and $\omega_1 - \omega_2$. The emerging beam therefore will contain components with these frequencies. Thus, the effect of a low frequency beam (*e.g.*, ω_2 in the infra-red region) may be observed in the optical region by choosing ω_1 in the optical range.

In many substances, the refractive index of the substance increases as the intensity increases. Thus, the effective refractive index of the material is larger near the centre of the propagating laser beam so that the rays bend towards the beam axis. This is known as *self-focussing* and is again a consequence of non-linear optics. This property is utilized in fibre-optics communication.

Holography

An extremely interesting application of lasers is to *holography*, *i.e.*, the production of the whole or complete, 3-dimensional picture of an object. It is based on the reconstruction of the electromagnetic fields reflected by the object.

Preparation of the photographic plate: Consider the electromagnetic field of a laser beam reflected by an object. For simplicity, the reflected beam is assumed to be a plane wave. This wave with amplitude R is allowed to interfere with a reference laser beam of amplitude A, at an angle θ, on a photographic plate [Fig. 6.5 (*a*)]. The exposed plate registers the interference fringes and is processed to give a hologram. The intensity registered is [Fig. 6.5 (*a*)]

$$\begin{aligned} I &= |A \exp(i2\pi(x - ct)/\lambda) + R \exp(i2\pi(\mathbf{n} \cdot \mathbf{r} - ct)/\lambda)|^2_{x=0} \\ &= A^2 + R^2 + 2AR \cos[2\pi y (\sin\theta)/\lambda] \end{aligned} \quad (6.91)$$

where it is assumed that A and R are real. The separation between the interference fringes is

$$d = \frac{\lambda}{\sin\theta} \quad (6.92)$$

Reconstruction of the wavefront: A similar reference beam is allowed to fall on the hologram which acts as a diffraction grating (grating separation $d = \lambda/\sin\theta$) and produces diffraction images. The angular separation between the central maximum and the first maximum on the two sides, is [see Fig. 6.5(*b*)]

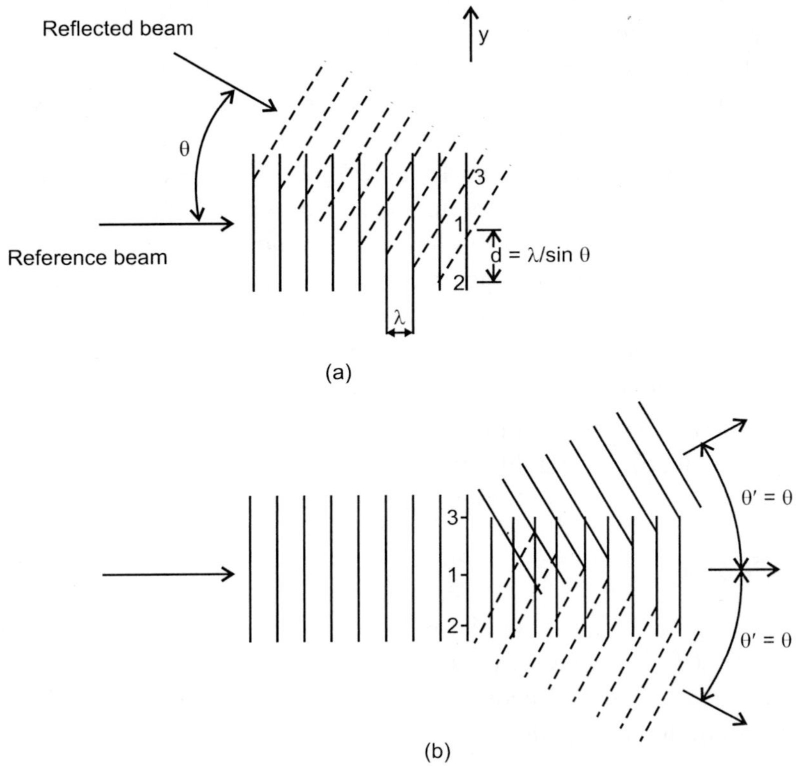

Fig. 6.5 (*a*) Interference of the plane wave object beam and the reference beam producing a hologram. (*b*) The hologram producing three components one of which is the same as the object beam.

$$\sin \theta' = \frac{\lambda}{d}$$
$$= \sin \theta \qquad (6.93)$$

or $\theta' = \theta$. Therefore, of the two maxima, one of the them has the same directionality as the beam reflected from the object had. The wavefront corresponding to this maximum is the same as that of the reflected beam and hence produces a three dimensional image. Analytically, the modulated amplitude is

$$B = IA \exp (i2\pi - ct)/\lambda) \big|_{x=0}$$
$$= A(A^2 + R^2) \exp (-i\omega t) + A^2R \exp [-i(\omega t + 2\pi y (\sin \theta)/\lambda)]$$
$$+ A^2R \exp [-i(\omega t - 2\pi y(\sin \theta)/\lambda)] \qquad (6.94)$$

where I given in Eq. (6.91) has been used. The first component corresponds to the central maximum, the second component corresponds to the first maximum along the direction of the original beam (it is to be noted that this wave moves downward as t increases), and the third component represents the other first maximum (the wave moves upward as t increases).

If the object beam is a diverging beam (as in normally the case), it is easily seen that the maximum intensity points 2 and 3 both move up with respect to 1 (see Fig. 6.6). As a consequence, the lower beam will also be diverging and will appear to start from the object, whereas the upper beam will converge to a real image of the object. This is seen by constructing wavefronts with circles of radius b with centre at point 1, radius $b \mp \lambda$ with centre of point 2, radius $b \pm \lambda$ with centre at point 3. The upper signs correspond to the lower beam and the lower signs correspond to the upper beam. It may also be noted that in viewing a hologram, the reference beam may have a frequency different from that of the beam used for recording the image. If the wavelength of the second reference beam is longer, the image observed will be magnified.

Holograms are very useful in studying the conditions at different levels by focussing the microscope at different planes of the reconstructed image, *e.g.,* in the investigation of sizes and distribution of particles, mechanical strains, etc.

Laser Cooling

In 1985, the group of, S. Chu and co-workers (among them Ashkin and J.E. Bjorkholm) at Bell Laboratories, Holmdel, NJ, reported success in cooling a dilute vapour of about 10^5 neutral sodium atoms in a volume of 0.2 cm^3 to a temperature of about 0.2 mK.

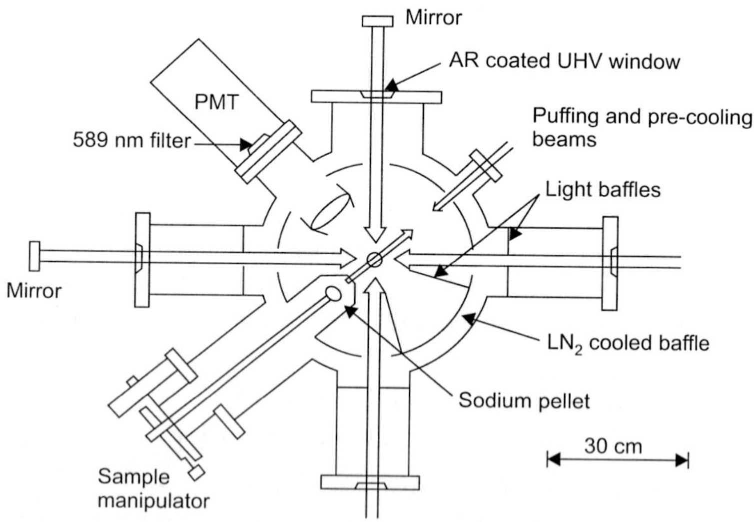

Fig. 6.6 Schematic drawing of the vacuum chamber, intersecting laser beams and atomic beam used for the Doppler cooling experiment. The laser beams enter the UHV windows vertically and horizontally.

The development of methods to cool dilute vapours of trapped atoms has made it possible to construct atomic clocks useful for precise timekeeping, *e.g.*, in connection with navigation in space and the exploration of the solar system. Another application using laser cooling is the development of atomic interferometers in which the de Broglie wavelength of slow atoms is used for interferometric measurements with ultrahigh precision, *e.g.*, the acceleration of gravity. It has become possible to develop instruments for atom optics to achieve atomic lithography. The atomic beams may be used to form nanometer structures on surfaces, *e.g.*, for electronic components. The recent observation of a Bose-Einstein condensation in a dilute atomic gas is also achieved using laser cooling and trapping.

(*Source: http://www.nobelprize.org/nobel_prizes/physics/laureates/1997/back.html*)

6.7 SOME EXPERIMENTAL METHODS

In this section, some experiments which are important for the study of atomic and molecular properties are described.

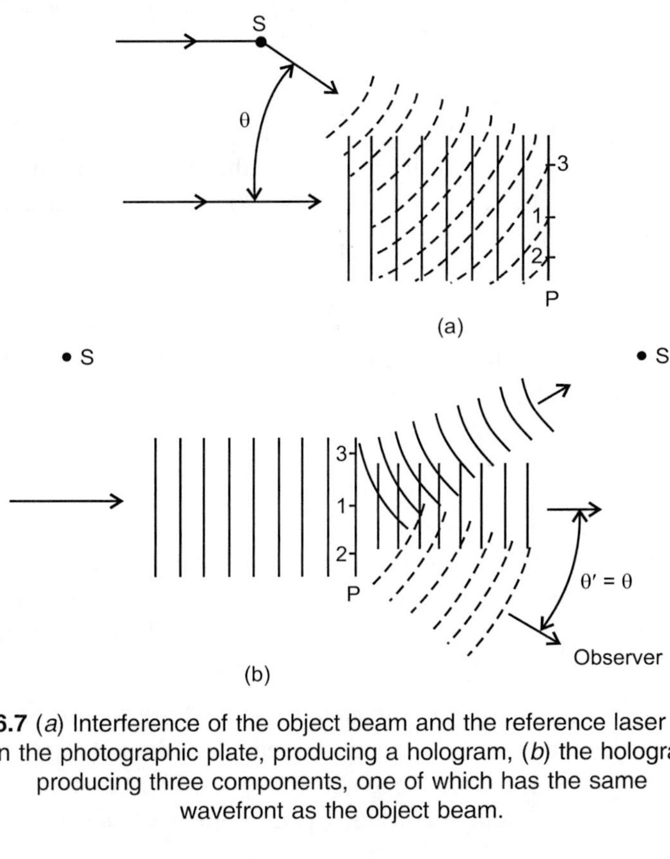

Fig. 6.7 (*a*) Interference of the object beam and the reference laser beam, on the photographic plate, producing a hologram, (*b*) the hologram producing three components, one of which has the same wavefront as the object beam.

Magnetic Resonance Experiments

These experiments depend on the interaction of an atomic or a nuclear system that has a nonzero magnetic moment, with a static magnetic field and a radiation field. The static field splits the levels into different components, and the radiation field with a particular frequency, called the *resonance frequency*, induces transitions between these levels. Analysis of these transitions leads to information about the magnetic moment of the system. The resonance experiments are described as *electron paramagnetic resonance* (abbreviated as epr) when applied to electronic magnetic moments, and as *nuclear magnetic resonance* (abbreviated as nmr) when applied to nuclear magnetic moments.

When an atom with nonzero electronic magnetic moment is placed in a magnetic field, each level with total angular momentum quantum number J splits into $2J + 1$ Zeeman levels [see Eq. (6.14)] with the energy difference between the energy levels being

$$\Delta E = \frac{e\hbar}{2m} gB \tag{6.95}$$

If now a radiation field of frequency ω is incident, then transitions between the different states take place with the absorption or emission of a photon, if

$$\hbar\omega = \frac{e\hbar}{2m} gB \tag{6.96}$$

They can be discussed along the same lines as in Sec. 6.3, except that the interaction energy is due to the interaction of the magnetic moment with the time-dependent magnetic field associated with the radiation. These transitions, known as *magnetic dipole transitions*, satisfy the selection rule

$$\Delta M_J = \pm 1, 0 \tag{6.97}$$

so that, within the Zeeman multiplets, the allowed transitions are only between adjacent levels. The frequency of the inducing radiation for $B = 10^4$ G (*i.e.*, 1 Wb/m^2), is of the order of

$$\omega \sim \frac{eB}{m} = 10^{11} \text{ rad/s} \tag{6.98}$$

which corresponds to $\lambda \approx 2$ cm, and is in the microwave frequency range. It may be recollected that radiation induces emission and absorption with equal probability [Eq. (6.52)]. Since at thermal equilibrium, there are more atoms in the lower state [see Eq. (6.82)], there is a net absorption of energy by the atoms.

In practice, the paramagnetic substance is placed inside a resonance cavity suspended between the poles of an electromagnet. Radiation of a given frequency is transmitted by a waveguide, made to interact with the substance, and is collected by a receiver and recorded. In the course of the experiment, the magnetic field produced by the electromagnet is gradually increased. As the value of B passes through the critical value satisfying Eq. (6.96), an intense absorption is observed. Since the experiments are done mostly for crystalline or liquid paramagnetic substances in which the energy levels of the atoms are perturbed by the internal fields, the resonance relation in Eq. (6.96) varies slightly from atom to atom. As a result, the absorption curve for the intensity of the transmitted radiation, as a function of B, has a finite width.

The nucleus of an atom has a magnetic moment given in Eq. (4.65),

$$\mu_N = g_N \frac{e}{m_p}\mathbf{I} \tag{6.99}$$

which is about 1000 times smaller than the magnetic moment of the electrons. Under the influence of the external field, the nuclear level splits into $(2I + 1)$ Zeeman levels and resonance absorption is observed for the frequency ω,

$$\hbar\omega = \frac{e\hbar}{m_p} g_N B \tag{6.100}$$

For $B = 10^4$ G, this corresponds to a frequency of about 10^8 rad/s which is in the radio-frequency range. Nuclear magnetic resonance is especially useful for the study of atoms and molecules which may have zero electronic magnetic dipole moment. These atoms generally have a nonzero nuclear magnetic moment and can be studied by the nuclear magnetic resonance techniques.

The *epr* and *nmr* techniques can be used for identifying the presence of certain elements, for determining the environment of the electron or the nucleus (by noting the shift in the resonance frequency due to the environment), and also for accurate measurement of magnetic fields.

Atomic and Molecular Beam Experiments

While magnetic resonance experiments are almost universal in their applications, their accuracy is limited by the fact that they are based on the differential population of nearby levels at thermal equilibrium and on the measurements of changes in the radiation intensity. If the material is available in the form of atomic or molecular beams, more accurate beam experiments can be performed.

Atomic and molecular-beam experiments are refinements of the Stern-Gerlach experiment (Sec. 4.2) due to Rabi, incorporating the observation of magnetic resonance. For simplicity, consider a beam of particles with the nuclear angular momentum characterized by $I = 1/2$ and an associated magnetic moment. In a typical set-up, the beam traverses three regions with magnetic fields B_1, B_2 and B_3 produced by magnets 1, 2 and 3 respectively (see Fig. 6.7). The first field B_1 is inhomogeneous with the gradient as shown, and splits the beam into two components with $M_I = \pm 1/2$ one of which, say with $M_I = -1/2$ is eliminated at the wall, while the second component with $M_I = 1/2$ moves along the trajectory shown. The second field B_2 is homogeneous and introduces an energy difference

$$\Delta E = \frac{e\hbar}{m_p} g_N B_2 \tag{6.101}$$

between the energy levels. In this region, there is also a radiation field of radio frequency ω (ω ~ 10^8 rad/s). If ω satisfies the resonance condition

$$\hbar\omega = \frac{e\hbar}{m_p} g_N B_2 \tag{6.102}$$

some of particles will undergo resonant transition to the $M_I = -1/2$ state. The third field B_3 also is inhomogeneous but has a gradient opposite to that of B_1, which will remove the particles with $M_I = -1/2$, at the wall, while those with $M_I = 1/2$ pass along the trajectory shown and register in the detector.

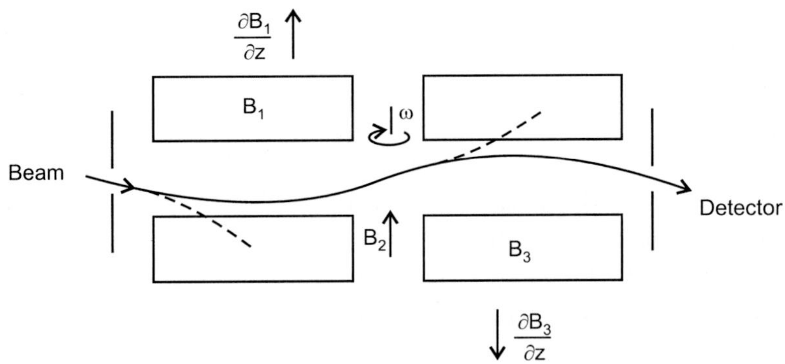

Fig. 6.8 Schematic diagram of the atomic/molecular beam resonance experiment. The dashed lines indicate the components removed.

In the actual experiment, the frequency ω is held fixed and the field B_2 is varied. When the resonance condition in Eq. (6.102) is satisfied, some of the particles undergo transition to the $M_I = -1/2$ state and are removed at the wall, which reduces the recorded beam intensity. The value of the field B_2 at which the minimum beam intensity is recorded can be used to calculate the value of g_N and hence the magnetic moment of the particles. For example, the reduction in intensity is observed for ^{31}P, at $B = 10^4$ G and $\omega = 1.08 \times 10^8$ rad/s which gives a value of $g_N = 1.13$.

The application of nuclear magnetic resonance best known to the general public is magnetic resonance imaging (MRI) for medical diagnosis and magnetic resonance microscopy in research settings, however, it is also widely used in chemical studies, notably in NMR spectroscopy such as proton NMR, carbon-13 NMR, deuterium NMR and phosphorus-31 NMR. Biochemical information can also be obtained from living tissue (*e.g.* human brain tumors) (see Fig. 6.9) with the technique known as *in vivo* magnetic resonance spectroscopy or chemical shift NMR Microscopy.

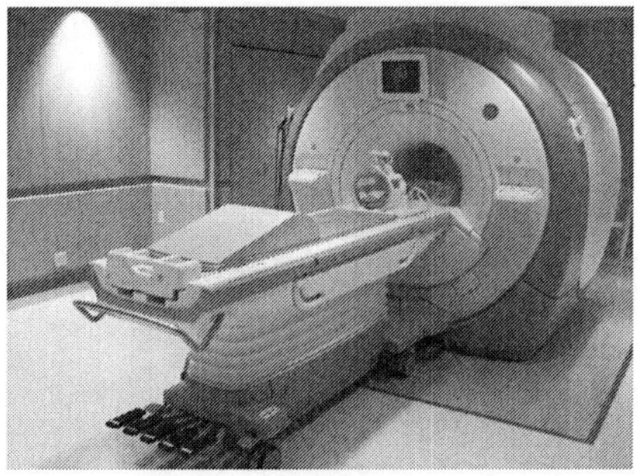

Fig. 6.9 Medical MRI.

Raman Effect

It was observed in 1928, by Raman and Krishnan, and simultaneously by Landsberg and Mandelshtam, that the spectrum of light scattered by gases, liquids and crystals, contains apart from the unshifted original frequency ω, new lines whose frequencies are given by

$$\omega' = \omega \pm \omega_1 \tag{6.103}$$

This is known as *Raman* effect, or more descriptively, as *combination scattering of light*.

The process of scattering of radiation may be regarded as being made up of absorption of the incoming photon and emission of the outgoing photon. If, as a result, the final state of the atom or molecule is the same as the initial state, the frequency of the photon is unchanged, giving the unshifted line. This process is known as *Rayleigh scattering*. On the other hand, if the final state of the atom or molecule is different, the process is an inelastic scattering of the photon and the frequency ω′ of the final photon is given by the energy conservation relation

$$\hbar\omega + E' = \mathrm{E} + \hbar\omega \tag{6.104}$$

or

$$\omega' = \omega + \frac{E - E'}{\hbar} \tag{6.105}$$

The shifted frequency is less than the original frequency if $E < E'$ and the corresponding lines are called the *Stokes lines*. It is more than the original frequency if $E > E'$ and the associated lines are called the *anti-Stokes lines*. At ordinary temperatures, there are more particles in the lower energy states, so that there are more transitions with $E_1 \rightarrow E_2$ than those with $E_2 \rightarrow E_1$, $E_2 > E_1$. Therefore, anti-Stokes lines are generally fainter (in some cases not even observable) than the Stokes lines. The anti-Stokes lines increase in intensity as the temperature is raised since this will increase the relative population of the higher energy states.

Since Raman effect is a two-step process, the selection rules can be deduced from those for the two separate steps. In particular, the selection rules for the transition between the rotational states of molecules, are $\Delta J = \pm 1$ for emission or absorption of photons, and hence Raman effect is observed for transitions with

$$\Delta J = \pm 2, 0 \tag{6.106}$$

For purely rotational transitions, only the $\Delta J = \pm 2$ transitions need be considered ($\Delta J = 0$ does not involve changes in energy). The change in the energy in the case of diatomic molecules, is given by

$$\Delta E = \pm \frac{\hbar^2}{2I} [J(J+1) - (J-2)(J-1)]$$

$$= \pm \frac{\hbar^2}{2I} (2J-1)\, J \geq 2 \tag{6.107}$$

where J refers to the higher state. For $J = 2$, $\Delta\omega = \pm 3\hbar^2/I$, for $J = 3$, $\Delta\omega = \pm 5\hbar^2/I$, for $J = 4$, $\Delta\omega = \pm 7\hbar^2/I$, etc. These lines are illustrated in Fig. 6.8. It is instructive to compare them with the equi-spaced rotational levels in absorption spectra [see Fig. 5.13(*a*)].

For transitions which involve changes in the vibrational states, ΔJ can be 0 or 2. These involve larger changes in energy and hence anti-Stokes lines are generally very faint. The frequency shift for a change in the vibrational state but with $\Delta J = 0$, corresponds to the missing central line in the vibrational-rotational spectrum. The spacing of the $\Delta J = 2$ lines about the $\Delta J = 0$ is given by Eq. (6.107). Raman spectra, involving changes in the vibrational states, provide useful information about the structure of the molecules.

Raman spectra are characteristic of the molecules (and atoms) and are extremely useful in the analysis of the complicated mixtures of molecules, especially of organic molecules. They are also important in the determination of the rotational and vibrational levels, and in the analysis of the structures of the molecules.

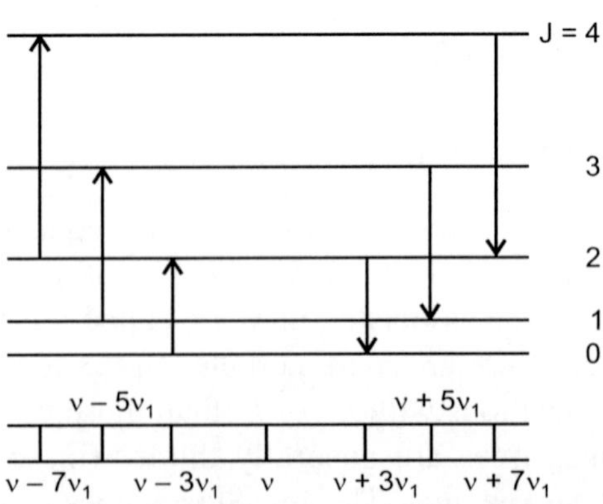

Fig. 6.9 The Raman spectrum for transitions within the rotational levels, ν_1 is the spacing of the rotational levels (see Fig. 5.13).

Measurement of Lifetimes

The measurement of lifetimes of excited atoms and molecules, being of the order of 10^{-8} s, is difficult. A few techniques of measuring short lifetimes are discussed here.

The lifetimes can be obtained by measuring the intensity of radiation from a collection of excited atoms, as a function of time. A voltage pulse is used to excite the atoms by electron bombardment. The pulse starts the multi-channel analyser in which channel n is active during the time $n\delta$ to $(n + 1)\delta$, δ being a time interval short compared with the lifetime τ of the atoms. The channel records the pulses produced by the photoelectrons generated by the radiation emitted. The pulse intensity is proportional to the number of excited atoms, and hence its time dependence allows us to calculate the lifetime [from Eq. (6.76)]. One of the difficulties is that the population in the decaying state may be continuously replenished by the particles in a higher excited state decaying to the lower excited state under consideration.

In another method for measuring lifetimes of excited ions, called the *beam-foil* technique, fast moving ions (accelerated by a potential difference) are excited by passing them through a thin foil. The intensity of radiation emitted as a function of the distance these excited ions travel, gives us information about the number of excited states as a function of time, and hence allows us to calculate the lifetime τ [from Eq. (6.76)].

An indirect method of calculating the natural lifetime is to measure the linewidth of the level, and use the relation $\tau = 1/\Delta\omega$ (essentially the uncertainty relation) to deduce the lifetime of the state. In this method, the *Doppler linewidth* and the *collision linewidth* (collisions affect the lifetime of a state), must be taken into account in isolating the natural linewidth from the total observed linewidth ($\Delta\omega$ used in the uncertainty relation is the natural line width).

6.8 EXAMPLES

The discussion in this chapter is now supplemented with some technical details and examples.

Example 1

Here, the proof of the important theorem stated in Sec. 6.2 is outlined. To prove the equality in Eq. (6.8), the z-axis is taken along the direction under consideration. Then it has to be proved that

$$\int \psi^*_{J,M_J} S_z \psi_{J,M_J'}\, d\tau = a \int \psi^*_{J,M_J} I_z \psi_{J,M_J'}\, d\tau \qquad (6.108)$$

for $\qquad [\mathbf{L}, \mathbf{S}] = 0, \mathbf{J} = \mathbf{L} + \mathbf{S}.$

Consider first the case $M_J \neq M_J'$. Since $[S_z, J_z] = 0$,

$$\int \psi^*_{J,M_J} (S_z J_z - J_z S_z) \psi_{J,\ M_J'} d\tau = 0 \tag{6.109}$$

This leads to (after integrating the second term by parts),

$$(M_J' - M_J)\hbar \int \psi^*_{J,M_J} S_z \psi_{J,M_J'} d\tau = 0 \tag{6.110}$$

or
$$\int \psi^*_{j,M_J} S_z \psi_{J,M'_J} d\tau = 0, \text{ for } M_J \neq M'_J \tag{6.111}$$

Since the right hand side of Eq. (6.108) is

$$aM_J'\hbar \int \psi^*_{J,M_j'} \psi_{J,M_J'} d\tau = 0, M_J \neq M_J' \tag{6.112}$$

the equation is satisfied for $M_J \neq M_J'$.

To prove Eq. (6.108) for $M_J = M_J'$, it is observed that

$$\int \psi^*_{J,M_J+1} S_z \psi_{J,M_J+1} d\tau - \int \psi^*_{J,M_J} S_z \psi_{J,M_J} d\tau = ah \tag{6.113}$$

where **a** is a constant independent of M_J. While this result is plausible in the sense that every increment of M_J (or J_z) may be expected to cause an increase in the average value of S_z, which depends only on the increment of M_J and not on M_J itself, it is quite difficult to prove it (see Ref. 22, p. 236). This result then leads to

$$\int \psi^*_{J,M_J} S_z \psi_{J,M_J} d\tau = a\hbar M_J + b \tag{6.114}$$

It is then noted that if all the angular momenta. **J, L** and **S** take opposite values, the average value of S_z also should change its sign:

$$\int \psi^*_{J,-M_J} S_z \psi_{J,-M_J} d\tau = -\int \psi^*_{J,M_J} S_z \psi_{J,M_J} d\tau \tag{6.115}$$

Substituting Eq. (6.114) in this relation give $b = 0$. Since $\hbar\ M_J$ is the eigenvalue of J_z, the required relation is obtained as

$$\int \psi^*_{J,M_J} S_z \psi_{J,M_J'} d\tau = a \int \psi^*_{J,M_J} J_z \psi_{J,M_J'} d\tau \tag{6.116}$$

in which both the sides are zero for $M_J \neq M_J'$. This proves the equality in Eq. (6.8).

Example 2

The interaction of an atom with an external constant electric field **E** in the z-direction, is obtained from Eq. (6.5):

$$H' = e\sum_i z_i |\mathbf{E}| \tag{6.117}$$

This interaction has the interesting new feature that it becomes indefinitely large and negative as $z_i \rightarrow -\infty$, so that the electrons in an atom can tunnel through the potential barrier and ultimately escape to infinity ($z \rightarrow -\infty$). Thus, there are no longer any true bound stages, and each level (including the ground state) acquires a linewidth due to the fact that it has a finite lifetime ($\Delta\omega \sim 1/\tau$). Also, the first-order energy shift given by Eq. (3.125) is zero for nondegenerates states,

$$\int \psi^*_n \, z_i \, \psi_n \, d\tau = 0 \tag{6.118}$$

Since z_i is odd and $|\psi_n|^2$ is even (nondegradable states are even or odd under parity). For degenerate states, which one encounters in the hydrogen atom, the problem is more complicated.

Consider the $2p$ and $2s$ states of the hydrogen atom. From Eq. (3.127), it is easy to show that the energies of the $m_i = \pm 1$ states are unperturbed. For the $m_l = 0$ states, the energy shifts are given by Eq. (3.128), with 1 standing for the $l = 1, m_l = 0$ state and 2 standing for the $l = 0, m_i = 0$ state. For this case, $V_{11} = V_{22} = 0$, so that $x = \pm 1$ and the energy shifts are

$$\Delta E = \pm e \,|\, \mathbf{E} \,|\, \int \psi_2^{(1)*} \, z \, \psi_2^{(2)} \, d\tau \tag{6.119}$$

Thus, for the hydrogen atom, in addition to acquiring linewidths, the spectral lines split into several components (in this discussion spin-obrit interaction has been neglected). *e.g.* the $n = 2 \rightarrow n = 1$ line splits into three components. This is known as Stark effect. As the strength of the electric field becomes large ($\gtrsim 10^7$ V/m), higher order corrections have to be included, and one has what is known as the *quadratic Stark effect* to distinguish it from the first order effect which is called the *linear* Stark effect.

Example 3

The Zeeman splitting for the hydrogen $2p \rightarrow 1s$ transitions is given by Eqs. (6.22) and (6.23).

For $B = 10^4$ G (1 Wb/m^2), $\Delta\omega_0 = 8.78 \times 10^{10}$ rad/s

so that

$$\Delta\lambda \approx (\pm 2/3, \pm 4/3)\,(0.0069)\ \text{Å for } 2\,{}^2P_{1/2} \rightarrow 1\,{}^2S_{1/2} \tag{6.120}$$

$$\approx (\pm 1/3, \pm 1, \pm 5/3)\,(0.0069)\ \text{Å for } 2\,{}^2P_{3/1} \rightarrow 1\,{}^2S_{1/2}$$

The shifts are observed to be very small.

Example 4

Here, the lifetime of the $2p$ state of the hydrogen state is calculated using Eq. (6.70).

Without any loss of generality, we assume that the atom is originally in the $l = 1, m_l = 0$ state. Using the wave functions given in Sec. 4.1,

$$|(\mathbf{r})_{1s,\,2p}| = \left| \int \frac{\exp(-r/a_1)}{(\pi a_1^3)^{1/2}}\, z\, \frac{r\cos\theta}{(32\,\pi\, a_1^5)^{1/2}} \exp(-r/2a_1)\, r^2 dr\, 2\pi d\cos\theta \right|$$

$$= \frac{2^{15/2}}{3^5} a_1 \approx 0.745\, a_1 \tag{6.121}$$

Substituting this in Eq. (6.70) and using the relation in Eq. (6.77) the lifetime τ of the $2p$ state ($\Delta E = \hbar\omega = 12.0$ eV) is

$$\tau \approx 1.6 \times 10^{-9} \text{ s} \tag{6.122}$$

which is in agreement with the experimental observation.

Example 5

The ratio of spontaneous transitions to stimulated transitions for particles in thermal equilibrium can be obtained from Eq. (6.63). Denoting the probability for spontaneous transitions by P_1 and that for stimulated transition by P_2, Eq. (6.63) reads

$$P_2 + P_1 = \exp(\hbar\,\omega_{nm'}/kT)\, P_2 \tag{6.123}$$

or

$$P_1/P_2 = \exp(\hbar\,\omega_{nm'}/kT) - 1 \tag{6.124}$$

For very low temperatures, the transitions are predominantly spontaneous but become predominantly stimulated for high temperatures. This is to be expected since the radiation density increases with temperature. For example, at room temperatures, the transitions between $2p$ and $2s$ states of the hydrogen atom ($\hbar\,\omega_{nm} \sim 10$ eV, $kT \sim 0.026$ eV) are predominantly spontaneous. However, in some of the hot stars, the surface temperatures are as high as 30000 K, so that there stimulated transitions also are important.

Example 6

Lasers provide an intense, collimated beam. To estimate the power, consider the original ruby laser which had a diameter of 1 cm and a length of 5 cm. If the ruby has about 10^{19} Cr atoms/cc, and all of them are excited, the total energy available is

$$E = 10^{19}\left(\frac{\pi}{4}5\right) \times (hv)$$

$$= 11.25 \text{ J} \tag{6.125}$$

If the pulse lasts for about 10^{-7} s, the power during this period is about 10^8 W.

The angular spread of the beam is

$$\alpha \sim 1.10 \times 10^{-5} \text{ rad} \tag{6.126}$$

In travelling a distance of 100 m, the beam will spread by about

$$\Delta x \sim 0.11 \text{ cm} \tag{6.127}$$

Example 7

If an excited state is replenished by decays from a higher excited state, the decay rate is not given by the simple exponential function in Eq. (6.76).

Consider three states with energies $E_0 < E_1 < E_2$ and decay probabilities A_{10}, A_{20} and A_{21}. Then the changes in $N_1(t)$ in time dt are

$$dN_1(t) = -A_{10}N_1(t)\,dt + A_{21}N_2(t)\,dt \tag{6.128}$$

$$dN_2(t) = -A_2N_2(t)\,dt \tag{6.129}$$

where $A_2 = A_{20} + A_{21}$. The solutions to these equations are

$$N_2(t) = \exp(-A_2 t)\,N_2(0)$$

$$N_1(t) = \exp(-A_{10} t)\,N_1(0) + \frac{A_{21}}{A_{10} - A_2}$$

$$(\exp[-A_2 t] - \exp[-A_{10} t])\,N_2(0) \tag{6.130}$$

It is observed that if $A_{21} N_2(0) > A_{10} N_1(0)$, $N_1(t)$ will increase for small t but will eventually start decreasing.

PROBLEMS

1. Obtain the energy levels of the $(1s)^2\ {}^1S_0$, $(1s)(2p)\ {}^1P_1$, and $(1s)(3d)\ {}^1D_2$ states of the helium atom in the presence of a magnetic field of strength 1 Wb/m^2. What are the shifts in Å for the allowed transitions, if $l_{P \to S}$ = 584.4 Å, $\lambda_{D \to P}$ = 6678 Å for the unperturbed states under consideration?
2. Describe the Zeeman patterns of the ${}^2D_{3/2}$ and ${}^2P_{3/2}$ states. Calculate the frequency shifts for the transitions ${}^2D_{3/2} \to {}^2P_{3/2}$ with $\Delta M_J = 0$, $\Delta M_J = 1$, and $\Delta M_J = -1$. If the sodium line with λ = 8195 Å corresponds to a ${}^2D_{3/2} \to {}^2P_{3/2}$ transition, what is the maximum shift in its wavelength when a magnetic field of 1 Wb/m^2 is introduced?
3. Obtain the shifts in the frequency for ${}^2D_{3/2} \to {}^2P_{1/2}$ transitions in the presence of a weak magnetic field.
4. What is the magnetic moment of sulphur in the ground state ${}^3P_2$? In the presence of a magnetic field of 2 Wb/m^2, what is the resonance frequency?
5. Indicate the energy levels of ${}^2P$ and ${}^2D$ states and the allowed transitions between them in the presence of a strong magnetic field.

6. For N^{14}, magnetic resonance is observed at a $B = 1$ Wb/m^2 and for a frequency of $\omega = 1.933 \times 10^7$ rad/s. What is the value of the Landé g-factor for N^{14} ($I = 1$ for N^{14})?

7. Interstellar space is occupied by atomic hydrogen. The $I = 1$ and $I = 0$ states are split by hyperfine interaction (the $I = 1$ state is higher than the $I = 0$ state), the energy separation being given by the 21 cm line. If interstellar space is characterized by a temperature of $T = 3$ K, what is the ratio at this temperature of spontaneous transitions to stimulated transitions between these states?

8. There are three states with energies $E_1 < E_2 < E_3$, whose populations are in thermal equilibrium. If an external radiation induces transitions between the states with energies E_1 and E_3 till they have equal populations, show that, in general, there is population inversion between the E_1, E_2 states or the E_2, E_3 states. What is the condition for the exception?

9. A helium-neon gas laser of 2.0 mW power, operates on 220 V-2.0 A power supply. What is the working efficiency of the laser?

10. HCl molecules are traversed by the Hg 2536.5 Å radiation. If the moment of inertia of HCl is 2.7×10^{-47} kg.m^2, what are the wavelengths of the first and second rotational Raman lines?

7

Quantum Statistics

Structures of the Chapter

It is clear from earlier discussions that a complete description of even one-electron atom is quite complicated. The complexity of the problems increases rapidly as the number of particles increases, so as to make a detailed solution of a many particle system almost impossibly difficult to obtain. However, as the number of particles becomes very large, say of the order of 10^{23} as in the case of macroscopic bodies, the very largeness of the degrees of freedom leads to the result that the average properties (*macroscopic* and in some cases, *microscopic*) of the system correspond to, statistically, the most probable behaviour of the system. This feature of many-particle systems forms the basis of the quantum statistical description of their properties.

The statistical energy distributions of a collection of particles are discussed first and then some important applications based on the most probable distributions are considered.

7.1 DISTINGUISHABLE ARRANGEMENTS

The basic assumption for obtaining the most probable statistical distribution is that *every physically distinct arrangement of particles in the various available states is equally likely to occur.* This implies that the most probable distribution is the one which has the largest number of distinguishable arrangements associated with it. Therefore, the procedure for determining the most probable statistical distribution involves two steps: (*i*) obtaining the number of distinguishable arrangements which give rise to the same distribution, and (*ii*) maximizing this number of arrangements with respect to different distributions. In this section, expressions for the number of distinguishable arrangements are obtained for a given distribution.

Consider a collection of N particles, which interact weakly with each other and with the wall. Since these particles interact only weakly, each particle will have a set of states with well-defined energies available to it. Let their possible energies be grouped into cells of sizes $\Delta\varepsilon_1$, $\Delta\varepsilon_2$..., $\Delta\varepsilon_i$ ..., with average energies ε_1, ε_2..., ε_i ..., respectively. These particles may belong to one of the following three classes of particles.

1. There are Q identical but distinguishable particles. These are classical particles whose trajectories may, in principle, be followed. There are f_i number of states and q_i number of particles in the i-th energy cell.
2. There are R identical, indistinguishable **bosons** with integral spin. The i-th energy cell has g_i number of states and r_i number of these bosons ($g_i \neq f_i$ in general).
3. There are S identical, indistinguishable fermions with half-integral spin. The ith energy cell has h_i number of states and s_i number of these fermions ($h_i \neq f_i$ or g_i, in general). It should be noted that no two of these fermions can be in the same state, so that $h_i \geq s_i$.

Our aim is to determine the number of different, distinguishable arrangements for each of these distributions. The problem is similar to that of determining the number of distinguishable ways in which Q identical balls can be placed in different boxes, so that there are q_1 balls in the first box with f_1 shelves, q_2 balls in the second box with f_2 shelves, etc. (each shelf corresponds to an energy level). In the case of classical particles, since they are distinguishable, the balls can be thought of as having different colours. In the case of bosons and fermions, the balls are identical in every way including their colour. However, for fermions, there is the additional restriction that at most one ball can be placed in each shelf.

Identical classical particles: We first determine the number of ways in which Q particles can be grouped into distinguishable sets of $q_1, q_2, .., q_i,...,$ particles and then the number of ways in which q_i particles can be distributed among f_i states of the i-th cell.

The first particle can be chosen in Q number of ways, the second in $(Q-1)$ ways, etc. However, since the different orders of choosing the same set of q_1 particles in the set lead to the same result, there are

$$P_1(q_1) = \frac{Q(Q-1)\cdots(Q-q_1+1)}{q_1\,!}$$

$$= \frac{Q\,!}{(Q-q_1)\,!\,q_1\,!} \tag{7.1}$$

number of ways choosing q_1 distinguishable particles from Q classical particles. Similarly, form the remaining $Q - q_1$ particles, q_2 distinguishable particles can be chosen in ways. Proceeding in this way, the total number of ways of choosing

$$P_2(q_2) = \frac{(Q-q_1)\,!}{(Q-q_1-q_2)\,!\,q_2\,!} \tag{7.2}$$

distinguishable sets of $q_1, q_2,..., q_i, ...$ particles from Q distinguishable particles is found to be

$$P_1(q_1)\,P_2(q_2)\,...\,P_i(q_i)\,... = \frac{Q\,!}{q_1\,!\,q_2\,!\cdots q_i\,!\cdots} \tag{7.3}$$

Now each of the q_i particles can occupy any one of the f_i states so that there are f_i^{qi} number of ways of distributing q_i distinguishable particles in f_i states. Thus the total number of distinguishable arrangements for the distribution of $q_1, q_2, ..., q_i, ...$ sets of distinguishable practices in $f_1, f_2, .., f_i, ...$ states is

$$P(q_i) = Q!\prod_{i=1}^{\infty}\frac{f_i^{q_i}}{q_i\,!} \tag{7.4}$$

Identical bosons: Since the particles are indistinguishable, there is only one way of grouping R particles into distinguishable sets of r_1, r_2,..., r_1, ... particles. Therefore, the total number of distinguishable arrangements is given by just the product of the number of ways in which r_i particles are distributed among g_i number of states.

For determining the number of ways in which r_i particles are distributed among g_i number of states, the states are regarded as being separated by portions. Since no partition is needed at the ends, $g_i - 1$ number of partitions is needed. Then the particles and the partitions are arranged in a row, *e.g.*

$$\times \times \,|\,|\, \times \,|\, \times \,|\, \times \times \times \tag{7.5}$$

where each × represents a particle, the vertical line represents a partition, and the arrangement shown represents 2, 0, 1, 1, 3 particles in five states (four partitions). The number of such distinguishable arrangements in the *i*-th cell is given by the number of different ways of arranging $(r_i + g_i - 1)$ objects of which r_i particles and $g_i - 1$ partitions belong to two groups of indistinguishable objects and is

$$P_i(r_i) = \frac{(r_i + g_i - 1)!}{r_i!(g_i - 1)!} \tag{7.6}$$

Therefore, the total number of distinguishable arrangements for the distribution of $r_1, r_2, ..., r_i, ...$ sets of bosons in $g_1, g_2, ... g_i, ...$ states is

$$P(r_i) = \prod_{i=1}^{\infty} \frac{(r_i + g_i - 1)!}{r_i!(g_i - 1)!} \tag{7.7}$$

Identical fermions: Here again, there is only one way of grouping S particles into distinguishable sets of $s_1, s_2, ... s_i, ...$ particles. For obtaining the number of ways of distributing s_i particles in h_i states, it is noted that each state can be occupied by at most one particle so that the states may be arranged in a row of h_i objects, indicating the occupation of each state, *e.g.*

$$0\,0 \times \times 0\,0 \tag{7.8}$$

where 0 indicates that the level is unoccupied, × indicates that the level is occupied by one particle, and the particular arrangement represents 0, 0, 1, 1, 0, 0 particles in the six energy levels. Therefore, the number of such distinguishable arrangements in the *i*-th cell is given by the number of different ways of arranging h_i objects of which s_i and $h_i - s_i$ belong to two groups of indistinguishable objects:

$$P(s_i) = \frac{h_i!}{s_i!(h_i - s_i)!} \tag{7.9}$$

Hence the total number of distinguishable arrangements for the distribution of $s_1, s_2, ..., s_i, ...$ sets of fermions in $h_1, h_2, ..., h_i ...$ states is

$$P(s_i) = \prod_{i=1}^{\infty} \frac{h_i!}{s_i!(h_i - s_i)!} \tag{7.10}$$

7.2 STATISTICAL DISTRIBUTIONS

The number of distinguishable arrangements for a given distribution of a mixture of the three classes of particles is

$$P(q_i, r_i, s_i) = (Q_i)\prod_i^{\infty}\left[\frac{f_i^{q_i}}{q_i\,!}\right]\left[\frac{(r_i+g_i-1)!}{r_i\,!(g_i-1)\,!}\right]\left[\frac{h_i\,!}{s_i\,!(h_i-s_i)\,!}\right] \quad (7.11)$$

The number of particles of each class should be conserved. Assuming that the total energy of the system is E,

$$\sum_i q_i = Q, \sum_i r_i = R, \sum_i s_i = S \quad (7.12)$$

$$\sum_i \varepsilon_i\,(q_i + r_i + s_i) = E \quad (7.13)$$

The most probable distribution corresponds to the maximum of $P\,(q_i, r_i, s_i)$, subject to the conditions (7.12) and (7.13).

In practice, it is more convenient to maximize $\ln P\,(q_i, r_i, s_i)$. The calculations are greatly simplified by using the following approximation (Stirling's formula):

$$\ln n\,! = \ln 2 + \ln 3 + ...+ \ln n$$

$$= \int_1^{n+1/2} \ln x\,dx + 0(1)$$

$$\approx n \ln n - n \quad (7.14)$$

where for large n only the first two leading terms have been retained. Then keeping only the leading terms gives

$$\ln P = Q \ln Q - Q + \sum_i (q_i \ln f_i - q_i \ln q_i + q_i)$$

$$+\sum_i [\,(r_i + g_i) \ln (r_i + g_i) - (r_i + g_i)$$

$$-r_i \ln r_i + r_i - g_i \ln g_i + g_i]$$

$$+\sum_i [\,h_i \ln h_i - h_i - s_i \ln s_i + s_i$$

$$-(h_i - s_i) \ln (h_i - s_i) + (h_i - s_i)] \quad (7.15)$$

At the maximum, this expression has to be stationaryfor small but arbitrary changes in q_i, r_i and s_i subject to the constraints (7.12) and (7.13). Taking the differential of ln P, one gets

$$\delta (\ln P) = \sum_{i=1}^{\infty} \delta q_i [\ln f_i - \ln q_i) + \sum_{i=1}^{\infty} \delta r_i [\ln (r_i - g_i) - \ln r_i]$$

$$+ \sum_{i=1}^{\infty} \delta s_i [\ln (h_i - s_i) - \ln s_i] = 0 \qquad (71.6)$$

subject to the conditions

$$\sum_{i=1}^{\infty} \delta q_i = 0, \sum_{i=1}^{\infty} \delta r_i = 0, \sum_{i=1}^{\infty} \delta s_i = 0, \qquad (7.17)$$

$$\sum_{i=1}^{\infty} \varepsilon_i (\delta q_i + \delta r_i + \delta s_i) = 0 \qquad (7.18)$$

Using the relations in Eq. (7.17) to eliminate δq_1, δr_1 and δs_1 in Eqs. (7.16) and (17.18) gives

$$\sum_{i=2}^{\infty} \delta q_i \ln \left(\frac{f_i q_1}{f_1 q_i} \right) + \sum_{i=2}^{\infty} \delta r_i \ln \left[\frac{(r_i + g_i) r_1}{(r_1 + g_1) r_i} \right]$$

$$+ \sum_{i=2}^{\infty} \delta s_i \ln \left[\frac{(h_i - s_i) s_1}{(h_1 - s_1) s_i} \right] = 0 \qquad (7.19)$$

$$+ \sum_{i=2}^{\infty} (\varepsilon_i - \varepsilon_1)(\delta q_i + \delta r_i + \delta s_i) = 0 \qquad (7.20)$$

From Eq. (7.20),

$$\delta q_2 = -\sum_{i=2}^{\infty} \frac{\varepsilon_i - \varepsilon}{\varepsilon_2 - \varepsilon_1} (\delta q_i + \delta r_i + \delta s_i) + \delta q_2 \qquad (7.21)$$

Using this in Eq. (7.19) to eliminate δq_2, then regarding δq_3, δq_4 ..., δr_2, δr_3, ...,δs_2, δs_3 ..., etc., as arbitrary variables and equating their coefficients to zero gives

$$\ln \frac{f_i q_1}{f_1 q_i} - \frac{\varepsilon_i - \varepsilon_1}{\varepsilon_2 - \varepsilon_1} \ln \frac{f_2 q_1}{f_1 q_2} = 0, i = 3, 4, ...,$$

$$\ln \frac{(r_i + g_i) r_1}{(r_1 + g_1) r_i} - \frac{\varepsilon_i - \varepsilon_1}{\varepsilon_2 - \varepsilon_1} \ln \frac{f_2 q_1}{f_1 q_2} = 0, i = 2, 3, \cdots \qquad (7.22)$$

$$\ln \frac{(h_i - s_i) s_1}{(h_1 - s_1) s_i} - \frac{\varepsilon_i - \varepsilon_1}{\varepsilon_2 - \varepsilon_1} \ln \frac{f_2 q_1}{f_1 q_2} = 0, i = 2.3, \cdots$$

These relations allow us to solve for the equilibrium distributions

$$q_i = \frac{q_1}{f_1} f_i \exp[-\beta(\varepsilon_i - \varepsilon_1)] \tag{7.23}$$

$$r_i = \frac{g_i}{\frac{r_1 + g_1}{r_1} \exp[\beta(\varepsilon_i - \varepsilon_1)] - 1} \tag{7.24}$$

$$s_i = \frac{h_i}{\frac{h_1 - s_1}{s_1} \exp[\beta(\varepsilon_i - \varepsilon_1) + 1} \tag{7.25}$$

where $$\beta = \frac{1}{\varepsilon_2 - \varepsilon_1} \ln \frac{f_2 q_1}{f_1 q_2} \tag{7.26}$$

These equations are identified for $q_1\, q_2$, r_1 and s_1, so that they are valid for all *i*. It is often the case that the number of bosons is not restricted. In this case, δr_1 also is an independent variable. Following the same steps as before gives instead of Eq. (7.24),

$$r_i = \frac{g_i}{\exp(\beta\varepsilon_i) - 1}, \sum_i r_i = \text{unrestricted} \tag{7.27}$$

In order to determine f_i, g_i and h_i, for the translational levels, it is assumed that the system is in a cubic box of length *I*, (the results are valid for other shapes as well *e.g.*, rectangular shape) for which the energy levels are [see Eq. (3.173)]

$$E = \frac{\hbar^2 \pi^2}{2ml^2}(n_x^2 + n_y^2 + b_z^2), n_x = 1, 2, \cdots \text{etc.} \tag{7.28}$$

Since every set of positive, nonzero integers (n_x, n_y, n_z) is associated with a state, the number of states in the absence of internal degrees of freedom is approximately equal to the volume in the first octant of the *n*-space. Therefore, the expression for f_i is

$$f_i = \frac{V}{4\hbar^2 \hbar^3}(2m)^{3/2} \in_i^{1/2} \Delta\varepsilon_i \tag{7.29}$$

and similar expressions for g_i and h_i. From this, the total number Q of the distinguishable particles and their total energy can be obtained as

$$Q = \Sigma q_i$$

$$= V\left(\frac{m}{2\pi\hbar^2 \beta}\right)^{3/2} \frac{q_1}{f_1} \exp(\beta\varepsilon_1) \tag{7.30}$$

$$E = \Sigma q_i \varepsilon_i$$

$$= V\left(\frac{m}{2\pi\hbar^2\beta}\right)^{3/2}\frac{q_1}{f_1}\left(\frac{3}{2\beta}\right)\exp(\beta\varepsilon_1) \quad (7.31)$$

It then follows that average energy is

$$\bar{\varepsilon} = \frac{3}{2\beta} \quad (7.32)$$

Since $\bar{\varepsilon}$ is equal to (3/2) kT for the classical particles where T is the absolute temperature it follows that

$$\beta = \frac{1}{kT} \quad (7.33)$$

In terms of T, the various distributions can be written as

$$q = f_i \exp(-\alpha_1 - \varepsilon_i/kT) \text{ Maxwell-Boltzmann,} \quad (7.34)$$

$$r_i = \frac{g_i}{\exp(\alpha_2 + \varepsilon_i/kT) - 1} \text{ Bose-Einstein,} \quad (7.35)$$

$$s_i = \frac{h_i}{\exp(a_3 + \varepsilon_i/kT) + 1} \text{ Fermi- Dirac} \quad (7.36)$$

These are known as the Maxwell-Boltzmann (for distinguishable particles), Bose-Einstein (for bosons), and Fermi-Dirac (for fermions) distributions, respectively. The constant α_1, α_2 and α_3 are determined from the conditions in Eq. (7.12) for the number of particles.

The following properties of the distributions may be noted:

1. For large ε_i/kT when $\exp(\alpha_{2,3} + \varepsilon_i/kT) >> 1$, all the three distributions have the some form, *i.e.* that of the Maxwell-Boltzmann distribution.
2. For distinguishable particles, the quantity q_i/f_i called the particle index, satisfies the relation

$$\frac{q_i/f_i}{q_i/f_j} = \exp[-(\varepsilon_i = \varepsilon_j)/kT] \quad (7.37)$$

   which implies that there is always a greater tendency for the particle to occupy a lower energy state than a higher energy state. This tendency is observed for bosons and fermions as well [see Eqs. (7.35) and (7.36)]. Equation (7.37) is valid for bosons and fermions also, if the 1 in the denominator of Eqs. (7.35) and (7.36) can be neglected.
3. It is noted that when the total number of bosons is unrestricted, $\alpha_2 = 0$ and the distribution of the bosons is given by Eq. (7.27). In this case, it is

observed that the bosons have a tendency to bunch together at low energies [see Fig. 7.1 (*a*)]. Also, the number of bosons increases as T increases, at all energies.

4. For the Fermi-Dirac distribution, s_i/h_i is the probability for a state to be occupied and is seen to be less than one for all ε_i as is required by the Pauli exclusion principle. In general, ε_3 is negative and it is convenient to write

$$s_i = \frac{h_i}{\exp[\varepsilon_i - \varepsilon_f)/kT] + 1} \tag{7.38}$$

For $T \to 0$, $s_i/h_i = 1$ for $\varepsilon_i < \varepsilon_f$ and $s_i/h_i = 0$ for $\varepsilon_i > \varepsilon_f$. This means that fermions occupy the lowest energy states available, subject to the exclusion principle. For finite but small T, $s_i/h_i \approx 1$ for $(\varepsilon_i - \varepsilon_f)/kT \ll -1$, and $s_i/h_i \approx 0$ for $(\varepsilon_i - \varepsilon_f)/kT \gg 1$. The quantity ε_f is called the Fermi energy (which depends on T), and it plays an important role in the behaviour of fermions. The distribution is illustrated in Fig. 7.1 (*b*).

5. In principle, every system of particles which interact weakly with each other, is described by either the Bose-Einstein or the Fermi-Dirac distribution. However, if the particles are localized (at the lattice points for example) and their wave functions do not overlap, they can be taken as being distinguishable (distinguished by the region of localization). In such cases, Maxwell-Boltzmann distribution can be applied to describe the system.

In what follows, some important physical properties of different systems are deduced using the statistical distributions given in Sec. 7.2

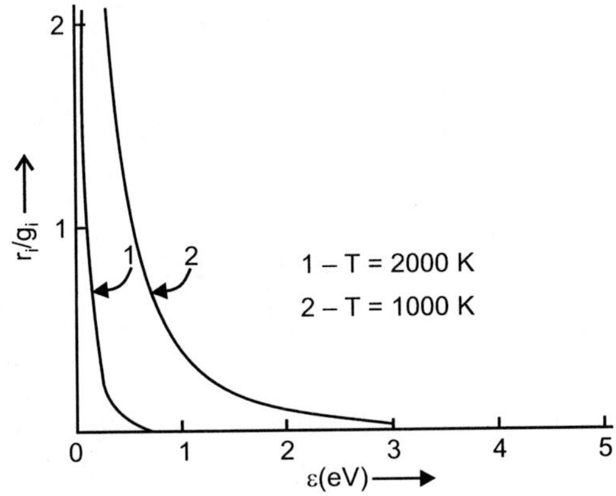

(a)

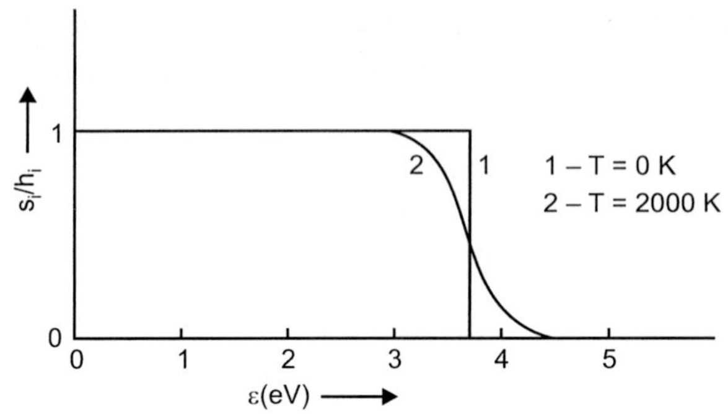

(b)

Fig. 7.1 (*a*) Particle index for bosons with $\alpha_2 = 0$, (*b*) particle index for fermions with $\varepsilon_f = 3.7$ eV.

7.3 APPLICATIONS OF MAXWELL-BOLTZMANN DISTRIBUTION

In this section, the specific heats of gases and solids are discussed in terms of the Maxwell-Boltzmann distribution for distinguishable particles.

Specific Heats of Gases

As mentioned in Sec. 5.7, the spacing of the electronic energy levels of molecules is of the order of 5 eV, whereas that of the vibrational levels is of the order of 1 eV or less, and that of the rotational levels is of the order 10^{-3} eV. Hence it follows from Eq. (7.37) that at ordinary temperatures ($kT = 0.026$ eV. For $T = 300$ K), while many rotational levels are excited almost all the particles will be in the lowest electronic state. Therefore, the energy of each particle may be written as:

$$E = \frac{1}{2m}p^2 + \left(n + \frac{1}{2}\right)\hbar\omega + \frac{\hbar^2}{2I}J\,(J+1) \tag{7.39}$$

and the average energy as

$$E = \frac{\sum_i \exp(-E_i/kT)\,f_i\,E_i}{\sum_i \exp(-E_i/kT)\,f_i} \tag{7.40}$$

Since the total energy is a sum of the energies of different modes of excitation, it can be shown that

$$\bar{E} = \bar{E}_{\text{tr}} + \bar{E}_{\text{vib}} + \bar{E}_{\text{rot}} \tag{7.41}$$

where the average of each term is only over the states corresponding to that mode of excitation.

It was noted in Eq. (7.32) that $\bar{E}_{tr}$ is $\frac{3}{2}kT$. The average vibrational energy is

$$\bar{E}_{\text{vib}} = \frac{\sum_{n=0}^{\infty} \exp[-(n+1/2)\,\hbar\omega/kT]\,(n+1/2)\,\hbar\omega}{\sum_{n=0}^{\infty} \exp[-(n+1/2)\,\hbar\omega/kT]} \tag{7.42}$$

This expression can be evaluated by using Eq. (2.11) and gives

$$\bar{E}_{\text{vib}} = \frac{1}{2}\hbar\omega + \frac{\hbar\omega}{\exp(\hbar\omega/kT) - 1} \tag{7.43}$$

where the first term is called the *zero-point energy.* The average rotational energy is

$$\bar{E}_{\text{rot}} = \frac{\sum_{j=0}^{\infty} \exp(-aJ\,(J+1)/kT)\,aJ\,(J+1)\,(2J+1)}{\sum_{j=0}^{\infty} \exp(-aJ\,(J+1)/kT)\,(2J+1)} \tag{7.44}$$

where $a = \hbar^2/2I$ and $(2J + 1)$ is the degeneracy of the rotational levels.

For applying the above results to symmetric diatomic molecules, some changes have to be made in the expression for $\bar{E}_{\text{rot}}$. For example, only even-J states are allowed by Pauli's exclusion principle for the *para*-hydrogen molecules. In this case, closed expression is obtained for $\bar{E}_{\text{rot}}$ by separating out the $J = 0$ and $J = 2$ terms and converting the remaining sum into an integral. This is done by first replacing J by $2l$ so that the summation is over $l = 0, 1, 2, \ldots$ and then substituting $x = l\,(2l + 1)$ for which $\Delta x = (4l + 1)\,\Delta l$. This is leads to

$$\bar{E}_{\text{rot}} = -\frac{1}{F}\,\partial F/\partial(1/kT) \tag{7.45}$$

$$F \approx 1 + 5\exp(-6a/kT) + \int_{6}^{\infty} \exp(-2ax/kT)\,dx, \tag{7.46}$$

where F is the denominator in Eq. (7.44). The lower limit corresponds to $l = 3/2$. Carrying out the integration, we obtain for *para*-hydrogen,

$$\overline{E} = \frac{3}{2} kT + \frac{1}{2} \hbar\omega + \frac{\hbar\omega}{\exp(\hbar\omega/kT) - 1} - \frac{1}{F} \frac{\partial F}{\partial (1/kT)} \tag{7.47}$$

$$F \approx 1 + 5 \exp(-6a/kT) + \frac{kT}{2a} \exp(-12\, a/kT)$$

The values of ω and a are obtained from the spectrum of the hydrogen molecule, and have the values

$$\hbar\omega \approx 0.5454 \text{ eV} \tag{7.48}$$
$$a \approx 0.007\ 55 \text{ eV}$$

The specific heat of *para*-hydrogen obtained from Eq. (7.47)

$$C_v = N_{\text{Avo}} \frac{\partial \overline{E}}{\partial T} \tag{7.49}$$

is plotted in Fig. (7.2) and is in very good agreement with the experimental observations. It may be observed that the contribution to C_v from the rotational energy becomes appreciable at $T \gtrsim 75$ K which corresponds to $kT \gtrsim a$, while the contribution from the vibrational to $kT \gtrsim \hbar\omega$. Ordinary hydrogen is a mixture of *ortho-* and *para*-hydrogen, there being about 25% *para*-hydrogen at room temperature, $I = \frac{1}{2}$ for the hydrogen atom). The specific heat of the mixture is a statistical average of the specific heats of the components. Its behaviour is similar to that given in Fig. (7.2) except that the hump around $T \approx 150$ K is now absent.

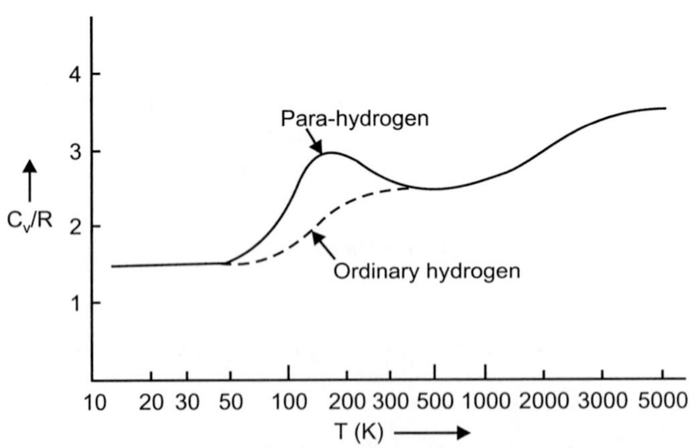

Fig. 7.2 The specific heat of *para*-hydrogen (solid line) and ordinary hydrogen (dashed line) at constant volume, as a function of absolute temperature.

Specific Heat of Solids

The specific heat of solids provides an important application of Maxwell-Boltzmann distribution.

It is assumed that the atoms of a solid are localized and perform simple harmonic motion about the equilibrium position. In the classical analysis, the number of states is taken to be proportional to $d^3 p\, d^3 r$, so that the average energy for the Maxwell-Boltzmann distribution is given by

$$\bar{E}_{cl} = \frac{\int \exp\left[-\left(\frac{1}{2m}p^2 + \frac{1}{2}br^2\right)\Big/kT\right]\left(\frac{1}{2m}p^2 + \frac{1}{2}br^2\right) d^3p\, d^3r}{\int \exp\left[-\left(\frac{1}{2m}p^2 + \frac{1}{2}br^2\right)\Big/kT\right] d^3p\, d^3r} \tag{7.50}$$

The expression can be evaluated by using the result

$$\int_0^\infty \exp(-ax^2)\, x^{2n}\, dx = (-1)^n \frac{d^n}{da^n}\int_0^\infty \exp(-ax^2)\, dx$$

$$= (-1)^n \frac{d^n}{da^n}\left(\frac{\pi^{1/2}}{2a^{1/2}}\right) \tag{7.51}$$

and comes out to be

$$\bar{E}_{cl} = 3kT \tag{7.52}$$

From this it follows that the specific heat is

$$C_{v,\,cl} = 3\,R \tag{7.53}$$

At about room temperatures and above, this result is in agreement with the experimental observations (law of *Dulong and Petit*). However, the experimental measurements (note that experimentally C_p is measured, though the difference $C_p - C_v$ is quite small for solids) show that, at low temperatures the specific heat rapidly decreases as T decreases and goes to zero as T approaches 0, K.

Einstein (1911) was the first person to appreciate that the low-temperature behaviour of the specific heat of solids is essentially governed by quantum properties of the system. He suggested that the energies of the oscillating atoms do not form a continuum. Instead, their allowed energies for oscillation in each direction, are

$$\varepsilon_n = nh\nu,\; n = 0, 1, 2, ... \tag{7.54}$$

where $\nu = 1/2\,\pi\,(b/m)^{1/2}$ is the natural frequency of the oscillator. Therefore, the average energy of the atom for oscillation in each direction is

$$\bar{\varepsilon} = \frac{\sum_{n=0}^{\infty} nh\nu\, e^{-nh\nu/kT}}{\sum_{n=0}^{\infty} e^{-nh\nu/kT}}$$

$$= \frac{h\nu}{e^{h\nu/kT} - 1} \tag{7.55}$$

which is the same expression encountered in Eq. (2.11) for Planck's oscillator. The specific heat of the system including oscillations in all the three direction is

$$C_v = 3\,N \frac{\partial}{\delta T} \overline{\varepsilon}$$

$$= 3\,R \left(\frac{h\nu}{kT}\right)^2 \frac{e^{h\nu/\lambda T}}{(e^{h\nu/kT} - 1)^2} \tag{7.56}$$

For large *T*, this expression reduced to the classical expression of 3*R* but at low temperatures it decreases rapidly and goes to zero ~ T^{-2} exp (– $h\nu/kT$) for $T \to 0$. Overall the expression describes the qualitative behaviour of specific heat quite well. However, experiments show that C_v goes to zero more gently, as T^3 near 0 K, and not as an exponential function. Still, the result clearly indicates that quantum oscillations govern the low temperature behaviour of the specific heat of solids.

An improved description of the specific heat of solids was given by Debye (1912) who observed that the motion of neighbouring atoms is correlated, and that the allowed frequencies of oscillation correspond to those of allowed standing elastic waves in the medium. The number of the allowed modes for the standing waves was calculated in Sec. 2.1 [see Eq. (2.8)], and is given by

$$dN_t(\nu) = \frac{8\,\pi\,V\nu^2\,d\nu}{v_t^3} \tag{7.57}$$

for the transverse modes (which correspond to the oscillations of atoms perpendicular to the direction of propagation of the waves—there are two independent directions of transverse oscillations), where v_t is the velocity of propagation for the transverse modes, and by

$$dN_l(\nu) = \frac{4\,\pi\,V\nu^2\,d\nu}{v_t^3} \tag{7.58}$$

for the longitudinal mode (which corresponds to the oscillation of atoms parallel to the direction of propagation of the waves), where v_l is the velocity of propagation for the longitudinal modes, *V* being the volume. However, since the medium of propagation consists of discrete atoms, Debye assumed that the total number of frequency modes is equal to the total number of degrees of freedom, *i.e.* $3N_{0,}$ N_0 being the number of atoms. This imposes an upper limit ν_m on the allowed frequencies,

$$3N_0 = 4\pi\,V\left(\frac{2}{v_t^3} + \frac{1}{v_l^3}\right) - \int_0^{\nu_m} \nu^2 d\nu$$

$$= \frac{4\pi V}{3}\left(\frac{2}{v_t^3} + \frac{1}{v_l^3}\right) v_m^3 \tag{7.59}$$

Since each mode is associated with an average energy given by Eq. (7.55), the total thermal energy is

$$E = 4\pi V\left(\frac{2}{v_t^3} + \frac{1}{v_l^3}\right) \int_0^{v_m} \frac{hv}{e^{hv/kT} - 1} v^2 dv \tag{7.60}$$

which in terms of Eq. (7.59) can be written as

$$E = \frac{9N_0}{v_m^3} \int_0^{v_m} \frac{hv}{e^{hv/kT} - 1} v^2 dv \tag{7.61}$$

Defining $x = hv/kT$ and $\theta = h^v m/k$ where θ is called the *Debye temperature*, the energy is

$$E = 9\, N_0\, kT \left(\frac{T}{\theta}\right)^3 \int_0^{\theta/T} \frac{x^3}{e^x - 1} dx \tag{7.62}$$

The molar specific heat C_v is given by $\partial E/\partial T$ for $N_0 = N_{\text{Avo.}}$

In the limit $T \to \infty$, $\theta/T \to 0$, the integral is $\frac{1}{3}(\theta/T)$, so that

$$C_v = 3R, \quad \text{for } \frac{\theta}{T} \ll 1 \tag{7.63}$$

which is the classical limit [Eq. (7.53)]. For $T \to 0$, $\frac{\theta}{T} \to \infty$, one can use

$$\int_0^{\infty} \frac{x^3}{e^x - 1} dx = \frac{\pi^4}{15} \tag{7.64}$$

to get

$$E = \frac{3}{5}\pi^4\, N_0\, kT\, (T/\theta)^3, \frac{\theta}{T} \gg 1 \tag{7.65}$$

and

$$C_v = \frac{12}{5}\pi^4\, R\, (T/\theta)^3, \frac{\theta}{T} \ll 1 \tag{7.66}$$

The model predicts that the specific heat at low temperatures is proportional to T^3, in agreement with the experimental observation.

The behaviour of C_v at other temperatures has to be evaluated numerically from the expression

$$C_v = 9R\left(\frac{T}{\theta}\right)^3 \int_0^{\theta/T} \frac{x^4\, e^x\, dx}{(e^x - 1)^2} \tag{7.67}$$

obtained from Eq. (7.61), and gives a universal curve as a function of θ/T (Fig. 7.3). The general agreement between theory band experiments is quite good, θ being about 100 K for lead, 160 K for sodium, 220 K for silver, 340 K for copper, 400 K for aluminium, 640 K for silicon, and about 1860 K for carbon (diamond). Some of the observed differences at intermediate temperatures can be explained by taking a more realistic spectrum for the allowed frequencies.

7.4. APPLICATIONS OF BOSE-EINSTEIN DISTRIBUTION

Bose-Einstein distribution describes the properties of bosons which may be massless, such as photons, or massive, *e.g.* ^{4}He. Some of their statistical properties are considered here.

Photon Gas

In Sec. 2.1, Planck's theory of blackbody radiation in terms of the allowed standing waves and the associated harmonic oscillators was discussed. A more modern and satisfactory description is in terms of the energy distribution of the photons regarded as massless bosons.

Since the number of photons is unrestricted, their distribution is given by Eq. (7.27),

$$r_i = \frac{g_i}{\exp(\beta\varepsilon_i) - 1} \tag{7.68}$$

where $\varepsilon_i = h\nu$. The number of energy levels is the same as the number of allowed standing waves given by Eq. (2.8), except that the standing waves in Eq. (2.5) are to be interpreted as the energy eigenstates of photons with energy eigenvalue $h\nu$. Therefore, g_i is

$$g_i = \frac{8\pi V}{c^3} \cdot \nu^2 d\nu \tag{7.69}$$

V being the volume. The energy density per unit volume is

$$U(\nu)\, d\nu = \left(\frac{8\pi h}{c^3}\right) \frac{\nu^3 d\nu}{e^{h\nu/kT} - 1} \tag{7.70}$$

which agrees with Planck's expression in Eq. (2.12)

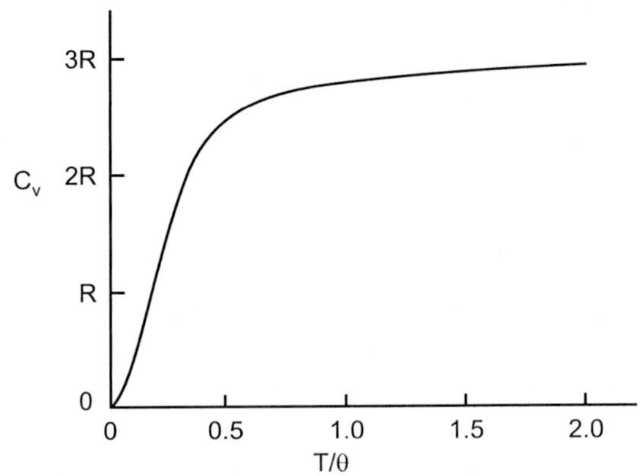

Fig. 7.3 The Debye specific heat as a function of T/θ, θ being the Debye temperature.

Photon Gas

As in the case of electromagnetic waves and the photons, the elastic waves in a solid have a quantum manifestation. The energy of these waves is in the form quanta called *phonons* each of which carries a quantum of energy $h\nu$ where ν is one of the allowed frequencies. These phonons are bosons, they interact with the atoms, they are absorbed and emitted, and their total energy of the thermal energy of the solid.

The number of phonons is unrestricted, so that their frequency distribution is given by Eq. (7.27),

$$r_i = \frac{g_i}{e^{\beta h\nu} - 1} \tag{7.71}$$

Phonons are transverse or longitudinal and the number of energy levels is given by Eqs. (7.57) and (7.58), with the upper limit ν_m for the frequency given by Eq. (7.59). Therefore, the total energy of the phonon gas is

$$E = 4\pi V\left(\frac{2}{v_t^3} + \frac{1}{v_l^3}\right)\int_0^{\nu_m} \frac{h\nu}{e^{h\nu/kT} - 1}\,\nu^2 d\nu \tag{7.72}$$

which is the same as the relation in Eq. (7.60).

Bose-Einstein Condensation

A gas with a given number of bosons whose mass is nonzero, shows remarkable quantum mechanical properties at low temperatures. In particular, it undergoes a phase transition, known as *Bose-Einstein condensation* which is of interest for two reasons. Firstly, it is an example which allows an exact mathematical treatment. Secondly, the observed changes in the properties of ^{4}He at $T = 2.17$ K can be explained in terms of Bose-Einstein condensation.

The distribution of a Bose-Einstein gas is given by Eq. (7.35) as

$$r_i = \frac{g_i}{\exp[(\varepsilon_i + \alpha_2)/kT] - 1} \tag{7.73}$$

where the constant α_2 is determined from the condition that the total number of particles is N. Here α_2 is different from that in Eq. (7.35)

$$N = \sum_i \left(\frac{g_i}{\exp[(\varepsilon_i + \alpha_2)/kT] - 1} \right) \tag{7.74}$$

The energies may be measured from the ground state energy which can be taken to be zero. This implies that since r_i is nonnegative, $\alpha_2 \geq 0$.

The number of energy levels is obtained from Eq. (3.173) by taking $l_x = l_y = l_z$. It is given by the volume element in the first octan to the n-space,

$$g_i = dn_x\, dn_y\, dn_z,\ n_x = 1, 2, ..., n_y = 1, 2..., n_z = 1, 2, ...,$$

$$= \frac{1}{2}\pi n^2 dn$$

$$= \frac{2\pi V(2m)^{3/2}}{h^3}\varepsilon^{1/2} d\varepsilon \tag{7.75}$$

where V is the volume. Therefore, using Eq. (7.74)

$$\frac{N}{V} = \frac{2\pi V(2m)^{3/2}}{h^3}\int_0^\infty \frac{\varepsilon^{1/2} d\varepsilon}{\exp[(\varepsilon + \alpha_2)/kT] - 1} \tag{7.76}$$

where the left-hand side is independent of temperature. It then follows that as T decreases, so does α_2, and the smallest value of T allowed by Eq. (7.76) is the one for which $\alpha_2 = 0$ (it should be noted that $\alpha_2 \geq 0$). This minimum value of T, called T_c, is given by

$$\frac{N}{V} = \frac{2\pi(2m)^{3/2}}{h^3}\int_0^\infty \frac{\varepsilon^{1/2} d\varepsilon}{\exp[(\varepsilon / kT_c) - 1} \tag{7.77}$$

Writing $x = \varepsilon/kT_c$, this relation reduces to

$$\frac{N}{V} = \frac{2}{\pi^{1/2}}\left(\frac{2\pi mk\, T_c}{h^2}\right)^{3/2}\int_0^\infty \frac{x^{1/2}\, dx}{e^x - 1}$$

$$= 2.612\left(\frac{2\pi mk\, T_c}{h^2}\right)^{3/2} \tag{7.78}$$

For $T < T_c$, Eq. (7.76) cannot be satisfied for $\alpha_2 \geq 0$. The reason for this difficulty is that the continuum expression in Eq, (7.75) for g_i is valid provided the population of no single level is significant. Now when T is sufficiently low, the particles will tend to occupy the ground state with $\varepsilon = 0$ which is not taken into account by the expression for g_i in Eq. (7.75) ($g_i = 0$ for $\varepsilon = 0$). This difficulty can be overcome by taking the ground state into account separately and using the continuum expression for g_i in Eq. (7.75) for the states with $\varepsilon > 0$. This leads to the more general expression:

$$N = \frac{1}{\exp(\alpha_2 / kT) - 1} + \frac{2\pi V(2m)^{3/2}}{h^3} \int_{\delta}^{\infty} \frac{\varepsilon^{1/2} d\varepsilon}{\exp[(\varepsilon + \alpha_2) / kT] - 1} \tag{7.79}$$

where δ is a small positive quantity.

For $T > T_c$ the-first term is small, *e.g.* at high temperatures one has Maxwell-Boltzmann distribution for which [using Eq. (7.76) without the unit term in the denominator]

$$\exp(\alpha_2/kT) = \left(\frac{2\pi mkT}{h^2}\right)^{3/2} \frac{V}{N} \gg 1$$

For $T < T_c$, α_2 is small but nonzero, and the nonsingular integral in Eq. (7.79) can be evaluated at $\alpha_2 = 0$. Using the variable $x = \varepsilon/kT$, Eqs. (7.79) and (7.78) given

$$N = N_0 + N(T/T_c)^{3/2},\ T \le T_c \tag{7.80}$$

where N_0 is the number of particles in the ground state. The fraction of particles in the ground state is

$$\frac{N_0}{N} = 1 - (T/T_c)^{3/2},\ T < T_c \tag{7.81}$$

and is shown in Fig. 7.4 (*a*). For T < T_c, a significant fraction of particles is in the ground state, and this occupation of the zero energy and zero momentum ground state is called Bose-Einstein condensation. The temperature T_c below which the condensation takes place is called he *condensation temperature.*

The particles in the ground state have zero energy and momentum, and hence do not contribute to the viscosity of the fluid. (Viscosity arises from the interaction between particles—viscous flow is accompanied by the excitation of vortices whose quantum is called a *roton.* The roton has a finite energy and hence cannot easily be excited at low temperatures.) These particles, being in the ground state, do not contribute to the total energy which therefore is obtained from the second term in Eq. (8.79) with $\alpha_2 = 0$, as

$$E = \frac{2\pi V(2m)^{3/2}}{h^3} \int_{\delta}^{\infty} \frac{\varepsilon^{3/2} d\varepsilon}{e^{\varepsilon/kT} - 1} \tag{7.82}$$

Using $x = \varepsilon/kT$, this expression comes out to be

$$E = 0.77\, Nk \frac{T^{5/2}}{T_c^{3/2}}\ T < T_c \tag{7.83}$$

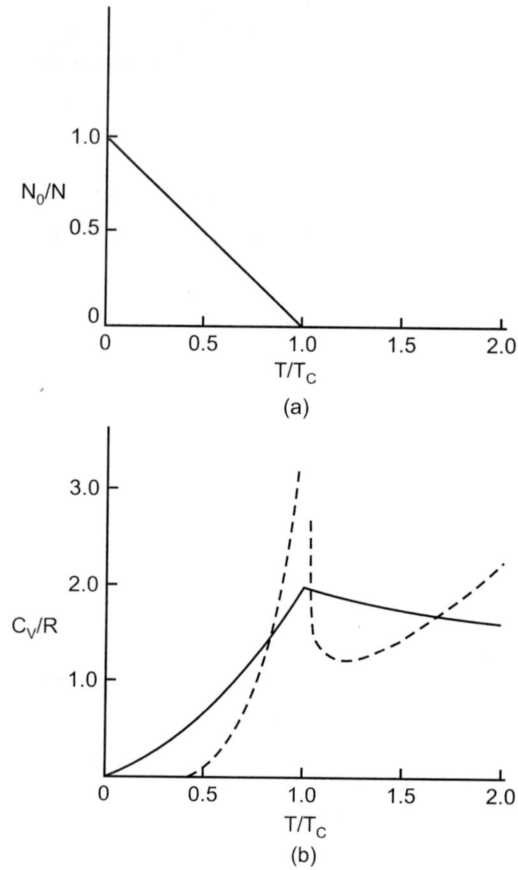

Fig. 7.4 (*a*) Fraction of particles in the ground state, (*b*) specific heat as a function of temperature. The solid line is for Bose-Einstein condensation and the dashed line is the experimental curve with $T_c = 2.17$ K.

Therefore, the specific heat is given by

$$C_v = 1.93\ R\left(\frac{T}{T_c}\right)^{3/2}, T < T_c \tag{7.84}$$

which at $T = T_c$ is greater than the classical value of 3/2 R. Detailed calculations show that C_v is continuous at $T = T_c$ but has a kink there, *i.e.* its derivative is discontinuous at $T = T_c$ [see Fig. 7.4 (*b*)].

Some examples of transitions which are possible manifestations Bose-Einstin condensation are discussed here.

Liquid ^{4}*He*: The phase transition of liquid ^{4}He at $T = 2.7$ K provided an interesting illustration of Bose-Einstein condensation. The properties of liquid helium (^{4}He liquefies at 4.2 K under normal conditions) show dramatic changes at 2.17 K. Above 2.17 K, it behaves like a normal liquid and is known as *helium I.* Below this temperature, it acquires some unusual properties, *e.g.* it flows through capillaries without any apparent viscosity. This form is known as *helium II* and many of its properties can be described by regarding it as a mixture of two fluids, one a normal fluid and the other a superfluid which has no viscosity. This mixture is similar to a Bose-Einstein gas with some condensation, the superfluid corresponding to the particles in the ground state. This would explain the zero viscosity. The identification of the two phenomena is further strengthened by the observation that the specific heat of ^{4}He also shows a singular behaviour at 2.17 K. The observed specific heat has the shape of λ [see Fig. 7.4 (*b*)] and hence the transition is called a *λ-transition* while the transition temperature is called the λ-point. It should be noted however that careful experiments indicate that the specific heat has a logarithmic infinity at the λ-point T_λ. This however may be due to the fact that the particles considered in Bose-Einstein condensation were noninteracting which is certainly no the case for the atoms of liquid helium. Finally, using V = 27.6 cm^3/mole for liquid helium in Eq. (7.78), one obtains T_c = 3.13 K compared with T_λ = 2.17 K. These observations strongly suggest that the λ-transition is a form of Bose-Einstein condensation.

Liquid ^{3}He: Helium has an isotope ^{3}He which is a fermion (it has 2 protons, 1 neutron and 2 electrons) and which liquifies at 3.2 K. It is found that ^{3}He, though a fermion, undergoes a transition to the superfluid state at 2.6×10^{-3} K. This arises from the fact that two ^{3}He atoms interact with each other and produce a weakly-bound system at low temperatures. This bound system is a boson which can undergo a transition to the superfluid state.

Hydrogen: The atoms of about half of the elements are bosons, *i.e.* they obey Bose statistics. Even then, Bose-Einstein condensation is not a common phenomenon. The reason for this is that the condensation takes its simplest form only for an ideal gas in which the atoms do not interact with each other. In real atoms, the electromagnetic interaction tends to bind them and most substances go into the solid state long before the critical temperature for Bose-Einstein condensation is reached. Therefore, condensation is expected in only those systems where the interaction between the atoms is weak compared to the zero-point energy of the atoms, *e.g.* in helium. An interesting possibility that is being currently considered is the Bose-Einstein condensation of atomic hydrogen [see Silvera and Walraven, *Sc. Am.* **246**, 1, 56 (1982)]. It is true that under ordinary conditions, the interaction between the hydrogen atoms is quite strong and binds them into molecules in which the spins of the two electrons are antiparallel. However, if the atoms with parallel electron spins are isolated, for example, by using strong inhomogeneous magnetic fields. Pauli's exclusion principle prevents the electrons from having overlapping wave functions. As a consequence, the force between these atoms is mostly repulsive and the atoms do not bind. The experiments show that atomic hydrogen with parallel electron spins, remains a gas at temperatures as low as 0.08 K. One may therefore be able to observe Bose-Einstein condensation in atomic hydrogen. The critical temperature at which the condensation takes place depends on the density and is given by Eq. (7.78). For example, at $\rho = 10^{24}$ m^{-3}, the predicted critical temperature is 0.016 K. While the densities of atomic hydrogen achieved in the laboratory, are as yet not sufficiently high to observe the condensation, they are only one or two orders of magnitude lower than those at which Bose-Einstein condensation is predicted to occur. An observation of Bose-Einstein condensation in atomic hydrogen would be an exciting, unambiguous demonstration of quantum properties of a collection of bosons.

Superfluidity in neutron stars: There are interesting speculations that superfluidity may be occurring in neutron stars. Neutron stars are thought to be the end products of stars whose masses are between 4 M_s and 20 M_s, M_s being the mass of the sun. They are very dense with an average density of about $10^{14} - 10^{15}$ g/cm^3, have a radius of about 10 km, and are primarily made up of neutrons. Under suitable conditions two neutrons, which are fermions, may form weakly-bound states. The bound system which is a boson, may undergo Bose condensation to become a superfluid. It may be expected that since the densities of the neutron stars are so large, the transition temperature there would also be very high.

Bose-Einstein Condensation

In the gas phase, the Bose-Einstein condensate (BEC) remained an unverified theoretical prediction for many years. In 1995 the research groups of Eric Cornell and Carl Weiman of JILA, at the University of Colorado at Boulder, produced the first such condensate experimentally.

Condensation happens when several gas molecules come together and form a liquid. It all happens because of loss of energy. Gases are really excited atoms. When they lose energy, they slow down and begin to collect. They can collect into one drop. Water condenses on the lid of a pot when water is boiled. It cools on the metal and becomes a liquid again. One would then have a condensate.

If a sufficiently dense gas of cold atoms can be produced without condensation into liquid state, the matter wavelengths of the particles will be of the same order of magnitude as the distance between them. It is at that point that the different waves of matter can 'sense' one another and co-ordinate their state, and this is Bose-Einstein condensation. It is sometimes said that a "superatom" arises since the whole complex is described by one single wave function exactly as in a single atom. We can also speak of *coherent matter* in the same way as of *coherent light* in the case of a laser.

This was achieved with alkali atoms. For rubidium with mass number 87, ^{87}Rb, and sodium with its single stable isotope ^{23}Na, which both have integer atomic spin, weak repulsive forces arise between the atoms in each case. BEC occurs if the density, expressed as the number of atoms in a λ-sided cube exceeds 2.6. The atoms for realistic densities must move very slowly, at speeds of the order of a few millimetres per second. This corresponds to temperatures of the order of 100 nK (nanokelvin), *i.e.* a tenth of a millionth of a degree above absolute zero.

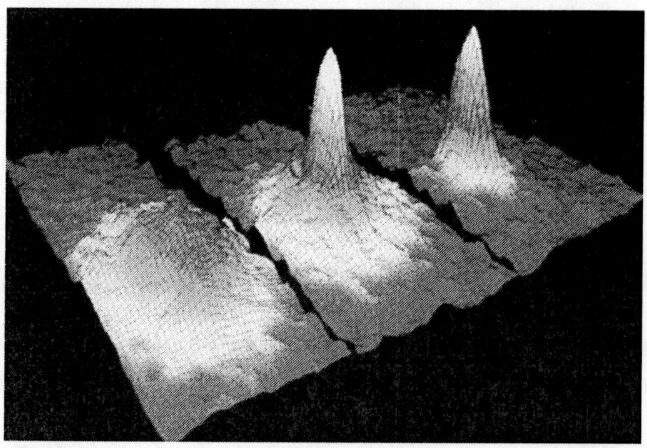

Fig. 7.5 Successive occurrence of Bose-Einstein condensation in rubidium. From left to right is shown the atomic distribution in the cloud just prior to condensation, at the start of condensation and after full condensation. High peaks correspond to a large number of atoms. Silhouettes of the expanding atom cloud were recorded 6 ms after switching off the confining forces of the atom trap.

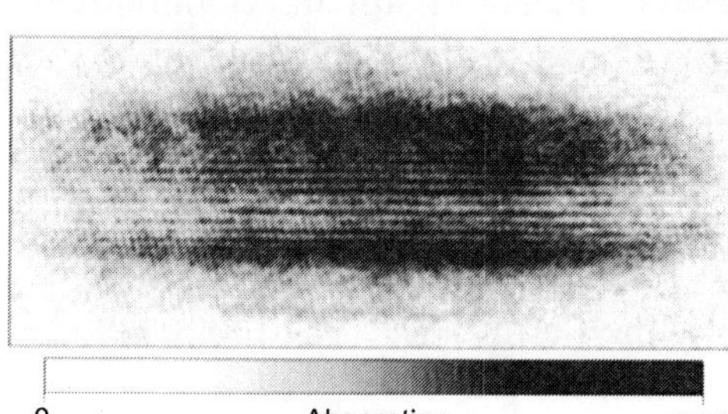

Fig. 7.6 Pattern of interference between two overlapping Bose-Einstein condensates of sodium atoms. The image was made in absorption. Matter-wave interferences have a periodicity of 15 micrometer. The recording shows that the atoms of the two condensates were fully co-ordinated.

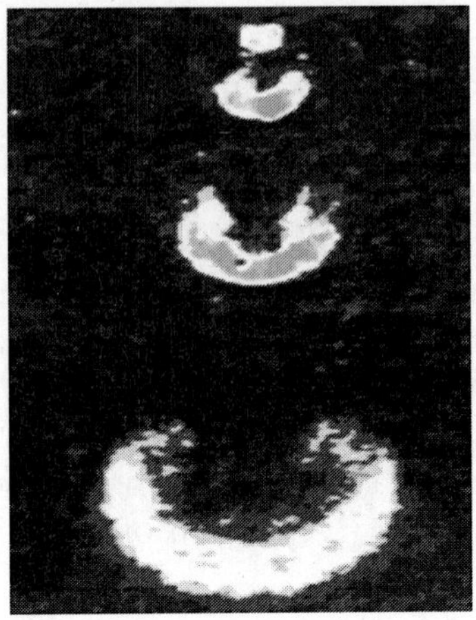

Fig. 7.7 Repeated release from the trap of parts of a Bose-Einstein condensate of sodium atoms. Pulses of coherent matter fall in the gravitational field—the phenomenon can be seen as an atom laser effect. The real size of the picture is 2.5 mm × 5 mm.

(*Source: http://www.nobelprize.org/nobel_prizes/physics/laureates/2001/public.html*)

7.5 APPLICATIONS OF FERMI-DIRAC DISTRIBUTION

The Fermi-Dirac distribution is dominated by the property that the occupation index s_i/h_i [Eq. (7.38)] is less than or equal to 1 at all temperatures, since no two fermions can occupy the same state. This provides a very useful framework for the description of several properties of metals in terms of what are known as *conduction electrons.*

Free-Electron Theory of Metals

According to this theory of metals (Pauli and Sommerfeld, 1927), the weakly-bound valence electrons become detached from the atom and move around freely. As a first approximation, the detailed interaction of the electrons with the lattice points (*i.e.* the ions) and with each other may be neglected, band the free electrons regarded as being in an average, constant potential in the macroscopic volume of the metal. Since the electrons are fermions, their distribution is given by the Fermi-Dirac distribution as

$$s_i = \frac{h_i}{e^{(\varepsilon - \varepsilon f)/kT} + 1} \tag{7.85}$$

where $\varepsilon_f(T)$ is the Fermi energy. The number of states h_i is the same as in Eq. (7.75), except for a factor of 2 to take into account the two spin states of the electron, giving

$$h_i = \frac{4\pi V\,(2m)^{3/2}}{h^3}\,\varepsilon^{1/2}\,d\varepsilon \tag{7.86}$$

The density of electrons as a function of energy, is then given by

$$\frac{dN\,(\varepsilon)}{d\varepsilon} = \left[\frac{4\pi V\,(2m)^{3/2}}{h^3}\right]\frac{\varepsilon^{1/2}}{e^{(\varepsilon-\varepsilon_f)/kT}+1} \tag{7.87}$$

and is shown in Fig. (7.8). The Fermi energy is determined from the condition that the total number of particles is *N, i.e.*

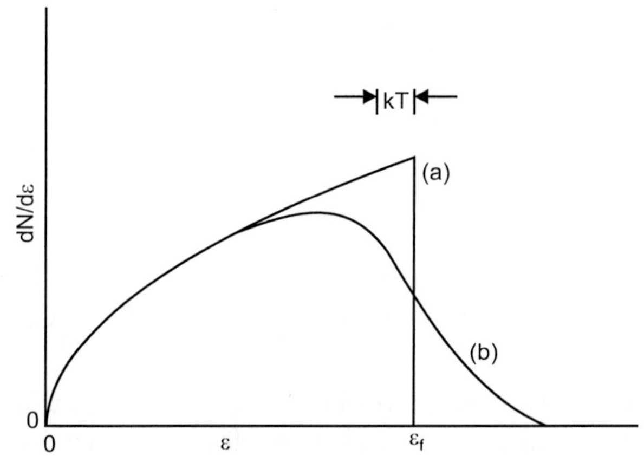

Fig. 7.8 Density of levels as function of ε, (*a*) for $T = 0$, (*b*) $kT = 0.1\ \varepsilon_f(0)$.

$$N = \frac{4\pi V\,(2m)^{3/2}}{h^3}\int_0^\infty \frac{\varepsilon^{1/2}d\varepsilon}{\exp[(\varepsilon-\varepsilon_f)/kT]+1} \tag{7.88}$$

At $T = 0$,

$$N = \frac{4\pi V\,(2m)^{3/2}}{3h^3}\int_0^{\varepsilon_f} \varepsilon^{1/2}d\varepsilon \tag{7.89}$$

$$= \frac{8\pi V\,(2m)^{3/2}}{3h^3}\,\varepsilon_f^{\;3/2}$$

which gives

$$\varepsilon_f(0) = \frac{h^2}{2m}\left(\frac{3N}{8\pi V}\right)^{2/3} \tag{7.90}$$

A slightly involved calculation gives for $kT \ll \varepsilon_f(0)$

$$\varepsilon_f(T) \approx \varepsilon_f(0)\left[1-\frac{\pi^2}{12}\left(\frac{kT}{\varepsilon_f(0)}\right)^2\right] \tag{7.91}$$

For metals, $N/V \approx 5 \times 10^{22}$ cm^{-3} for which Eq. (7.89) implies $\varepsilon_f = 4.5$ eV. The actual value of $\varepsilon_f(0)$ for some of the metals is 4.7 eV for Li, 2.1 eV for K, 7.0 eV for Cu, and 5.5 eV for Au. This means that the approximation in Eq. (7.91) is adequate for most purposes ($kT \approx 0.026$ eV at $T = 300$ K). For $kT \ll \varepsilon_f$ most of the electrons are in the lowest energy states allowed by Pauli's exclusion principle, and the electron gas is said to be degenerate (completely degenerate at $T = 0$). It is interesting to note that because of the exclusion principle, the average energy of the electron gas is quite substantial even at $T = 0$:

$$\bar{\varepsilon}(0) = \frac{\int_0^{\varepsilon_f} \varepsilon\varepsilon^{1/2} d\varepsilon}{\int_0^{\varepsilon_f} \varepsilon^{1/2} de}$$

$$= \frac{3}{5}\varepsilon_f \tag{7.92}$$

which is of the order of a few eV (compare with $kT \approx 0.0226$ eV at room temperature).

Specific heat of meals: An interesting property of the specific heat of metals is that it is described quite well by the **Debye theory**. **Since the Debye theory** includes only the phonon contributions, *i.e.* lattice vibrations, this implies that the contribution from the free electrons to the specific heat of metals is small. This is explained by the fact that, unlike the phonons, the free electrons satisfy Fermi-Dirac statistics. When the temperature T is increased, only a few electrons in the range $|\varepsilon - \varepsilon_f| \approx kT$ are excited to the higher energy states (see Fig. 7.8). It is only these electrons that contribute to the specific heat, as a result of which the contribution of the electron gas to the specific heat is quite small. Roughly speaking, it is seen from Fig. (7.8) that the number of electrons which are excited is $kT \left.\frac{dN}{d\varepsilon}\right|_{\varepsilon=\varepsilon_f}$ and their energy increases by an amount of about $2kT$. Therefore, the total energy of the systems is given by

$$E(T) \approx E(0) + 2k^2 T^2 \left.\frac{dN}{de}\right|_{\varepsilon=\varepsilon_f} \tag{7.93}$$

and the heat capacity per mole of the electron gas is

$$C_v^{el} \approx 4k^2 T \left.\frac{dN}{d\varepsilon}\right|_{\varepsilon=\varepsilon_f} \tag{7.94}$$

$$\approx 3R\left(\frac{kT}{\varepsilon_f}\right)$$

Here, we have used Eq. (7.87) for $dN/d\varepsilon$ and Eq. (7.89). A more detailed calculation gives

$$C_v^{el} = \frac{\pi^2}{2} R\left(\frac{kT}{\varepsilon_f}\right) \tag{7.95}$$

Since kT/ε_f is quite small at ordinary temperatures, the electronic specific heat also is small and the total specific heat is described quite well by the Debye theory. It should, however, be noted that the Debye specific heat at low temperatures is proportional to $R(T/\theta)^3$ [see Eq. (7.66)] so that at sufficiently low temperatures the electronic specific heat becomes dominant. At low temperatures, the total specific heat is given by

$$C_v = \frac{12}{5}\pi^4 R\,(T/\theta)^3 + \frac{\pi^2}{2} R\left(\frac{kT}{\varepsilon_f}\right) \tag{7.96}$$

and the observed nonzero limit of C_v/T as $T \to 0$, for metals such as copper, indicates the presence of the linear electronic contribution. Experimentally, in the case of copper, C_v/T for $T \to 0$ is about 0.7×10^{-3} J/mol/K^2 whereas the value predicted for copper ($\varepsilon_f \approx 7$ eV), by Eq. (7.96), is about 0.54×10^{-3} J/mol/K^2. The difference is a measure of the deviation of the model from the real situation.

Electrical and thermal conductivities: Some general characteristics of the electrical and thermal conductivities of metals can be discussed in terms of the free-electron theory of metals. This discussion will be based on the assumptions that (*i*) the conducting electrons move with the velocity $v_f = (2\varepsilon_f/m)^{1/2}$, which is reasonable since most of the conducting electrons will be in states close to the Fermi level, (*ii*) the electrons have a mean free path of λ and that they carry information over a distance of λ ($\lambda \approx 500$ Å) .

In the presence of an external electric field **E**, the electrons acquire an average drift velocity **v** which is equal to half of the average acceleration $\varepsilon\mathbf{E}/m$ multiplied by the interval λ/v_f between two collisions. Therefore, the current is $\frac{1}{2} en\,(e\lambda\mathbf{E}/mv_f)$ where n is the electron density. This satisfies Ohm's law since v_f, being large, is essentially independent of **E**. The electrical conductivity is then

$$\sigma = e^2 n\lambda/2mv_f \tag{7.97}$$

For calculating thermal conductivity, it is noted that since the electrons carry information over a distance of λ, the energy carried across an area by the electrons is $\varepsilon \pm (1/2)\lambda\left(\frac{\partial \varepsilon}{\partial x}\right)$ in opposite directions. Therefore, if $\frac{1}{3}n$ electrons are assumed to have a velocity perpendicular to the area, the net energy transferred across a unit area, per unit time, is

$$\frac{dQ}{dt} = -\frac{1}{6} n v_f \left(\lambda \frac{\partial \varepsilon}{\partial x}\right) \tag{7.98}$$

(where the negative sign indicates that the energy is transferred in a direction opposite to the gradient). From this relation, the thermal conductivity is obtained by writing $\partial\varepsilon/\partial x$ as $(\partial\varepsilon/\partial T)$ $(\partial T/\partial x)$ which leads to the coefficient of thermal conductivity K,

$$K = \frac{1}{6} n v_f \lambda \frac{\partial \varepsilon}{\partial T} \tag{7.99}$$

Since $\partial\varepsilon/2T$ is the specific heat per electron, using Eq. (7.95) gives

$$K = \frac{\pi^2 n k^2 \lambda T}{6m\, v_f} \tag{7.100}$$

It follows from Eqs. (7.97) and (7.100) that

$$\frac{K}{\sigma T} = L$$

$$= \frac{\pi^2}{3}\left(\frac{k}{e}\right)^2 \tag{7.101}$$

which is the same for all metals. This relation is known as **Wiedemann-Franz** law. The constant L, know as *Lorenz number*, has a value of 2.45×10^{-8} JΩ/s K, while the experimental values of $K/\sigma T$ for some of the metals at 0°C are 2.31×10^{-8} for Ag, 2.47×10^{-8} for Pb and 2.19×10^{-8} for Na.

While it is obvious that the free electrons are responsible for transporting charge, it is suggested by the validity of the Wiedemann-Franz law that the free electrons play a dominant role in the transfer of energy as well, in preference to the phonons. It is also noted that the thermal conductivity of metals is in general greater than that of insulators, sometimes by as much as two orders of magnitude. It is therefore reasonable to say that most of thermal conductivity in meals is due to the free electron gas.

Thermionic emission: When a metal is heated, electrons are emitted from the surface. Thermionic emisson can be studied by subjecting the electrons to a small potential difference and analysing the thermionic emission current as a function of temperature.

The electrons in a metal may be regarded as particles in a potential well with barrier at the boundary. The barrier arises from the fact that when an electron tries to escape from the surface, its image in the surface, being of opposite charge, pulls it back. In order to escape, the electrons must then have a minimum energy ϕ above the Fermi level. This energy ϕ is known as the *work function,* and usually has a value of the order of a few eV, *e.g.* 2.3 eV for Na, about 4.5 eV for Cu, etc.

An electron that is emitted must satisfy the condition (the metal surface is taken to be perpendicular to the z-direction)

$$\frac{p_z^2}{2m} \geq \varepsilon_f + \phi \tag{7.102}$$

Since the current at a point is $\mathbf{v}\,\rho$, ρ being the charge density, the amount of charge emitted by a unit area, per unit time is

$$j = e\int v_z\, dN \tag{7.103}$$

Here dN is the number of electrons per unit volume, with momentum between $\mathbf{p}$ and $\mathbf{p} + d\mathbf{p}$. It is equal to s_i/V, where s_i is given in Eq. (7.85) and h_l in Eq. (7.86). Using the relation $p^2 = 2m\varepsilon$, $2\pi\,(2m)^{3/2}\,\varepsilon^{1/2}\,d\varepsilon$ is replaced in h_i by $4\pi\, p^2\, dp$ or $dp_x\, dp_y\, dp_z$. It is then integrated over only positive p_z to give

$$j = \frac{8e}{h^3}\int_0^\infty dp_x \int_0^\infty dp_y \int_{[2m(\varepsilon_1+\phi)]^{1/2}}^\infty dp_z \left(\frac{p_z}{m}\right)\frac{1}{\exp[(\varepsilon-\varepsilon_f)/kT]+1} \tag{7.104}$$

Since ϕ is generally of the order of a few eV, the 1 in the denominator can be ignored to get

$$j = \frac{8e}{h^3}\int_0^\infty dp_x e^{-px^2/2mkT}\int_0^\infty dp_y e^{-py^2/2mkT}\int_{[2m(\varepsilon_f+\phi)]^{1/2}}^\infty dp_z \left(\frac{p_z}{m}\right)\exp\left[-\left(\frac{p_z^2}{2m}-\varepsilon_f\right)\Big/ kT\right]$$

$$= \frac{4\pi}{h^3}\, me\, k^2 T^2\, e^{-\phi/kT} \tag{7.105}$$

This is known as the Richardson-Dushman equation and is generally written as

$$j = AT^2 \exp\,(-\phi/kT) \tag{7.106}$$

where A has the value 1.2×10^6 A/m^2/K^2. It is in good agreement with the experiments provided (*i*) the constant A is modified to take into account the possibility that the electron may be reflected when it comes across a change in the potential near the surface, (*ii*) ϕ varies with temperature, with the crystal direction and with surface impurities. The experimental values of A are usually though not always, smaller than the one predicted by Eq. (7.105), *e.g.* 0.4×10^6 for Cr, 0.30×10^6 for Ni, etc. in MKS units

7.6 SUPERCONDUCTIVITY

Superconductivity is an interesting phenomenon in which electrons, which are fermions, behave like bosons. The reason for this is that under some special conditions, pairs of electrons form weakly-bound states which exhibit properties of Bose systems.

When the temperature of some metals, semiconductors and alloys is lowered to a few degrees kelvin, the electrical resistance of the material suddenly drops to zero [see Fig. 7.9 (*a*)]. The substance is then said to have become a *superconductor* and the temperature T_c at which the transition takes place is known as the *critical* or *transition temperature, e.g.* $T_c = 0.015$ K for tungsten, 3.72 K for tin, 9.3 K for niobium, and the highest known value 23.2 K for the Nb_3 Ge alloy. The transition to a superconducting state is quite sharp for a pure and physically perfect specimen. In some cases it has been observed to occur within a temperature range of 10^{-5} K. However, for impure or physically imperfect specimens, the transition may be over a range as large as 0.1 K or more.

Superconductivity has not been observed in all substances. In particular, it has not been detected in alkali metals, ferromagnetic substances, and relatively good conductors of electricity such as Ag, Cu, Au. Matthias has pointed out that superconductivity occurs only in substances which have an average of two to eight valence electrons per atom. Also, a small atomic volume is favourable for superconductivity. It is worth noting that an alloy may be a superconductor even if it is composed of two metals which themselves are not superconductors, *e.g.* Bi-Pd.

Some of the important properties of superconductors are the following:

1. The current in the superconductors persists for a very long time. This is demonstrated by placing a loop of the superconductor in a magnetic field, lowering its temperature below T_c and then removing the field.

   The current which is set up is found to persist over a period longer than two years without any attenuation.
2. The magnetic field does not penetrate into the body of the superconductor (permeability $\mu = 0$). This property, known as the *Meissner effect,* is the fundamental characterization of superconductivity. However, when the magnetic field B is greater than a critical value $B_c(T)$ [see Fig. 7.9 (*b*)], the superconductor becomes a normal conductor [$B_c(T)$ is zero at $T = T_c$ and has the largest value at $T = 0$].

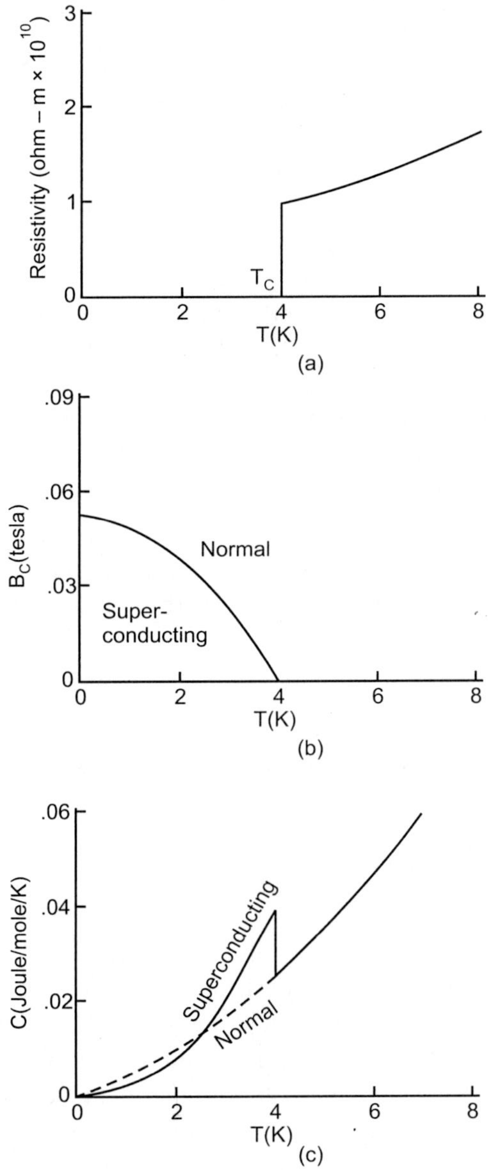

Fig. 7.9 (*a*) Resistivity of a superconductor as a function of *T*, (*b*) the critical magnetic field at which superconductivity disappears, (*c*) specific heat in superconducting and normal states.

3. When the current through the superconductor is increased beyond a critical value $I_c(T)$, the superconductor again becomes a normal conductor [$I_c(T) = 0$ at $T = T_c$].
4. The specific heat of the material shows an abrupt change at $T = T_c$, jumping to a larger value for $T < T_c$ [see Fig. 7.9 (*c*)].

The theory of superconductivity was given by Barden, Cooper and Schrieffer, and is known as the *BCS* theory. In this theory the electrons experience a special kind of mutual attraction which at large distances dominates over the Coulomb repulsion between them. It is the lattice of the material which provides the necessary medium for producing the attractive forces. An electron moving in the metal disturbs the lattice, producing phonons (quanta of vibrational motion). These phonons may be absorbed by another electron which may, at the microscopic level, be far away (about 10^{-6} m) from the first electron. In effect, the two electrons interact and the electron-electron interaction energy due to a phonon exchange is negative. If at low temperatures this attraction exceeds the Coulombic repulsion, the two electrons form a weakly bound state called a *Cooper pair* (a Cooper pair has an energy of about 10^{-3} eV in superconductors). If such pairs are created, the conductor becomes a superconductor.

Cooper pairs are spin-zero bosons and hence can be in the same state. They are in the ground state and are described by a wave function which extends over the entire body of the metal. In a sense, this is quantum mechanics on a macroscopic scale.

The Energy Gap

The specific heat of a superconductor at *very low* temperatures, is observed to be of the form

$$C_v = A\left(\frac{T}{\theta}\right)^3 + ae^{-b/kT} \tag{7.107}$$

where the first term is the lattice specific heat [Eq. (7.67)]. The second term is the electronic contribution to specific heat. The exponential form of this term suggests the presence of an energy gap [see Eq. (7.82) which indicates that if there is an energy gap Δ, the leading behaviour of the total energy of bosons at low temperatures is given by $E \sim e^{-\Delta/kT}$]. The energy gap comes from the binding energy of Cooper pairs. Since energy is required to break the bond in Cooper pairs, a sharp jump in the specific heat is observed at the transition temperature. Furthermore, when $T \sim T_c$, there is a substantial number of electrons in the normal state so that the energy gap is

$$\Delta \sim kT_c \tag{7.108}$$

The detailed BCS theory gives the binding energy of the Cooper pair as

$$\begin{aligned} E_b(T) &\approx 3.5\, kT_c \quad \text{for } T \to 0 \\ &\to 0 \quad \text{for } T \to T_c \end{aligned} \tag{7.109}$$

and this is the energy gap between the ground state and the dissociated state. For $T_c = 4$K, and energy gap of the order of 3×10^{-4} eV is obtained. The smallness of the energy gap is the reason why superconductivity is a low-temperature phenomenon.

The attenuation of a current implies a change in the state of conducting electrons. Since the ground state is separated by the energy gap Δ, the excitation of the Cooper pairs is not possible at low velocity [corresponding to currents less than $I_c(T)$]. This implies motion without friction and therefore, superconductivity of the metal (or alloy).

Magnetic Properties

The magnetic properties of superconductors are quite complicated. In the class of superconductors known as type *I* superconductors (which includes most of the elemental superconductors), the magnetic field is excluded from the body of the superconductors for $B < B_c(T)$ showing perfect **Meissner** effect. However, the Meissner effect disappears for $B > B_c(T)$.

For type II superconductors, an example of which is lead-indium alloy, perfect Meissner effect occurs for $B < B_1(T)$, but only a partial exclusion of the field for $B_1(T) < B < B_2(T)$, and a complete penetration of the field for $B > B_2(T)$. The reason for this behaviour is that for $B(T)$ between $B_1(T)$ and $B_2(T)$, the material is in a mixed state. A close investigation of the specimen shows the presence of small circular regions in the normal state, called *vortices* or *fluxoids*. They are surrounded by large regions which are in the superconducting state. It is the presence of both the states which gives rise to partial penetration of the field. Materials with high critical temperatures tend to fall in the class of the type II superconductors.

Since usually $B_2(T) >> B_c(T)$, carefully-prepared type II superconductors are used for the manufacture of high-field magnets which require almost no power input and little cooling. Technology based on superconductors would receive a major boost if superconductivity could be produced at higher temperatures, say at liquid nitrogen temperature ($T_b = 77.4$ K). This possibility has been considered recently.

There are two additional properties which are of interest, quantization of flux enclosed by a superconductor and Josephson junctions, which are discussed briefly.

Quantization of Flux

Consider a superconducting loop in which a current is circulating. The current generates a magnetic field whose flux across the area enclosed by the loop is quantized.

The wave function of the superconducting particles satisfies the equation [see Eq. (6.1)].

$$\left[\frac{1}{2m}(-i\hbar\nabla - q\mathbf{A})^2 + V\right]\psi = E\psi \tag{7.110}$$

with q being the charge of the particles, and is given by

$$\psi(\mathbf{r}) = \exp\left[i\frac{q}{\hbar}\int_{\mathbf{r}_o}^{r} A.\,d\,\mathbf{l}\right]\phi(\mathbf{r}) \tag{7.111}$$

where the integral involved is a line integral and $\phi(\mathbf{r})$ satisfies Eq. (7.110) in the absence of the field, *i.e.* for $\mathbf{A} = 0$. Now, the wave function must have the same phase even after going around the entire loop, *i.e.*

$$\frac{q}{\hbar}\oint \mathbf{A}\,.\,d\,\mathbf{l} = 2n\pi,\ n = 0, \pm 1, \pm 2,... \tag{7.112}$$

where the integration is along the entire loop. Using Stokes theorem

$$\frac{q}{\hbar}\oint \mathbf{A}\,.\,d\,\mathbf{l} = \frac{q}{\hbar}\int \nabla\times\mathbf{A}.\,d\mathbf{s}$$

$$= 2n\pi,\ n = 0, \pm 1, \pm 2,... \tag{7.113}$$

where the surface integral is across the surface enclosed by the loop. Since $\nabla \times \mathbf{A} = \mathbf{B}$, this leads to the relation for flux ϕ,

$$\phi = \int \mathbf{B}.d\,\mathbf{S}$$

$$= \frac{h}{q}n, n = 0, \pm 1, \pm 2,... \tag{7.114}$$

Thus the flux takes only quantum values of integral multiples of h/q. The value of $h/e = 4 \times 10^{-15}$ Wb is quite small but is macroscopically detectable. The quantized flux was observed experimentally by Deaver and Fairbank and independently by Doll and Näbauer (1961). The observed flux was found to be integral multiples of h/q with $q = -2e$. This is an additional confirmation of the BCS theory according to which it is the Cooper pairs, with charge $-2e$ each, that are the carriers of current in superconductors.

Josephson Junctions

The discovery of Josephson junctions has made the direct macroscopic measurement of the ratio $\hbar/e$ possible.

Consider two pieces of a superconductor separated by a thin layer of an insulator (about 20 Å in thickness). If now a voltage V is applied to the superconductors, an ac current j flows across the junction. This effect, known as *Josephson effect,* is due to the tunnelling of the superconducting electrons across the insulator. In the process they emit microwave radiation of angular frequency ω such that

$$\hbar\omega = |q|V \tag{7.115}$$

where $q = -2e$ is the charge of a Cooper pair. It was also found that when the junction was illuminated by a radiation of frequency ω, and the potential V was varied, the current across the junction showed a jump whenever the condition

$$n\hbar\omega = 2e\,V,\; n = \text{an integer,} \tag{7.116}$$

was satisfied. Thus knowing ω and V, it was possible to obtain an accurate measurement of $\hbar/e$.

For obtaining the current across the junction, let ψ_1 and ψ_2 be the solutions of the Schrödinger equation for particles on the two sides of the insulator, with Hamiltonian H_0 in the absence of an external radiation. The approximate form of these wave functions in the region of the junction, is

$$\psi_1 = \exp\,[-(x+a)k_1]\;\exp\,(-\frac{i}{\hbar}Et) \tag{7.117}$$

$$\psi_2 = \exp[-(a-x)\,k_2]\;\exp\,[-\frac{i}{\hbar}(E+qV)t] \tag{7.118}$$

where V is the potential across the junction (see Fig. 7.7). If now radiation of frequency ω is incident on the junction, it may be assumed that a superposition of ψ_1 and $\psi_{2,}$

$$\psi = b_1\,(t)\,\psi_1 + b_2\,(t)\,\psi_2 \tag{7.119}$$

is an approximate solution of the Schrödinger

$$i\hbar\frac{\partial\psi}{\partial t} = (\,H_0 + H_1\,)\,\psi \tag{7.120}$$

Here H_1 is the interaction with the external radiation which is taken to be v cos ωt at one end (say the first) of the insulator and zero at the other. Multiplying Eq. (7.120) successively by ψ_1 and y_2 and integrating, one obtains (after neglecting the overlapping terms).

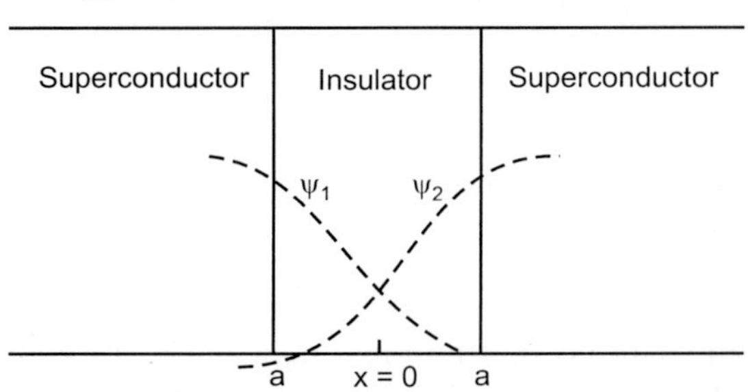

Fig. 7.10 The **Josephson** junction and the associated wave functions.

$$i\hbar \frac{\partial b_1(t)}{\partial t} = v(\cos \omega t)\, b_1(t) \tag{7.121}$$

$$i\hbar \frac{\partial b_2(t)}{\partial t} = 0 \tag{7.122}$$

These equations are fairly easy to solve. However, it is more instructive to solve them perturbatively. Assuming that v is small, we replace the $b_1(t)$ on the right hand side by $b_1(0)$ and integrate the two sides to get

$$b_1(t) \approx b_1(0) - \frac{i}{\hbar\omega} v(\sin \omega t)\, b_1(0) \tag{7.123}$$

$$b_2(t) = b_2(0) \tag{7.124}$$

Now the quantum mechanical generalization of current is

$$\mathbf{j} = \mathrm{Re}\, \frac{q}{m} \psi^* \mathbf{p}\, \psi$$

$$= -\,\mathrm{Re}\left(\frac{i\hbar q}{m} \psi^* \nabla \psi \right) \tag{7.125}$$

Substituting Eq. (7.119) for ψ with $b_1(t)$ and $b_2(t)$ given by Eqs. (7.123) and (7.124), the current across the junction is

$$j \approx j_0 \left[\sin\left(\frac{qVt}{\hbar} + \delta_0 \right) \right] - \frac{v}{\hbar\omega} (\sin \omega t) \cos\left(\frac{qVt}{\hbar} + \delta_0 \right) \Bigg] \tag{7.126}$$

where j_0 and δ_0 are constants. It is interesting to note that the ac current persists even in the absence of external radiation. On taking the time average, the contribution of the second term is nonzero if

$$\frac{|qV|}{\hbar} = \omega \tag{7.127}$$

If higher order perturbations are included, there are nonzero contributions to the current, for higher harmonics as well,

$$\frac{|qV|}{\hbar} = n\omega,\ n = 1, 2, \ldots \tag{7.128}$$

in conformity with the result in Eq. (7.116) for $q = -2e$. For $n = 1$, $v = 4.836 \times 10^{11}\ Vs^{-1}$ where V is in millivolts. Since V is usually of the order or several millivolts, the Josephson frequency is in the microwave range.

One of the most important applications of the Josephson effect is the determination of the fundamental constant $e/\hbar$ which occurs in Eq. (7.128) with

$q = -2e$. It has been possible to determine this ratio to an accuracy of a few parts in 10^6. It is to be noted that Josephson effect is very sensitive to the magnetic field. The use of this property has led to many important applications of Josephson junctions, singly or in combinations.

Around 1986 Georg Bednorz and Karl Müller, working at IBM in Zurich, discovered that certain semiconducting oxides became superconducting at 35 K, then considered a relatively high temperature. In particular, the lanthanum barium copper oxides, an oxygen deficient perovskite-related material, proved promising.

Later on, Maw-Kuen Wu and his graduate students, Ashburn and in 1987, and Paul Chu and his students around same time discovered YBCO has a T_c of 93 K. (The first samples were $Y_{1.2}Ba_{08}CuO_4$.) Their work led to a rapid succession of new high temperature superconducting materials, ushering in a new era in material science and chemistry.

YBCO was the first material to become superconducting above 77 K, the boiling point of liquid nitrogen. All materials developed before 1986 became superconducting only at temperatures near the boiling points of liquid helium ($T_h = 4.2$ K) or liquid hydrogen ($T_b = 20.28$ K) — the highest being Nb_3Ge at 23 K. The significance of the discovery of YBCO is the much lower cost of refrigerant used to cool the material to below the critical temperature.

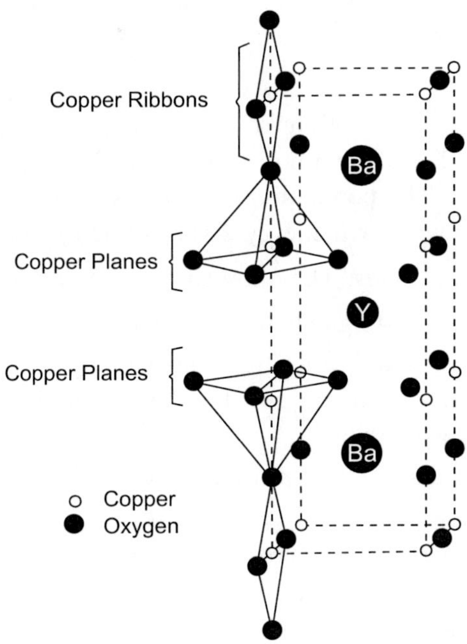

Fig. 7.11 Perovskite structure of $Y_1Ba_2Cu_30_7/Y_1Ba_2Cu_4Os$.

There is no widely accepted theory to explain their properties. Cuprate superconductors (and other unconventional superconductors) differ in many important ways from conventional superconductors, such as elemental mercury or lead, which are adequately explained by the BCS theory. There also has been much debate as to high-temperature superconductivity coexisting with magnetic ordering in YBCO, iron-based superconductors, several other exotic superconductors, and the search continues for other families of materials. HTS are Type-II superconductors, which allow magnetic fields to penetrate their interior in quantized units of flux, meaning that much higher magnetic fields are required to suppress superconductivity. The layered structure also gives a directional dependence to the magnetic field response.

Several commercial applications of high temperature superconducting materials have been realized. For example, superconducting materials are finding use as magnets in magnetic resonance imaging, magnetic levitation, and Josephson junctions. (The most used material for power cables and magnets is BSCCO (bismuth strontium calcium copper oxide).

(*Source: Wikipedia*)

7.7 EXAMPLES

In this section, some examples which illustrate and extend the main ideas quantum statistics are discussed.

Example 1

Consider the statistical distributions of two identical particles among three sets of states $g_1 = 1$, $g_2 = 2$, $g_3 = 1$ with energies 0, ε, 2ε, respectively (both the g_2 states have energy ε). The populations of these sets are (n_1, n_2, n_3). The most probable distribution with total energy 2ε has to be found.

Distinguishable particles: The allowed population distributions are:

1. (1, 0, 1) has two distinguishable arrangements (A, 0, B), (B, 0, A)
2. (0, 2, 0) with four possible distinguishable arrangements (AB, 0), (0, A, B), (A, B) and (B, A) in the g_2 set, where A and B represent the two distinguishable particles.

Thus, the second distribution is twice as probable as the first distribution.

Bosons: For bosons, the (1, 0, 1) distribution has only one distinguishable arrangement while (0, 2, 0) has three distinguishable arrangements (AA, 0), (A, A) (0, AA) in the g_2 set. Therefore, the (0, 2, 0) is three times as probable as the (1, 0, 1) distribution.

Fermions: For fermions, of the two distribution (1, 0, 1) and (0, 2, 0), each has only one possible distinguishable arrangement (it should be recalled that the **Pauli principle** forbids more then one particle in each state). Thus, the two distributions are equally probable.

The number of distributions can be verified by Eqs. (7.4), (7.7) and (7.10).

Example 2

The extension of the classical distributions for the case of bound and ionized atoms in equilibrium is of special interest in astrophysics and plasma physics. The equilibrium distribution is obtained by using arguments similar to those used in Sec. 6.4 for obtaining the Einstein coefficients *A* and *B*.

The transitions in this case are

$$M_0 \rightleftharpoons M^+ + \bar{e} \tag{7.129}$$

where M_0 is the neutral atom and M^+ is its ion. In contrast to the discussion in Sec. 6.4, here the final states form a continuum. Let N_0 be the number of M_0 atoms and N_+ be the number of M^+ ions. The number of ionization transitions to a set of states f_i, is obtained from Eq. (6.60) as

$$N_{mn} = B_{mn}\, u(\omega)\, N_0 f_i$$

$$f_i = \frac{(2m)^{3/2}\, V}{4\pi^2\, \hbar^3}\, \varepsilon^{1/2}\, d\varepsilon \tag{7.130}$$

where f_i is given in Eq. (7.29) and $\hbar\omega = \varepsilon + E_I$, E_I being the ionization energy and ε is the energy of the electron. The number of reverse reactions, *i.e.* recombinations, is given by the first equation in Eq. (6.60) except that now the expression is also proportional to the number dN_e of electrons in f_i states

$$N_{mn} = (\, B_{nm}\, u\,(\omega) + A_{nm}\,)\, N_+ dN_e| \tag{7.131}$$

Equating N_{mn} and $N_{nm,}$ gives for $u\,(\omega)$

$$u\,(\omega) = \frac{A_{nm}}{B_{mn}\, f_i \dfrac{N_0}{N_+\, dN_e} - B_{nm}} \tag{7.132}$$

In analogy with Eq. (6.66) B_{nm} is taken to be equal to B_{nm} (this can be justified by more rigorous arguments). Comparing $u\,(\omega)$ with the expression in Eq. (66.5) gives

$$\frac{N_+\, dN_e}{N_0} = \exp\,[-\,(\varepsilon + E_I)/kT]\, f_i \tag{7.133}$$

Substituting for f_i and integrating over dN_e and ε, finally gives

$$\frac{n_+\, n_e}{n_0} = \left(\frac{2\pi mkT}{h^2}\right)^{3/2} \exp\,(-\,E_I\,/\,kT) \tag{7.134}$$

where $n_0 = N_0/V$, etc. This equation is known as the *Saha equation* (1920).

In deriving this relation, the degeneracy of states has been ignored. The degeneracy can be incorporated by multiplying the right-hand side by $g + g_e/g_0$ where g is the degeneracy of the appropriate state, in particular, $g_e = 2$ corresponding to the two spin states of the electron. As an illustration it is noted that if M_0 is the hydrogen atom in the ground state and M_+ is the proton, then $g_+ = 2$ and $g_0 = 4$ so that the degeneracy factor is 1. The Saha equations is very useful in plasma physics and also in astrophysics.

Example 3

As an applications of Maxwell-Boltzmann statistics, consider the ratio of *para*-hydrogen to *ortho*-hydrogen ordinary hydrogen at the room temperature. Since *ortho-hydrogen* has $I = 1$,

$$\frac{\mathbf{H}\,(para)}{\mathbf{H}\,(ortho)} = \frac{\sum_{J=0,2\ldots} (2J+1)\exp[-aJ(J+1)]}{3\sum_{j=1,3} (2J+1)\exp[-aJ\,(J+1)]} \tag{7.135}$$

where $a = 0.00755/kT$, kT being in eV. In evaluating the sum, the first term is separated out and the remaining sum is converted into an integral by replacing J by $2l$ and taking $x = l(2l+1)$. Therefore,

$$\frac{\mathbf{H}\,(para)}{\mathbf{H}\,(ortho)} \approx \frac{1 + \int_1^\infty e^{-2ax}\,dx}{3[3e^{-2ax} + \int_3^\infty e^{-2ax}\,dx]}$$

$$= \frac{2a + e^{-2a}}{3(6ae^{-2a} + e^{-6a})} \tag{7.136}$$

At $T = 27°C$, $a = 2.93$ and the ratio comes out to be 0.33 which is in very good agreement with experimental observations.

Example 4

Copper has an atomic weight of 63.5, a density of 8.9 g/cc, and $v_t = 2.32 \times 10^3$ m/s and $v_l = 4.76 \times 10^3$ m/s. Its Debye temperature is

$$\theta = hv_m/k$$

$$v_m = \left[\frac{9N_0}{4\pi V} \bigg/ \left(\frac{2}{v_t^3} + \frac{1}{v_l^3}\right)\right]^{1/3} \tag{7.137}$$

Since $N_0 = 6.02 \times 10^{23}$/g mol, and $V = (63.5/8.9)$ cc,

$$\nu_m = 7.1 \times 10^{12}\ \text{s}^{-1} \tag{7.138}$$

and hense $\theta = 341$ K which is in good agreement with the experimental value of 343 K.

From the Debye temperature one can estimate the specific heat at low temperatures from Eq. (7.66). For example, at $T = 30$ K

$$C_v \approx 0.16\,R, \left(\frac{T}{\theta} \approx 0.088\right) \tag{7.138a}$$

Example 5

For estimating the transition temperature T_c for ^{4}He, if $V = 27.6$ cm^3/mole

$$\frac{N}{V} = 2.18 \times 10^{28}\ \text{m}^{-3} \tag{7.139}$$

On substituting this in Eq. (7.78)

$$T_c = 3.13\ \text{K} \tag{7.140}$$

It may also be noted that α_2 is a small quantity for $T < T_c$. Using N_0 given in Eq. (7.81) gives

$$\frac{1}{\exp(\alpha_2/kT) - 1} \approx N\,[1 - (T/T_c)^{3/2}],\ T < T_c \tag{7.141}$$

which, on the expanding the exponential functions gives

$$\alpha_2 \approx \frac{kT}{N\,[1 - (T/T_c)^{3/2}]},\ T < T_c \tag{7.142}$$

Thus α_2 is very small for $T < T_c$ except when T is close to T_c.

Example 6

The electronic properties of Cu may be deduced by assuming that each atom contributes one free electron. The atomic weight of Cu is 63.54 and its density is 8.96 g/cc so that

$$\frac{N}{V} \approx 87.44 \times 10^{28}\ \text{m}^{-3}\text{g}$$

From Eq. (7.90), the Fermi energy at 0 K is

$$\varepsilon_f(0) \approx 7.0\ \text{eV} \tag{7.143}$$

The change in the Fermi energy T increases [see Eq. (7.91)] from 0 K to 300 K is very small, about $- 7.8\ 10^{-5}$ eV and hence $\varepsilon_f(T)$ can for most purposes be taken to be a constant.

The total specific heat of Cu at low temperatures, including the electronic and phonon contributions is [from Eq. (7.96)]

$$C_v \approx 4.935\, R\left(\frac{kT}{7}\right) + 2.34\times10^2\, R\left(\frac{T}{341}\right)^3 \tag{7.144}$$

where T is in kelvin and kT is in eV. The two contributions become comparable at $T \approx 3.21$. One may be therefore except that the linear term will be important for $T \lesssim 3$ K.

The mean free path λ can be estimated from Eq. (7.97). The conductivity for Cu is $5.82 \times 10^7/\Omega$ m, and $v_f = (2\varepsilon_f/m)^{1/2}$ is about 1.57×10^6 m/s. One then obtains

$$\lambda \approx 770 \text{ Å} \tag{7.145}$$

which is in reasonable agreement with the experimentally measured value of about 530 Å.

Example 7

A very interesting application of the Fermi-Dirac distribution is to white dwarfs and neutron stars, regarded as a degenerate gas of electrons and neutrons respectively. A very sketchy and approximate discussion of the main ideas is given here.

When a star contracts, a part of its gravitational energy escapes as radiation but the remainder is retained as kinetic energy. At equilibrium, there is the approximate relation

$$\frac{GM^2}{R} \approx N\,\varepsilon_f \tag{7.146}$$

where G is the gravitational constant, M is the mass of the star, R is its radius, N is the number of particles and ε_f is the Fermi kinetic energy of the particles which is of the same order of magnitude as the average kinetic energy. For the highly degenerate fermions, relativistic kinematics should be used and

$$\varepsilon_f = (pf^2c^2 + m^2 c^4)^{1/2} - mc^2 \tag{7.147}$$

where m is the mass of the degenerate particles. For obtaining p_f, Eq. (7.89) is written in the form

$$N = \frac{8\pi V}{3h^3}\, pf^3 \tag{7.148}$$

It is interesting to note that while the validity of Eq. (7.89) is limited to nonrelativistic situations, Eq. (7.148) is valid even for large velocities substituting these relations in Eq. (7.146), gives

$$\frac{GM^2}{R} = N\left[\left(\frac{3Nh^3}{8\pi V}\right)^{2/3} c^2 + m^2 c^4\right]^{1/2} - Nmc^2 \tag{7.149}$$

Furthermore, since $N = M/m_N$, m_N being the mass of the neutron or the proton, and $V = \frac{4\pi R^2}{3}$, this equation simplifies to

$$\frac{G^2 m_N^2 M^2}{R^2} + \frac{2Gm_N M\, mc^2}{R} = \left(\frac{9Mh^3 c^3}{32\pi^2 m_N}\right)^{2/3} \frac{1}{R^2} \tag{7.150}$$

Since R is positive, this implies the condition

$$\left(\frac{9Mh^3 c^3}{32\pi^2 m_N}\right)^{2/3} > G^2\, mN^2 M^2 \tag{7.151}$$

or

$$M < \frac{2}{2\pi}\left(\frac{1}{m_N^2}\right)\left(\frac{he}{2G}\right)^{3/2} \tag{7.152}$$

$$< 5\, M_{sun}$$

These calculations are only order of magnitude calculations. More refined calculations provide a somewhat lower upper bound, $< 3M_{sun.}$

This example demonstrates the importance of quantum distributions even on an astronomical scale.

Example 8

As noted before, when the temperature of a material is lowered, the specific heat shows a sudden increase when it becomes a superconductor at $T = T_c$. This is because Cooper pairs are formed below $T = T_c$ and some energy goes into breaking them. However, the specific heat of the superconducting material goes to zero faster than T as $T \to 0$, unlike the linear T behaviour expected for a free-electron gas, the reason being that the specific heat of the Cooper pairs goes to zero faster than T as $T \to 0$.

PROBLEMS

1. Three identical particles with total energy 6ε are distributed among four energy levels with energies ε, 2ε, 3ε and 4ε of which the second level has a degeneracy of 3. What are the possible distributions if the particles are (*i*) distinguishable, (*ii*) bosons and (*iii*) fermions? Which is the most probable distribution in each case?

2. For a gas at temperature T, what is the most probable energy of a particle? What is the approximate fraction of particles which have more then ten times the average energy of the particles?
3. For producing Balmer absorption lines, a substantial number of the hydrogen atoms should be in the first excited state. What is the percentage of the hydrogen atoms which are in the first excited state in the sun's photosphere which is at a temperature of about 6000 K, taking into account the degeneracy of the states?
4. What is the fraction of the hydrogen molecules which are in the first excited vibrational state at 3000 K? What is the fraction in the second excited vibrational state?
5. Show that for $T \gg T_D = \theta$, the total energy of a solid is given by

$$E = 3RT_D\left[\frac{T}{T_D} - \frac{3}{8} + \frac{1}{20}\left(\frac{T}{T_D}\right)^{-1} + \ldots\right]$$

and the corresponding specific heat by

$$C_v = 3R\left[1 - \frac{1}{20}\left(\frac{T_D}{T}\right)^2 + \ldots\right]$$

The approximation is quite good even for $T \sim T_D$. Estimate the specific heat of copper at $T = 300$ K (T_D for copper is 343 K).
6. Obtain an expression for the energy of a 2-dimensional lattice. What is the specific heat of this lattice for $T \to \infty$ and for $T \to 0$? The situation is applicable for layer structures such as graphite whose specific heat at low temperatures is proportional to T^2.
7. For Cu, the lattice specific heat a low temperature has the behaviour for $C_v \sim 4.6 \times 10^{-5}\, T^3$ J/mol K. Estimate the Debye temperature for Cu.
8. Given that the Debye temperature for diamond is 1860 K, what is the specific heat of diamond at room temperature?
9. What is the number of excited phonons and the average energy per phonon at a given temperature? Show that the average goes to $\frac{2}{3}h\nu_{max}$ for $T \to \infty$ and is proportional to T for $T \to 0$.
10. Show that the average kinetic energy of the electrons emitted in thermionic emission is $2kT$, and that the average value of the square of velocity perpendicular to the surface is $2kT/m$.
11. Calculate the Fermi energy at 0 K, of silver (density 10.5 g/cc, atomic weight ≈ 107.87) and sodium (density 0.97 g/cc, atomic weight ≈ 22.99),

assuming one free electron per atom. What is the ratio of $\varepsilon_f(T)/\varepsilon_f(0)$ for these elements at $T = 300$ K?

12. Show that the electronic specific heat of Cu at room temperature is very small compared to its lattice specific heat.

13. If electrons are treated as distinguishable particles, at what temperature would they have an average energy of 5.5 eV (*i.e.* the Fermi energy of silver)?

14. Given that the Fermi energy of Cu is 7.0 eV at room temperature, what is the number of electrons per unit volume with energy greater than 8.0 eV?

15. Given that the electrical conductivity of aluminium is $3.55 \times 10^7\ \Omega^{-1}\ m^{-1}$ estimate its thermal conductivity at room temperature.

16. What is the minimum frequency of radiation which can break apart Cooper pairs in niobium?

17. A microwave radiation of frequency 10^{10} Hz is incident on a Josephson junction. What is the minimum voltage across the junction for which a jump in the current is observed?

8

Solid State Physics

Structures of the Chapter

The description of the properties of solids is an important application of the quantum principles to a collection of a large number of particles arranged in order. Many of the solids are in a crystalline form, associated with a lattice. A lattice is obtained by an infinite, regular repetition, in space, of basic units. In a perfect crystal, the atoms are arranged according to a pattern in each unit, and the pattern is repeated for all the units. Our concern will be mainly with such crystals, *e.g.*, sodium chloride, diamond, etc. This leaves out some important substances, such as amorphous solids which have some short-range order but no long-range order (*e.g.*, glass, amorphous semiconductors etc.), and others like wood, plastics, etc. which have huge molecules. Furthermore, most crystalline materials are usually a collection of small crystals packed together, and even the single crystals may have some disorder introduced by an impurity or a missing atom. These disorders play an important role in determining their properties.

The forces which bind the atoms in solids, and the structure of solids are considered first. The energy levels in solids are significant and often decisive in determining their electromagnetic properties, and lead to important applications. After considering the band structure of energy levels in solids, in particular, semiconductors, some aspects of magnetic and dielectric properties of solids are discussed.

8.1 BINDING FORCES IN SOLIDS

The binding forces between the atoms in a solid are electromagnetic in origin. It is the valence electrons of the constituent atoms and the quantum effects which are important in determining the details of these forces. The binding forces are similar to those in molecules though the many-particle effects bring in some new aspects. We briefly discuss the different types of bonds in solids. In a real solid, more than one of these bonds may contribute to the binding of the atoms.

Ionic Bonds

As in the case of molecules, these bonds are formed when it is energetically favourable for a valence electron to be transferred from one atom to another resulting in a net electrostatic attraction between the ions. However, in the solid there are many more ions and the electrostatic interaction of an ion with all the other ions should be taken into account. This is done by writing the electrostatic energy per ion pair as

$$V_{el} = -\frac{\alpha e^2}{4\pi\varepsilon_0 R} \tag{8.1}$$

where R is the distance between the nearest neighbours and α is called the *Madelung constant*. For example, in the case of NaCl, KCl, etc. (but not CsCl), the crystal structure is what is called *face centred cubic* structure (Fig. 8.1) with an ion pair associated with each lattice point. The electrostatic energy per ion pair is obtained by writing it as a sum of terms representing the interaction of a given ion with the nearest neighbours, the next nearest neighbours, etc.

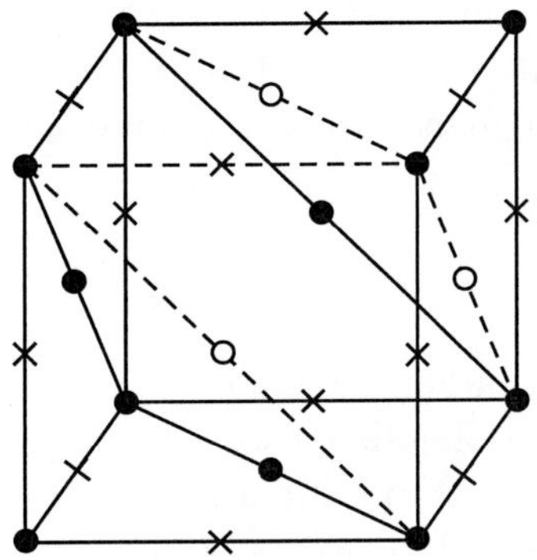

Fig. 8.1 The KCl crystal structure (*fcc*) with K^+ denoted by circles and Cl^- by crosses.

$$V_{el} = -\frac{e^2}{4\pi\varepsilon_0 R}(6 - 12/2^{1/2} + 8/3^{1/2} - 3 + ...)$$

$$\approx -\frac{e^2}{4\pi\varepsilon_0 R}(1.75) \tag{8.2}$$

where R is the distance between the nearest neighbours and the Madelung constant turns out to be $\alpha \approx 1.75$.

To get an idea about the details of the ionic bonds, consider solid KCl. As in the case of the KCl molecule, here also it takes an energy of 0.54 eV to transfer one electron from the K atom to the Cl atom (see Sec. 5.6). Representing the van der Waals repulsion, which becomes important when the electronic wave functions overlap, by b/R^n, the energy per ion pair is given by

$$E_{\text{pair}} = 0.54 - \frac{14.4\alpha}{R} + \frac{b}{R^n} \tag{8.3}$$

where $\alpha \approx 1.75$, R is in Å and the energy is in eV (the energy in Eq. (8.3) is with respect to the energy of the isolated neutral atoms, taken to be zero). The energy per pair, E_{pair}, is known as *cohesive energy per ion pair*. The equilibrium condition for the ions is

$$\frac{dE}{dR} = 0 \tag{8.4}$$

which implies

$$E_{\text{pair}} = 0.54 - \frac{14.4\,\alpha}{R_0}\left(1 - \frac{1}{n}\right) \tag{8.5}$$

R_0 being the equilibrium separation. From a knowledge of $R_0 \approx 3.14$ Å and $E_{\text{pair}} \approx -6.67$ eV, one gets $n \approx 10$ which is in approximate agreement with what is expected from a detailed theoretical analysis.

The ionic bind is quite strong, the binding per pair of atoms being about 5 eV. This leads to rather high melting temperatures for ionic crystals, *e.g.* 801°C for NaCl.

Covalent Bonds

Covalent bonds discussed earlier in the context of molecules (see Sec. 5.6), are important in the formation of solids also. These bonds arise when two atoms find it energetically favourable to share their electrons (this is particularly true for identical atoms such as two Cl atoms). In such cases the shared electrons are found preferentially between the atoms, with opposite spins as required by Pauli's principle, providing each atom with a complete shell. These bonds are especially important in group IV elements with half-filled shells, *e.g.* C, Si, Ge, etc. which can accommodate eight electrons in their outermost shells. These atoms can form four covalent bonds which are directional. Every atom may be considered to be at the centre of a tetrahedron, sharing an electron with each of the four nearest neighbours which are at the four corners of the tetrahedron. Since a covalent bond involves the sharing of two electrons, one from each atom, this allows the atoms to have closed shells with eight electrons.

Covalent bonds are usually quite strong (binding energy is a few eV per bond) and directional. As a consequence, crystals with covalent bonds are hard but brittle, and have a high melting point. This is especially so for diamond, which has a cohesive energy of about 7 eV per atom. It is the hardest material known, and has a melting point of more than 3550°C.

If the bonds are between different types of atoms, the electrons may spend more time with one of the atoms, so that the bonds are partially ionic and partially covalent. An important example of this is ZnS (zinc blende) which also has tetrahedral structure but with different atoms at the centre and the corners, *e.g.* Zn at the centre and *S* at the corners. In this case, the bonds are partially ionic and partially covalent.

Metallic Bonds

As was pointed out in the discussion of the free-electron theory of metals, the valence electrons in metallic atoms, being loosely bound, escape from the atom. These essentially free electrons provide a medium of negative charge which helps to bind the positive ions. That this leads to a lower energy state can be seen from the following arguments.

An electron in an isolated atom is confined to a small volume around the nucleus. This confinement gives rise to an uncertainty in the momentum, $\Delta p \sim \hbar/r$, where r is the radius of the atom. Consequently, the electron has a fairly substantial amount of kinetic energy, of the order of several eV. However, in the crystalline, metallic state, the electrons are essentially free to be anywhere in the entire crystal. As a result, there is a considerable reduction in their kinetic energy. This is the source of metallic bonding.

The bond between two metallic atoms is somewhat weaker than ionic or covalent bonds. This leads to relatively low melting points, for example 63°C for K. However, the cohesive energy of metals is fairly large since each valence electron interacts with several ions. The metallic bonds are not directional which allows the planes of atoms to slide over each other quite easily. Hence, metals are found to be ductile and malleable rather than brittle. The existence of essentially free electrons gives rise to high electrical and thermal conductivity for metals.

Van der Waals Bonds

When two neutral atoms approach each other, there is an attractive force between them because of the induced electric dipole moments (see Sec. 5.6). These forces, through weak, are important for atoms which do not form ionic, covalent or metallic bonds with each other. They lead to the solidification of inert gases He, Ne, Ar, Kr, Xe, and some organic molecules, such as methane. The binding due to van der Waals forces is usually weak which results in a low melting point, *e.g.* melting point of Ar is – 189°C, for these solids. They are also soft, electrically insulating and generally insoluble.

Hydrogen Bonds

The hydrogen atom has only one electron and would normally form a covalent bond with only one other atom. However, if the other atom is strongly electronegative, the electron may be transferred to the electronegative atom. The remaining proton being small in size, about 10^{-15} m compared to the usual atomic size of 10^{-10} m, can bind only two neighbouring negative ions (the two spheres would be glued together by the proton in-between). This gives rise to what is known as the *hydrogen bond* which connects only two atoms.

The hydrogen bond is important in the formation of ice and in the polymerization of hydrogen fluoride.

It may be noted that in a real crystal, the actual binding is due to a mixture of different types of bonds though one of them may be predominant.

8.2 CRYSTAL STRUCTURES

It is convenient to discuss the structures of crystals in terms of space lattices. A *space lattice* is an array of lines, which divides the space into identical volumes. These volumes fill the space completely and are known as *unit cells*. The intersections of the lines are the *lattice points* and each lattice point is usually associated with an atom or a group of atoms. Therefore may also be atoms at the centres of the cells or the faces of the cells. Obviously, the choice of a unit cell is not unique. It is usually dictated by convenience. For example, in the two dimensional space lattice shown in Fig. 8.2, the unit cell may be taken to be *ABCD* or *ABEC* though it may be more convenient to choose *ABCD*. The essential requirement is that repetition (or equivalently, translations) of the unit cell should cover the entire space lattice. A *primitive cell* is the unit cell with the smallest volume. The choice of a primitive cell also in not unique. It is usually chosen by convenience or tradition.

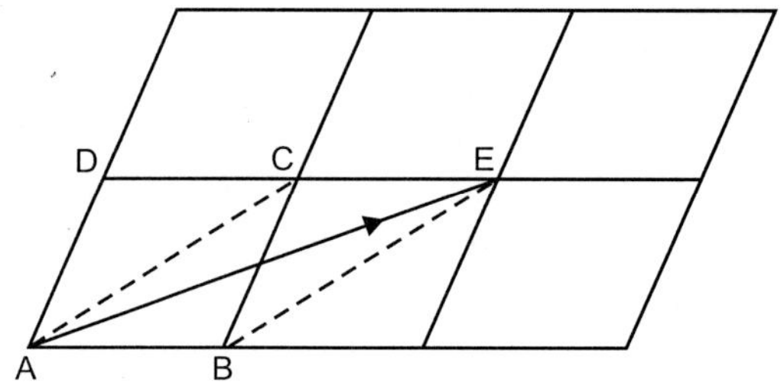

Fig. 8.2 A two-dimensional space lattice. Unit cell may be taken to be *ABCD* or *ABEC*. With *A* as the origin, the lattice point *E* is represented by the lattice vector 2*a* + *b*.

The important characteristic of a space lattice is that every lattice point has an identical surrounding. This severely constraints the possible space lattices. It was sown by Bravais (1848) that there are only 14 space lattices (Fig. 8.3). It is important to appreciate that while there are only 14 space lattices, there are a very large number of crystal structures since different patterns of atoms can be associated with a given lattice point.

A few details of some of the more common lattices are discussed here. It may be noted that the centres of atoms in various lattices are located at the corners or the centres indicated, and the atoms in some sense can be regarded as touching the nearest neighbours. The number of nearest neighbours is called the *coordination number*, and give an indication of the closeness of the packing of the atoms. Another quantity of interest is the *packing fraction*, which is the fraction of the available volume occupied by the atoms. With the assumption that the atoms are hard spheres and that the nearest neighbours touch each other, we can deduce the packing fraction for a given crystal pattern.

Simple cube: The simple cubic structure is perhaps the simplest possible form, tough polonium is the only element which has this structure. The reason for this is that the cubic structure is rather an open form with a considerable amount of empty space.

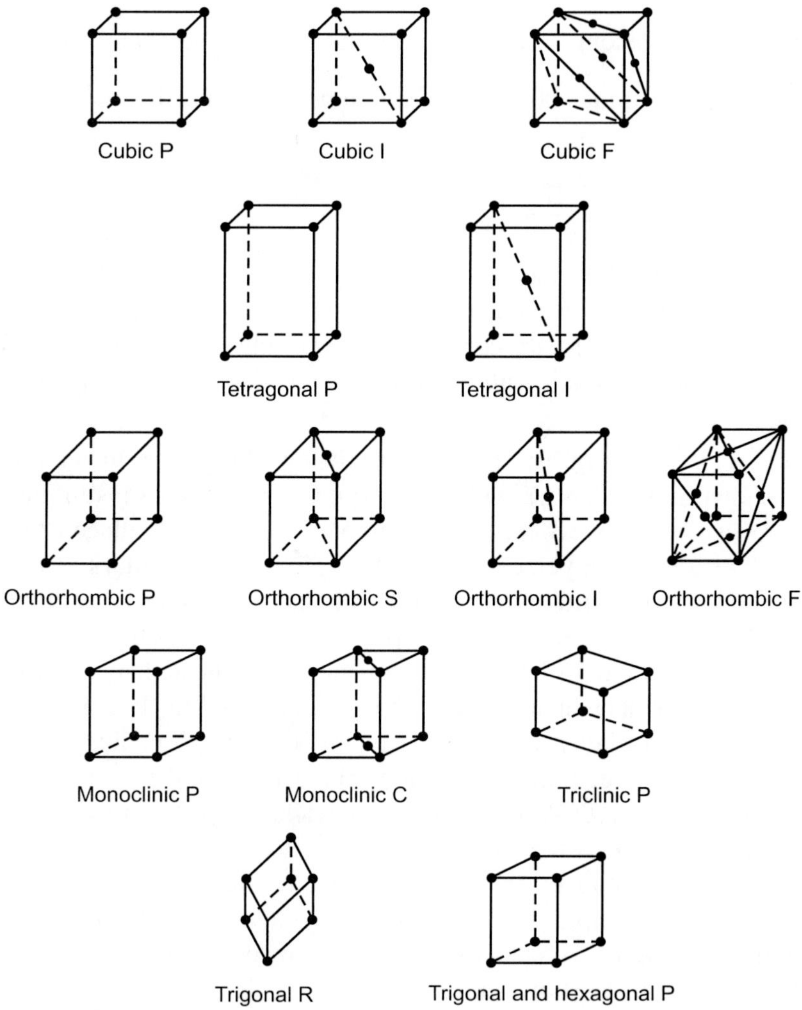

Fig. 8.3 The 14 Bravais or space lattices; *P* cells are primitive, *I* cells are body-centred, *F* cells are face-centred, *C* cells are base-centred, and *R* cells are rhombohedral.

The coordination number for the simple cubic structure is 6. For this structure, each atom at the corner is shared by eight cubes so that the volume occupied by the atoms in a cube is 4/3 πR^3 (R is the radius of the atom).

Since the length of the cube is $2R$, packing fraction f is

$$f = \frac{\pi}{6} \approx 52\% \tag{8.6}$$

Body centred cube: A closer packing is provided by the body centered cubic (bcc) lattice. Many elements crystallize in this form, *e.g.,* Fe, Cr, Mo, W, Li, K, Na, Rb, Cs, etc. The coordination number for these structures is 8. There are two atoms which belong to each cell so that the volume occupied is $\frac{8\pi}{3}R^3$. Since the diagonal of the cube is $4R$, the volume of the cube is $(4R/3^{1/2})^3$ and the packing fraction f is

$$f = \frac{3^{1/2}\pi}{8} \approx 68\% \tag{8.7}$$

It should be noted that some binary compounds such as CsCl, have a body centred cubic structure, with one type of atoms (*e.g.*, Cl) at the centres, and the other type of atoms (*e.g.*,) at the corners. However, since these atoms are different the lattice is a simple cubic lattice with a pair of Cs and Cl atoms associated with each corner.

Close-packed structures: The most efficient packing of atoms in a plane is the one shown in Fig. 8.4(*a*), with each atom touching six atoms in the plane. There are two ways in which the successive layers can be placed over one another. The second layer can have atoms in positions marked as *B* (or equivalently *C*). The third layer may have atoms in positions *A* leading to the hexagonal close-packed (hcp) structure [Fig. 8.4(*b*)]. The positions of the atoms in the successive layers of this structure may be indicated by *ABABAB*... . Examples of elements which have this structure are Cd, Mg, Ti, Zn, etc. On the other hand, the third layer may have atoms in positions *C* leading to the face-centred cubic (fcc) structure. The positions of the atoms in the successive layers may be indicated by *ABCABC*... . The close packed layers of the face-centred cubic structure, are in the body diagonal planes [Fig. 8.4(*c*)]. Some examples of elements that have the face-centred cubic structure, are Cu, Ag, Au, Al, Pd, and Pt. An interesting example is that of NaCl, KCl, etc. (Fig. 8.1) which consist of two interpenetrating fcc sublattices, one of them made up of Na^+ or K^+ ions and the other of Cl^- ions, as shown in Fig. 8.1.

Both the close-packed structures, the hexagonal close-packed structure and the face-centred cubic structure have a coordination number of 12 and have the highest packing fraction. The packing fraction for the fcc structure is obtained by noting that four atoms belong to each cell and occupy a volume of $\frac{16\pi}{3}R^3$. The length of a side of the cube is $2^{3/2}R$ so that the packing fraction f is

$$f = \frac{\pi}{3(2^{1/2})} \approx 74\% \tag{8.8}$$

For the hcp structure, six atoms belong to each cell and occupy a volume of $8\pi R^3$. The area of the base of the hexagon is 6 $(3^{1/2})$ R^2 while the height of the hexagon is $4(2/3)^{1/2}$ R, from which the packing fraction is again found to be

$$f = \frac{\pi}{3(2^{1/2})} \approx 74\% \tag{8.9}$$

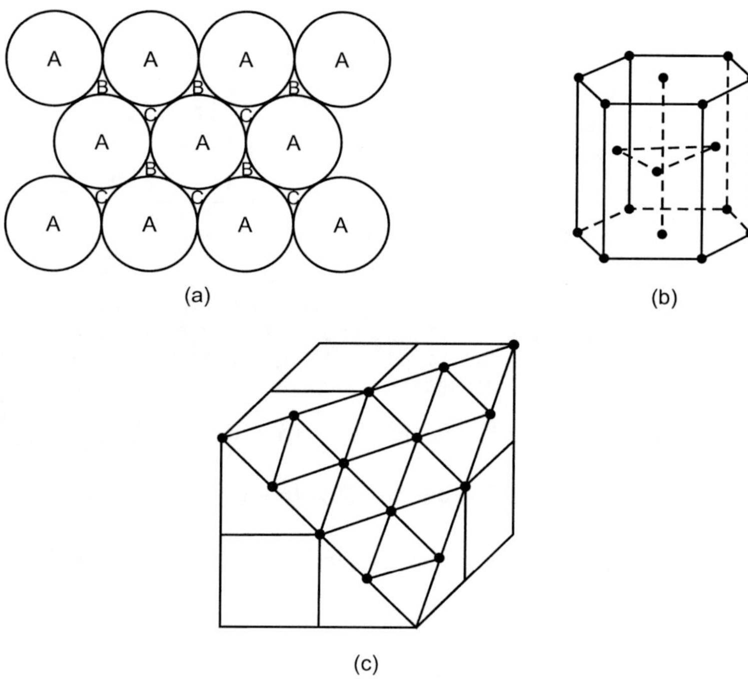

Fig. 8.4 Close-packed structures. (*a*) different layers, (*b*) hexagonal close-packed, (*c*) body-diagonal plane of face centred cube.

Equal packing fractions for the hcp and fcc structures is to be expected since both of them have close-packing. In each specific case, that structure which has a lower total energy is chosen.

The diamond structure: The diamond structure is of considerable importance since many practically important elements belonging to group IV, such as C, Si and Ge, crystallize in this form. It may be regarded as a superposition of an fcc lattice and another fcc lattice obtained by translating the first lattice along the body diagonal by a quarter of the diagonal [see Fig. 8.5 (*a*). The structure however is not close-packed since a corner atom of one face-centred cube is not in contact with an atom in a face with a corner atom of the second cube. A more useful concept is to regard each atom of one cube (*e.g.*, a corner of the second cube) as being at the centre of a tetrahedron formed by the four nearest neighbours which belong to the other cube [see Fig. 8.5(*b*)]. This also brings out the fact that each atom in this structure usually forms four covalent bonds with its four nearest neighbours. Some binary compounds such as ZnS also crystallize in the diamond structure, with Zn atoms forming one fcc and S forming the other fcc. In these cases, the bonds are partly ionic and partly covalent. Other compounds with this structure are SiC, CdS, InSb, GeP, etc.

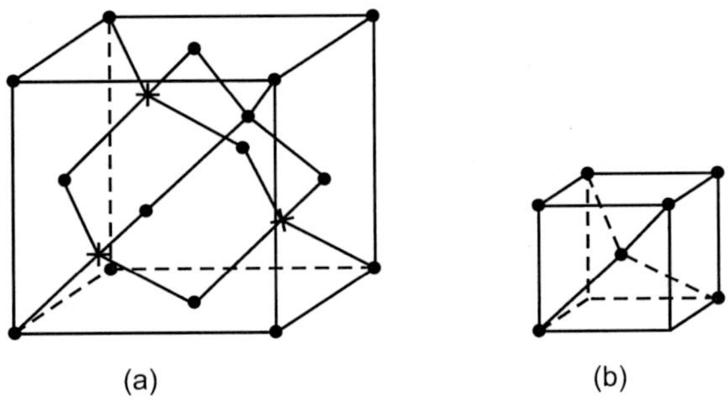

Fig. 8.5 Diamond structure, (*a*) circles represent an fcc lattice and crosses represent the fcc lattice obtained by translating the first lattice along the body diagonal by 1/4 the diagonal, (*b*) a typical tetrahedron in the diamond structure.

The coordination number for the diamond structure is 4. The packing fraction may be obtained by noting that eight atoms belong to the cell (4 are inside, $\frac{1}{2} \times 6$ in the faces and $\frac{1}{8} \times 8$ at the corners) and that the diagonal is equal to $8R$. This leads to a packing fraction f of

$$f = \frac{32\pi}{3} R^3 / (8R/3^{1/2})^3$$

$$= \frac{3^{1/2}\pi}{16} \approx 0.34 \tag{8.10}$$

which means that the packing in diamond structure is rather loose.

Directions and Planes in Crystals

Let the sides of the unit cell which is taken to be a parallelopiped, be designated by the vectors **a**, **b** and **c**. Let these vectors intersect at a corner A which is taken to be the origin. The position vector of any lattice point may be represented by a linear combination of a, b and c, as

$$r = y'\mathbf{a} + b'\mathbf{b} + w'\mathbf{c} \tag{8.11}$$

where u', v' and w' are integers. For example, the point E in the two dimensional lattice in Fig. 8.2 is represented by $2\mathbf{a} + \mathbf{b}$. The direction of the vector is then represented by a set of numbers $[u, v, w]$ which is obtained by multiplying u', v' and w' by the lowest common denominator. If some of the components are negative, this is indicated by a bar over the number, *e.g.,* if u is negative, the direction is given by $[|\bar{u}|, v, w]$.

A crystal plane is characterized by the intercepts u', $\mathbf{a}$, v' $\mathbf{b}$ and w' $\mathbf{c}$ along the three axes. The reciprocals $1/u'$, $1/v'$ and $1/w'$ are reduced to the simplest integers h, k, l by multiplying the reciprocals by their lowest common denominator. The plane is then denoted by the set of numbers (h, k, l), called the *Miller indices*. If some of the intercepts are along the negative axes, this is indicated by a bar over the corresponding index, *e.g.,* if u' is negative, the Miller indices are $(|\bar{h}|, k, l)$. It is clear from this that parallel planes (with the corresponding intercepts having the same sign) are represented by the same Miller indices. When an intercept is at infinity, the corresponding Miller index is zero. For example, (1, 0, 0) represents a plane parallel to the yz plane, with a positive intercept along the x-axis, while a plane with intercepts $-2\mathbf{a}$, $3\mathbf{b}$ and parallel to the z-axis, is denoted by the Miller indices $(\bar{3}, 2, 0)$.

Diffraction by a Lattice

Crystal structures are determined experimentally by analysing the diffraction of x-rays by a crystal. Since x-rays have a wavelength of about 1 Å, the atoms in a crystal serve as a grating and produce diffraction maxima. The measurement of the positions and intensities of these maxima gives information about the crystal structure. The conditions for the maxima were obtained in terms of the Bragg condition in Eq. (2.38) by considering the superposition of scattering from different planes. This is an over-simplified picture. Here a more rigorous and complete derivation of the maxima is presented by looking at the superposition of scattered waves from different atoms.

Consider a radiation described by $B \exp[i(\mathbf{k}_0.r - \omega t)]$, $|\mathbf{k}_0| = 2\pi/\lambda$, incident on a lattice with atoms at points

$$r_n = u\mathbf{a} + v\mathbf{b} + wc, \quad u = 0, \ldots, n_1; \quad v = 0, \ldots, n_2; \quad w = 0, \ldots, n_3 \tag{8.12}$$

The scattered wave will have the same wavelength as the incident beam, but will propagate in some other direction. Since the scattered beam has the same phase as the incident beam at the point of scattering, it is given by

$$A = \sum_n B' \exp[i\mathbf{k}_0.\mathbf{r}_n] \exp[i\mathbf{k}.(\mathbf{r} - \mathbf{r}_n)],$$

$$|\mathbf{k}| = |\mathbf{k}_0| = \frac{2\pi}{\lambda}$$

$$= B' \exp[i\mathbf{k}\,.\,\mathbf{r}] \sum_{u,\,v,\,w} \exp[i\mathbf{q}\,.\,(u\mathbf{a} + v\mathbf{b} + w\mathbf{c})],$$

$$\mathbf{q} = \mathbf{k}_0 - \mathbf{k} \tag{8.13}$$

The summation leads to

$$|A| = \left| B' \left(\frac{\sin[\frac{1}{2}\mathbf{a}\,.\,\mathbf{q}(n_1+1)]}{\sin(\frac{1}{2}\mathbf{a}\,.\,\mathbf{q})} \right) \left(\frac{\sin[\frac{1}{2}\mathbf{b}\,.\,\mathbf{q}(n_2+1)]}{\sin(\frac{1}{2}\mathbf{b}\,.\,\mathbf{q})} \right) \times \left(\frac{\sin[\frac{1}{2}\mathbf{c}\,.\,\mathbf{q}(n_3+1)]}{\sin(\frac{1}{2}\mathbf{c}\,.\,\mathbf{q})} \right) \right| \tag{8.14}$$

This amplitude is large, proportional to $B'(n_1 + 1)(n_2 + 1)(n_3 + 1)$, if

$$\begin{aligned} \frac{1}{2}\mathbf{a}\,.\,\mathbf{q} &= l_1\pi \\ \frac{1}{2}\mathbf{b}\,.\,\mathbf{q} &= l_2\pi \\ \frac{1}{2}\mathbf{c}\,.\,\mathbf{q} &= l_3\pi \end{aligned} \tag{8.15}$$

where l_1, l_2 and l_3 are integers not all simultaneously equal to zero.

The above conditions for maxima are expressed conveniently in terms of what is known as the *reciprocal lattice*. The reciprocal lattice is defined by a unit cell with basic vectors $2\pi\dfrac{\mathbf{b}\times\mathbf{c}}{\mathbf{a}\,.\,\mathbf{b}\times\mathbf{c}}, 2\pi\dfrac{\mathbf{c}\times\mathbf{a}}{\mathbf{b}\,.\,\mathbf{c}\times\mathbf{a}}, 2\pi\dfrac{\mathbf{a}\times\mathbf{b}}{\mathbf{c}\,.\,\mathbf{a}\times\mathbf{b}}$. Then, Eq. (8.15) implies that a maximum is observed if

$$\begin{aligned} \mathbf{q} &= \mathbf{k}_0 - \mathbf{k} \\ &= l_1\left(2\pi\frac{\mathbf{b}\times\mathbf{c}}{\mathbf{a}\,.\,\mathbf{b}\times\mathbf{c}}\right) + l_2\left(2\pi\frac{\mathbf{c}\times\mathbf{a}}{\mathbf{b}\,.\,\mathbf{c}\times\mathbf{a}}\right) + l_3\left(2\pi\frac{\mathbf{a}\times\mathbf{b}}{\mathbf{c}\,.\,\mathbf{a}\times\mathbf{b}}\right) \end{aligned}$$

$$|\mathbf{k}| = |\mathbf{k}_0| = \frac{2\pi}{\lambda} \tag{8.16}$$

where l_1, l_2 and l_3 are integers but $|\mathbf{k}_0 - \mathbf{k}| \neq 0$. Thus, the condition for a diffraction maximum is that the associated $\mathbf{q}$ vector is a reciprocal lattice vector.

It is seen that $\mathbf{q}$ is perpendicular to $\mathbf{a}/l_1 - \mathbf{b}/l_2$ and $\mathbf{b}/l_2 - \mathbf{c}/l_3$. Hence it is perpendicular to a plane in the lattice space which has intercepts $\mathbf{a}/l_1$, $\mathbf{b}/l_2$, $\mathbf{c}/l_3$ along the three axes. This plane has Miller indices $(l_1/n, l_2/n, l_3/n)$ where n is the highest common factor of l_1, l_2, l_3. For the situation in which **a**, **b**, **c** are orthogonal the distance between these planes is (see Example 2 of Sec. 8.8)

$$d = \left(\frac{l_1^2}{n^2 a^2} + \frac{l_2^2}{n^2 b^2} + \frac{l_3^2}{n^2 c^2} \right)^{-1/2} \tag{8.17}$$

Also the magnitude of $|\mathbf{q}|$ is

$$|\mathbf{q}| = 2\,|\mathbf{k}| \sin\theta$$

$$= 2\pi \left(\frac{l_1^2}{a^2} + \frac{l_2^2}{b^2} + \frac{l_3^2}{c_2} \right)^{1/2} \tag{8.18}$$

where 2θ is the angle between $\mathbf{k}_0$ and $\mathbf{k}$. Using $|\mathbf{k}| = 2\pi/\lambda$, gives the Bragg condition

$$2d \sin\theta = n\lambda \tag{8.19}$$

There are two general diffraction methods used for studying crystal structure. In the *method of Laue,* a single crystal is illuminated by x-radiation with a continuous spectrum. Since the Bragg condition or equivalently Eq. (8.16) will be satisfied for some wavelengths, diffraction maxima will appear on the photographic plate. This method is useful for determining the orientation of crystal planes. In the second method, known as the *powder method*, many small crystals bound together in a wire or rod are illuminated by an x-ray of a given frequency. Some of the small crystals will have the correct orientation to produce diffraction maxima. The diffraction angles and the intensities of the maxima provide information about the cell structure and dimensions. It may be noted that structures can also be deduced from the diffraction of neutron and electron beams (see Sec. 2.4). The considerations of x-ray diffraction can be extended to these cases if for wavelength we use the corresponding de Broglie wavelength of the particle.

8.3 BAND THEORY OF SOLIDS

While the free-electron theory of metals is able to explain several electromagnetic and thermal properties of metals, its validity is restricted. It is not able to expeain the fact that the Hall coefficient (which essentially gives the sign of the charge carriers, see Example 3 in Sec. 8.8) of alkaline earth metals (Be, Mg, Ca, Sr, etc. which are divalent metals) is positive. Nor can it explain the electomagnetic properties of insulators and semiconductors, *e.g.* increase in conductivity with temperature of semiconductors, insulators becoming good conductors in the presence of electromagnetic radiation. The description of these properties requires a more detailed knowledge of the energy levels of the electrons in solids. In particular, the interaction of the electrons with the ions, which gives rise to *energy bands,* should be included. Energy bands are central to the understanding of the properties of solids. They can be discussed in terms of (*i*) tight-binding approximation, (*ii*) nearly free electron approximation. These are applicable in different situations but lead to qualitatively similar results. Brief descriptions of these approximations are given here.

Tight Binding Approximation

In this approximation, the crystal is formed by bringing together N atoms. When the atoms are separated by a large distance, their energy levels are the same *i.e.* degenerate. As they come close together, these levels are perturbed by interactions between the atoms and the degeneracy is removed. Thus, there will

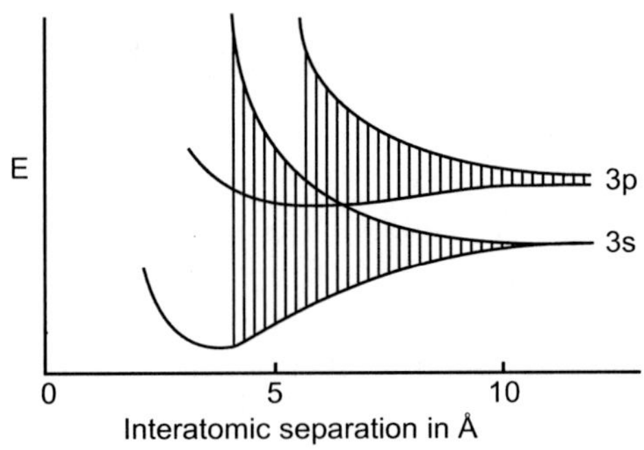

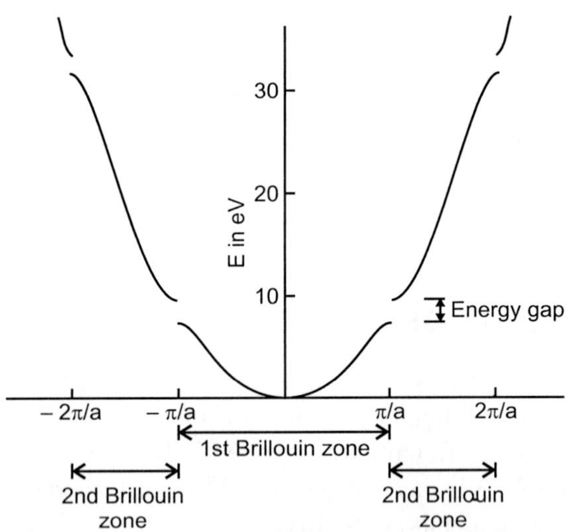

Fig. 8.6 (*a*) Energy bands in the tight binding approximation, (*b*) Brillouin zones and energy bands in the nearly free electron approximation. $a = 2\text{Å}$, $V_1 = 1$ eV.

be $2N$ levels for an s state and $6N$ levels for a p state (two spin states for each electron). Since **N** is very large, one gets bands of closely-spaced energy levels, called *energy bands* which in general are separated by some energy gaps. The spread for the inner-lying levels will be small since they are not greatly influenced by the presence of other atoms, and it will be larger for the outer levels [sec Fig. 8.6 (*a*)]. Since the spread is determined by the structure of each atom and the interatomic distance, the density of levels in a band will increase with the number of atoms (keeping the interatomic distance constant). It may happen that some outer-lying bands overlap and this will have a profound influence on the properties of the crystal.

The tight-binding approximation is reliable mainly for narrow bands of low-lying energy levels, for which the effect of the interatomic interaction is small.

Nearly Free Electron Approximation

Alternatively, the electrons may be regarded as moving in a periodic potential which describes their interaction with the ions. Here, the interaction between the electrons is neglected. For simplicity, the influence of such an interaction is considered in the one-dimensional lattice.

A one-dimensional periodic potential with period a can be written as

$$V(x) = \frac{1}{2}\sum_{m} V_m \exp(2\pi\, mix/a),\ m = 0, \pm 1, \ldots \tag{8.20}$$

with $V_m = V_{-m}$. The effect of only the $|m| = 1$ term is considered, with

$$V_1(x) = V_1 \cos(2\pi\, x/a) \tag{8.21}$$

The wave function will satisfy the equation

$$-\frac{\hbar^2}{2m}\frac{\partial^2}{\partial x^2}\psi(x) + V_1\cos(2\pi\, x/a)\,\psi(x) = E\psi(x) \tag{8.22}$$

The general form of the solutions of such an equation is given by the *Bloch theorem* which states that $\psi(x)$ can be written in the form

$$\psi(x) = \exp(ikx)u_k(x) \tag{8.23}$$

where $u_k(x)$ is periodic with period a. If the number of lattice points is N, a periodic boundary condition is imposed on ψ that $\psi_k(Na) = \psi_k(0)$. This is equivalent to closing a linear chain and implies that the allowed values of k are

$$k = \frac{2\pi n}{Na},\ n = 0, \pm 1, \ldots \tag{8.24}$$

These allowed values of k are conveniently divided as follows, into what are known as the *Brillouin zones*:

$$k = \frac{2\pi n}{Na},\ n = 0, \pm 1, \pm 2, \ldots N/2, \text{ 1st Brillouin zone,}$$

$$n = -N/2, \pm (1 + N/2), \pm (2 + N/2), ..., N,$$

2nd Brillouin zone etc. (8.25)

where for simplicity of notation it has been assumed that N is an even number. It is important to note that each Brillouin zone has $2N$ allowed states (including a factor of 2 for the electron spin).

It is assumed that in the presence of the interaction, the wave function is a superposition of mainly two plane waves,

$$\psi(x) = A\frac{e^{ikx}}{L^{1/2}} + B\frac{e^{ik'x}}{L^{1/2}} \tag{8.26}$$

where for the sake of definiteness we take $k' \leq k$. We substitute this expression in Eq. (8.22), multiply the equation by e^{-ikx} or $e^{-ik'x}$ and integrate to obtain

$$\left.\begin{aligned} \frac{\hbar^2 k^2}{2m} A + \frac{1}{2} V_1 B = EA \\ \frac{\hbar^2 k'^2}{2m} B + \frac{1}{2} V_1 A = EB \end{aligned}\right| k - k' = \frac{2\pi}{a} \tag{8.27}$$

In these equations, the interaction connects only the states which satisfy the condition

$$k - k' = \frac{2\pi}{a}.$$

This is related to the fact that for scattering by a one-dimensional lattice, a Bragg maximum is obtained for precisely the same condition [see Eq. (8.15)]. The interaction which gives rise to the scatting, is also responsible for mixing the two plane-wave states in Eq. (8.26). Solving the two homogeneous equations gives

$$E = \frac{\hbar^2}{4m}(k^2 + k'^2) \pm \left[\frac{\hbar^4 (k^2 - k'^2)^2}{16\, m^2} + \frac{1}{4} V_1^2\right]^{1/2} \tag{8.28}$$

$$\frac{B}{A} = \frac{2}{V_1}\left\{\frac{\hbar^2 (k'^2 - k^2)}{4m} \pm \left[\frac{\hbar^2 (k^2 - k'^2)^2}{16\, m^2} + \frac{1}{4} V_1^2\right]^{1/2}\right\}$$

where $k' = k - \dfrac{2\pi}{a}$. Clearly, the changes introduced, by the perturbation V_1 are significant mainly for $|k| \approx |k'|$ and hence for $k \approx \pi/a$. For $k < \pi/a$, the negative sign corresponds to $|A| \geq |B|$, which is therefore associated with the $0 \leq k \leq \pi/a$ branch of the solutions, while for $k > \pi/a$, the positive sign corresponds to

$|A| \approx |B|$, which is associated with the $\pi/a < k \leq 2\pi/a$ branch of the solutions. The most important feature of these branches is that there is an energy gap of V_1 at $k = \pi/a$ between the two Brillouin zones described by these branches, and there are $2N$ allowed states (including the negative k states) in each zone. This is qualitatively similar to the energy bands. It also follows from Eq. (8.28) that since $k - k' = 2\pi/a$, $\partial E/\partial k = 0$ at $k = \pi/a$. The energy bands are illustrated in Fig. 8.6(b). The gap at $k = 2\pi/a$ comes from the V_2 term etc.

In the case of the three dimensional problem the condition in Eq. (8.27) is modified to read $\mathbf{k} - \mathbf{k}' = \mathbf{q}$ where q is a reciprocal lattice vector defined in Eq. (8.16). The corresponding Brillouin zones are obtained form the requirement that $|\mathbf{k}| = |\mathbf{k}'|$ at the boundaries. This leads to the condition $2\mathbf{k} \cdot \mathbf{q} - \mathbf{q}^2 = 0$ which defines the boundaries as some plane perpendicular to the reciprocal lattice vectors $\mathbf{q}$. The perturbations of energies at the edges of the Brillouin zones lead to distorted equal-energy surfaces in the k-space. In particular, the electrons in a crystal occupy the lowest energy levels at $T = 0$ K, subject to the Pauli principle. The surface of the region of all occupied states in the k-space is called the *Fermi surface.* Since the energy distortions are prominent mainly near the surfaces of the Brillouin zones, the nearness of the Fermi surface to the surfaces of the Brillouin zones, and its shape are of importance for the understanding of the properties of electrons in crystals in general and metals in particular.

Another property of the energy bands worth noting is the density of states. The free particle energy density is proportional to $E^{1/2}$ [see Eq. (7.29)]. However, the periodic potential distorts the energy levels in such a way that there are no energy levels in the energy gaps and the density of states goes to zero at the bottom and the top of the energy band.

For obtaining information about the energy bands and the density of states, x-rays are used to knock out electrons in the energy bands. An analysis of the x-rays emitted when electrons for higher energy bands undergo transitions to these vacant levels, provides information about the energy bands and the density of states.

Effective Mass

A useful idea in the band theory of solids is that of the *effective mass* of an electron in a solid. One is led to this idea in an effort to simulate an electron in a periodic potential by a free electron but with an effective mass.

The energy of a free electron is given by

$$E = \frac{\hbar^2 k^2}{2m} \tag{8.29}$$

Therefore its mass may be defined as

$$m = \frac{\hbar^2}{(\partial^2 E / \partial k^2)} \tag{8.30}$$

This definition may be extended to apply to a particle in a periodic potential so that the effective mass of an electron in a one-dimensional crystal is

$$m^* = \frac{\hbar^2}{(\partial^2 E / \partial k^2)} \tag{8.31}$$

Here, however, m^* is a function of k. As can be seen from Fig. 8.6(b), m^* is positive near the bottom of each zone, negative near the top of each zone and is infinite at the point of inflection. The large effective mass can be interpreted as being due to the strong binding force between the electron and the lattice for some k values, which makes it difficult to move the electron. The negative mass may be interpreted in terms of the Bragg reflection when k is close to π/a, $2\pi/a$, etc. on account of which a force in one direction, because of reflection, leads to a gain of momentum in the opposite direction.

A detailed analysis shows that when an external electric field **E** is applied, the acceleration of the electron is given by

$$a = -e|\mathbf{E}|\left(\frac{1}{\hbar^2}\frac{\partial^2 E}{\partial k^2}\right) \tag{8.32}$$

which again simulates a free particle motion with the effective mass m^* given in Eq. (8.31). For a three dimensional crystal, the anisotropy is taken into account in the relation

$$\sum_j m^*_{ij}\, a_j = -e\,E_i \tag{8.33}$$

$$m^*_{ij} = \frac{\hbar^2}{(\partial^2 E / \partial k_i \partial k_j)}$$

The concept of effective mass provides a satisfactory description of the charge carriers in crystals. In normal circumstances, the conduction of current is by the electrons, particularly in the case of crystals in which an energy band is only partially filled (*e.g.*, alkali metals for which the band is only half-filled). On the other hand, consider a band which is nearly full except for a few vacancies near the top of the band. This situation of a full band with the vacancies in the negative charge, negative mass states may be regarded as corresponding to the presence of positive charge, positive mass particles. These *hole* states with positive charge, also act as charge carriers. In elements like Be, Zn, Cd, etc. It is the hole states which are the dominant charge carriers and hence they have positive Hall coefficients (see Example 3 in Sec. 8.8).

Metals, Insulators and Semiconductors

The existence of energy bands, *i.e.*, the allowed bands consisting of allowed energy levels and forbidden bands which are the gaps between the allowed bands, provides a simple explanation for the general properties of metals, insulators and semiconductors.

Consider the order in which the energy levels of the energy bands are filled. To start with, the electrons in the inner shells fill the corresponding narrow energy bands, and are not influential in determining the general properties of the crystal. The relevant electrons are the valence electrons and they occupy what is known as the *valence band.* At 0 K, the valence electrons occupy the lower levels of the valence band. It is the position of the higher levels which determines the conductivity properties of solids.

There are three important cases of the available higher energy levels, which are shown in Fig. 8.7. In the first case the electrons fill only the lower half of the valence band at 0 K. This happens for example, in the case of sodium for which there are N electrons and $2N$ levels in the $3s$ band. At finite temperature some of these electrons are excited to higher energy levels. In these cases of partially filled valence bands, an external electric field will transfer some of these electrons to the nearby higher energy vacant states and in addition provide additional velocity in the direction of the field. Such crystals are good conductors of electricity and are *metals*. It is worth noting that partial filling of the valence band, which for metals is also the conduction band, is observed if there is an odd number of electrons in the valence shell, *e.g.*, sodium, or if the energy bands overlap. Overlapping of bands is found, for example, in the case of Mg, Zn, etc. which are good conductors.

In the second case [Fig. 8.7(b)], the valence electrons completely fill the valence band which is separated by a large energy gap ΔE from the conduction band. For example, the covalent bonding in diamond splits the $2s$ and $2p$ levels into two bands (each of which is a mixture of $2s$ and $2p$ states) which are separated by an energy gap ΔE of about 6 eV, with the valence electrons filling the lower band. When an electric field is applied to such a crystal, there is no significant change in the states of the valence electrons since a transition to an available level requires an energy which is at least equal to the energy gap, which is about 6 eV in the case of diamond. Such solids are *insulators*. The description in terms of energy bands implies that if a radiation of high enough frequency is incident on an insulator, the electrons in the valence band may absorb the radiation and undergo transition to the conduction band. These excited electrons can easily change their velocity since many states are available to them, and act as efficient carriers of current. In this situation, an insulator can become a good conductor and the effect is known as *photoconductivity*. It may be observed that even at finite temperatures only a very small number of electrons are in the conduction band (energy gap ΔE is very large compared to kT), and an insulator remains a poor conductor of electric current.

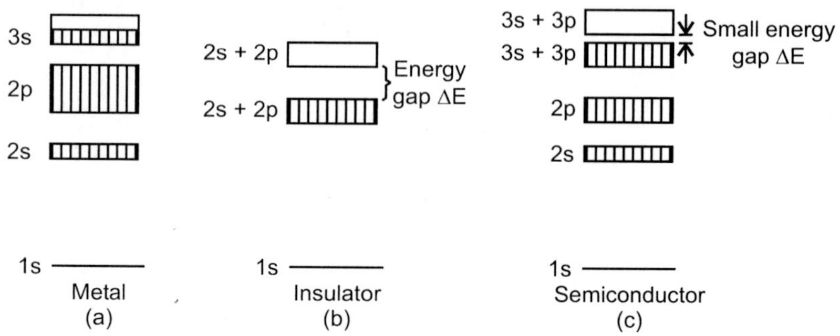

Fig. 8.7 Schematic illustration of the energy bands for (*a*) metals, (*b*) insulators with an energy gap, and (*c*) semiconductors with a small gap.

The third case [Fig. 8.7(*c*)] is qualitatively similar to that of insulators except that the energy gap between the conduction band and the valence band is much smaller, 1.1 eV for is and 0.7 eV for Ge. At 0 K, all the electrons are in the valence band and the conduction band is empty, and the solid behaves like an insulator. However, at room temperatures, an appreciable number of electrons are excited to the conduction band ($kT \approx 0.026$ eV compared to the energy gap which is about 1 eV). These electrons can carry charge. Simultaneously, the electrons in the valence band can undergo transitions to the vacant states left behind by the transitions to the conduction band. Effectively, the holes (or the vacancies) serve as carriers of positive charge. The conductivity of these solids lies between those of metals and insulators, and they are known as *semiconductors*.

An important characteristic which distinguishes metals from semiconductors is the temperature dependence of their conductivities. As the temperature is raised, more and more phonons are excited, which can scatter electrons and hence reduce their mobility. Therefore the conductivity of metals generally decreases as temperature increases. However, in the case of semiconductors, the decrease in the mobility is more than compensated by the increase in the number of carriers, electrons as well as holes. As a result, the conductivity of semiconductors increases (at moderate temperatures) as temperature increases.

8.4 SEMICONDUCTORS

As mentioned before, semiconductors are crystals whose valence band is completely filled but which have a small energy gap ($\Delta E \sim 1$ eV) between the conduction band and the valence band. Their conductivity is in-between that of metals (~ $10^8\ \Omega^{-1}\ m^{-1}$) and that of insulators (~ $10^{-11}\ \Omega^{-1}\ m^{-1}$), and increases with temperature. Because of the narrowness of the energy gap and the proximity of the energy levels of the impurity to the valence and conduction bands, semiconductors have rather striking electronic properties which make them very useful in the development of sophisticated electronic equipment. Here the positions and populations of the semiconductor energy levels which determine their electronic properties are discussed.

It is useful to classify semiconductors into two categories. The class of semiconductors which are pure, such as silicon, germanium (which are group IV elements), GaAs, PbS, etc., are known as *intrinsic* semiconductors. In the second class of semiconductors known as *extrinsic* (or *impurity*) semiconductors, the properties of the semiconductors are modified by the introduction of carefully controlled amounts of impurities.

To see how the impurities affect the properties of semiconductors, consider the specific examples of silicon and germanium. These are group IV elements which have diamond structure in which each atom has a covalent bond with each of the four nearest neighbours at the corners of a tetrahedron. If a small amount of a group V element, such as phosphorus, arsenic or antimony, is introduced during the formation of the crystal, the group V atom will take the place of one of the group IV atoms and form four covalent bonds with the nearest neighbours. However, since it has five electrons in the valence shell, the fifth electron is only weakly bound to the atom. They *i.e.*, the 'fifth' electrons occupy localized energy levels which are just below the conduction band [see Fig. 8.8(*a*)]. The electrons in these levels are easily excited to the states in the conduction band and serve as current carriers. Since group V atoms donate electrons for conduction they are known as *donors*, and the new energy levels just below the conduction band as *donor levels*. The charge carriers in this case being negatively charged, the corresponding extrinsic semiconductors are known as n-type semiconductors. Alternatively, if a small amount of a group III element such as boron, aluminium or indium is introduced, the group III element will form only three covalent bonds with the nearest neighbours. Thus, there is a vacancy or a hole associated with each of these atoms. Since an electron in these states would be fairly tightly bound, the vacant states provide localized energy levels which lie just above the valence band [see Fig. 8.8(*b*)]. The neighbouring electrons can easily be transferred to these levels, as a result of which holes are created in the valence band. Since the states near the top of the band have negative mass (see Sec. 8.3), these holes behave as positive mass, positive charge carriers. The group III impurity atoms are known as *acceptors* and the new energy levels just above the valence band as *acceptor levels*. The charge carriers in this case being positively charged, the corresponding extrinsic semiconductors are known as p-type semiconductors.

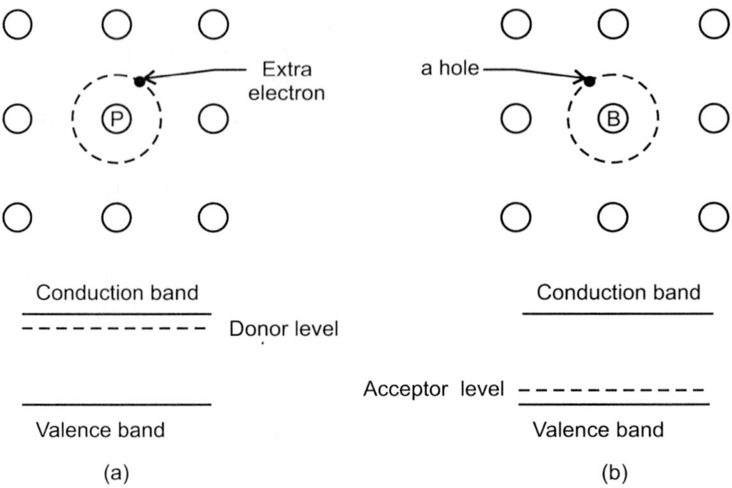

Fig. 8.8 Extrinsic semiconductors (*a*) *n*-type with *P* as the donor atom, (*b*) *p*-type with *B* as the acceptor atom.

The electronic properties of the semiconductors are influenced by the positions of the Fermi energy and the concentrations of the charge carriers.

ε_f for Intrinsic Semiconductors

In an intrinsic semiconductor, every electron transferred to the conduction band leaves behind a hole. Therefore, the total number of electrons in the conduction band is equal to the total number of holes in the valence band.

For calculating the total number of electrons in the conduction band, the number of states in the conduction band, per unit volume is taken to be [see Eq. (7.86)]

$$dN_c = \frac{4\pi(2m_e^*)^{3/2}}{h^3}(\varepsilon - \varepsilon_c)^{1/2}\, d\varepsilon \tag{8.34}$$

Here m_e^* is the effective mass of the electrons in the conduction band and ε_c is the lowest energy in the conduction band. Therefore, the number of electrons in the conduction band, per unit volume, is

$$n_c = \frac{4\pi(2m_e^*)^{3/2}}{h^3}\int_{\varepsilon_c}^{\infty}\frac{(\varepsilon - \varepsilon_c)^{1/2}\, d\varepsilon}{e^{(\varepsilon - \varepsilon_f)/kT} + 1} \tag{8.35}$$

Assuming that $(\varepsilon_c - \varepsilon_f) \gg kT$, the unit term in the denominator can be neglected and the integral evaluated [substitute $(\varepsilon - \varepsilon_c) = x^2$]. This then gives

$$n_c = \frac{2(2\pi m_e^* kT)^{3/2}}{h^3}\exp[(\varepsilon_f - \varepsilon_c)/kT] \tag{8.36}$$

For obtaining the number of holes in the valence band it is noted that the probability that a state is not occupied by an electron, is

$$P_h = 1 - \frac{1}{\exp[(\varepsilon - \varepsilon_f)/kT] + 1}$$

$$= \frac{1}{\exp[(\varepsilon_f - \varepsilon)/kT] + 1} \tag{8.37}$$

In analogy with Eq. (8.34), the number of hole states in the valence band per unit volume, is taken to be

$$dN_v = \frac{4\pi(2m_h^*)^{3/2}}{h^3}(\varepsilon_v - \varepsilon)^{1/2}\, d\varepsilon \tag{8.38}$$

where m_h^* is the effective mass of the holes in the valence band and ε_v is the highest energy in the valence band. The number of holes in the valence band, per unit volume, is

$$n_h = \frac{4\pi(2m_h^*)^{3/2}}{h^3}\int_{-\infty}^{\varepsilon_v} \frac{(\varepsilon_v - \varepsilon)^{1/2}\, d\varepsilon}{\exp[(\varepsilon_f - \varepsilon)/kT] + 1} \tag{8.39}$$

Assuming that $(\varepsilon_f - \varepsilon_v) \gg kT$, the unit term in the denominator can be neglected, and this leads to

$$n_h = \frac{2(2\pi m_h^* kT)^{3/2}}{h^3}\exp[(\varepsilon_v - \varepsilon_f)/kT] \tag{8.40}$$

It is interesting to note that

$$n_e n_h = \frac{4(4\pi^2 m_e^* m_h^*)^{3/2}}{h^6}(kT)^3 \exp[(\varepsilon_v - \varepsilon_c)/kT] \tag{8.41}$$

which is independent of ε_f but depends on the energy gap $(\varepsilon_c - \varepsilon_v)$.

For an intrinsic semiconductor, $n_e = n_h$ so that

$$(m_e^*)^{3/2} \exp[(\varepsilon_f - \varepsilon_c)/kT] = (m_h^*)^{3/2} \exp[(\varepsilon_v - \varepsilon_f)/kT] \tag{8.42}$$

or
$$\varepsilon_f = \frac{1}{2}(\varepsilon_c + \varepsilon_v) + \frac{3}{4} kT \ln(m_h^*/m_e^*) \tag{8.43}$$

At $T = 0$, the Fermi energy lies halfway between the valence and conduction bands. For finite temperatures, m_h^* is usually greater than m_e^*. However, since $\varepsilon_c - \varepsilon_v \approx 1$ eV and $kT \approx 0.026$ eV at room temperature, ε_f increases but slowly with temperature. The expression in Eq. (8.43) justifies the assumption that $(\varepsilon_c - \varepsilon_f) \gg kT$ and $(\varepsilon_f - \varepsilon_v) \gg kT$ at ordinary temperatures.

The knowledge of ε_f allows us to calculate the number of carrier electrons and holes, and the conductivity of intrinsic semiconductors. Substituting for ε_f in Eqs. (8.36) and (8.40), gives

$$n_e = n_h = \frac{2(2\pi kT)^{3/2}(m_e^* m_h^*)^{3/4}}{h^3} \exp[(\varepsilon_v - \varepsilon_c)/2kT] \tag{8.44}$$

For deducing conductivity, it is noted that the current density j is given by

$$j = e(n_e \bar{v}_e + n_h \bar{v}_h) \tag{8.45}$$

where $\bar{v}_e$ and $\bar{v}_h$ are the magnitudes of the average drift velocities of the electrons and holes, respectively. The conductivity of the semiconductor is therefore given by

$$\sigma = e(n_e\mu_e + n_h\mu_h) \tag{8.46}$$

where the mobilities μ_e and μ_h are defined by

$$\mu_e = \bar{v}_e/E, \mu_h = \bar{v}_h/E \tag{8.47}$$

E being the magnitude of the electric field. Substituting the expressions for the carrier densities,

$$\sigma = e(\mu_e + \mu_h) \frac{2(2\pi kT)^{3/2}(m_e^* m_h^*)^{3/4}}{h^3} \exp[(\varepsilon_v - \varepsilon_c)/2kT] \tag{8.48}$$

which leads to

$$\ln \sigma = -\left(\frac{\varepsilon_c - \varepsilon_v}{2k}\right)\frac{1}{T} + \frac{3}{2}\ln T + c \tag{8.49}$$

where c is a constant. Here, it has been assumed that the mobilities are independent of T. Actually, they do vary as a function of temperature. However, the main variation is due to the $1/T$ term and a plot of ln σ as a function of $1/T$ gives an approximate straight line. The slope of the straight line gives an estimation of the energy gap $(\varepsilon_c - \varepsilon_v)$ of the semiconductor.

It may be noted $m_e^* \approx 0.25\ m$, $m_h^* \approx 0.3\ m$ for Si and $m_e^* \approx m_h^* \approx 0.1\ m$ for Ge, m being the electron mass. These values imply that at a temperature of 300 K, the carrier concentrations are about 2.3×10^{15} m^{-3} for Si and about 10^{18} m^{-3} for Ge. The intrinsic conductivity at this temperature has the values of about 10^{-4} $(\Omega\ \text{m})^{-1}$ for Si and about 0.1 $(\Omega\ \text{m})^{-1}$ for Ge.

ε_f for Extrinsic Semiconductors

In an extrinsic semiconductor, the donor or acceptor levels play an important role in the determination of the Fermi energy and the conductivity of the semiconductor.

Consider an n-type semiconductor, with N_d number of donors per unit volume. Then the number of vacancies per unit volume, in the donor levels of energy ε_d is

$$n_d = \left[1 - \frac{1}{\exp[(\varepsilon_d - \varepsilon_f)/kT] + 1}\right] N_d \tag{8.50}$$

From the condition that the number of electrons in the conduction band is equal to the total number of vacancies in the donor levels and the valence band, one gets

$$c_0(m_e{*}T)^{3/2} \exp[(\varepsilon_f - \varepsilon_c)/kT] = c_0(m_h{*}T)^{3/2} \exp[(\varepsilon_v - \varepsilon_f)/kT)$$

$$+ \frac{N_d}{\exp[(\varepsilon_f - \varepsilon_d)/kT] + 1} \tag{8.51}$$

where $c_0 = 2(2\pi k)^{3/2}/h^3$. Now $\varepsilon_c - \varepsilon_d \approx 0.01$ eV for Ge and about 0.045 eV for Si, and for the cases of practical interest N_d is of the order of 10^{22} m^{-3}. So, at ordinary temperatures, most of the electrons in the conduction band are from the donor levels. For $T \to 0$, the unit term in the denominator can be neglected giving

$$c_0\,(m_e{*}T)^{3/2} \exp[(\varepsilon_f - \varepsilon_c)/kT] \approx N_d \exp[(\varepsilon_d - \varepsilon_f)/kT] \tag{8.52}$$

which leads to

$$\varepsilon_f = \frac{1}{2}(\varepsilon_d + \varepsilon_v) + \frac{1}{2} kT \ln\left[\frac{N_d}{c_0(m_e{*}T)^{3/2}}\right] \tag{8.53}$$

where $c_0 = 2(2\pi k)^{3/2}/h^3$. At $T = 0$, the Fermi level lies halfway between ε_c and ε_d. At room temperature, ε_f is below ε_d for the cases of interest and most of the donor atoms are ionized. In this region the number of vacancies, *i.e.*, rhs of Eq. (8.51) can be taken to be N_d to get

$$\varepsilon_f = \varepsilon_c + kT \ln\left[\frac{N_d}{c_0(m_e{*}T)^{3/2}}\right], \quad (\varepsilon_d - \varepsilon_f) \gg kT \tag{8.54}$$

For example, in the case of Si doped with a donor impurity to the extent of 10^{22} m^{-3}, the Fermi energy at 300 K is $\varepsilon_f \approx (\varepsilon_c - 0.15)$ eV. At higher temperatures, a detailed analysis of Eq. (8.51) shows that ε_f tends to the value $\frac{1}{2}(\varepsilon_c + \varepsilon_v)$, *i.e.*, the value for the intrinsic semiconductor. The conductivity for n-type of semiconductors is mainly due to the electrons in the condition band (at not very high temperatures) and is given by

$$\sigma \approx e n_e \mu_e \tag{8.55}$$

which leads to

$$\ln \sigma \approx - \ln \{\exp [(\varepsilon_f - \varepsilon_d)/kT] + 1\} + c_1 \tag{8.56}$$

where c_1 is a constant (it is assumed that μ_e is temperature independent). Plotted as a function of $1/T$, ln σ has a negative slope for large $1/T$, *i.e.*, small $T(\varepsilon_f > \varepsilon_d)$, but flattens out for larger T once $(\varepsilon_d - e_f) \gg kT$. For very high temperatures, intrinsic conductivity begins to dominate and the expression for ln σ tends to that given in Eq. (8.49).

For a *p*-type semiconductor with N_a number of acceptors per unit volume the number of electrons in the acceptor levels, per unit volume, is

$$n_a = \frac{N_a}{\exp[(\varepsilon_a - \varepsilon_f)/kT] + 1} \tag{8.57}$$

Equating the total number of electrons in the conduction band and the acceptor levels with the number of holes in the valence band, and proceeding as before, gives

$$\varepsilon_f = \frac{1}{2}(\varepsilon_a + \varepsilon_v) - \frac{1}{2} kT \ln\left[\frac{N_a}{c_0 (m_h^* T)^{3/2}}\right] \text{for } T \to 0 \tag{8.58}$$

so that at T = 0, the Fermi level is half way between ε_a and ε_v. At room temperature, essentially all the acceptor levels are occupied and so

$$\varepsilon_f = \varepsilon_v - kT \ln\left[\frac{N_a}{c_0 (m_h^* T)^{3/2}}\right], (\varepsilon_f - \varepsilon_a) \gg kT \tag{8.59}$$

For Si doped with an acceptor impurity to an extent of 10^{22} m^{-3}, the Fermi energy at 300 K is given by $\varepsilon_f = (\varepsilon_v + 0.15)$ eV. At higher temperatures ε_f tends to the value of $\frac{1}{2}(\varepsilon_c + \varepsilon_v)$. The conductivity of *p*-type semiconductors is primarily due to the holes in the valence band, and therefore one has as in Eq. (8.56),

$$\ln \sigma = - \ln \{\exp [(\varepsilon_a - \varepsilon_f)/kT] + 1\} + c_2 \tag{8.60}$$

where c_2 is a constant. Plotted as a function of $1/T$, the behaviour of ln σ is similar to that for *n*-type semiconductors.

ε_f for *pn* Junctions

Junctions between *p*-type and *n*-type semiconductors play an important role in the development of semiconductor devices. A *pn* junction is a junction at the microscopic level between a *p*-type and an *n*-type semiconductor. Such junctions are developed by the diffusion of impurity atoms.

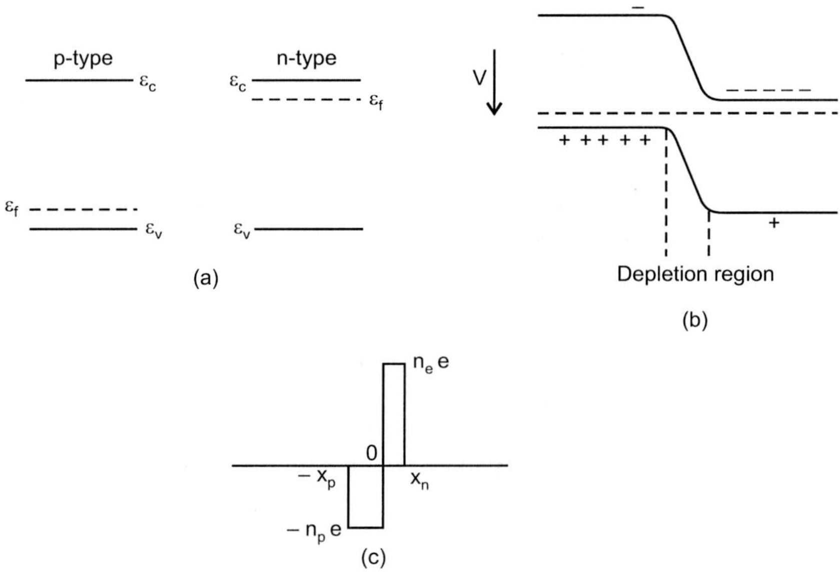

Fig. 8.9 The pn junction, (*a*) before equilibrium, (*b*) after equilibrium, and (*c*) charge density across the boundary.

The Fermi level of an *n*-type semiconductor is close to ε_c while that of a *p*-type semiconductor is close to ε_v (Fig. 8.9). Therefore, there are many more electrons in the conduction band of the *n*-type semiconductor and many more holes in the valence band of the *p*-type semiconductor. As a result, when a *pn* junction is formed, electrons diffuse from the *n*-type to the *p*-type semiconductor and occupy the vacant states there. Similarly, the holes diffuse from the *p*-type to the *n*-type semiconductor and allow the electrons to occupy their vacant states. As a result, there is a narrow depletion region at the boundary where there are no charge carriers. Instead, there is a thin layer of positive charge on the *n*-side (due to positive ions left behind) and a thin layer of negative charge on the *p*-side (due to extra electrons occupying the acceptor levels). This double layer of charges creates a potential difference across the junction which opposes the flow of electrons from the *n*-type to *p*-type and of holes from the *p*-type to *n*-type semiconductor. The flow of electrons and holes stops when the Fermi energy on the two sides has the same value (see Fig. 8.9). It must be appreciated that the shifting of the Fermi energy levels is due to the electric potential across the junction and that the relative positions of the various energy levels on the two sides, remain unchanged. The potential difference, across the boundary is equal to the difference in the Fermi levels of the separate *n*-type and *p*-type semiconductors, and is given by

$$V_0 = \frac{1}{2}(\varepsilon_c - \varepsilon_v + \varepsilon_d - \varepsilon_a) + \frac{1}{2}kT \ln\left[\frac{N_d N_a}{c_0^2 (m_e^* m_h^* T^2)^{3/2}}\right] \quad \text{for } T \to 0 \quad (8.61)$$

and $$V_0 = \varepsilon_c - \varepsilon_v + kT \ln \left[\frac{N_d N_a}{c_0^2 (m_e^* m_h^* T^2)^{3/2}} \right] \tag{8.62}$$

at room temperature.

The width of the depletion region can be estimated by the following model calculation. It is assumed that there is a width of x_p in the p-type and x_n in the n-type of semiconductor. Using Maxwell's equation $\nabla \cdot (\kappa\varepsilon_0 \mathbf{E}) = \rho$, we get

$$\kappa\varepsilon_0 E = \rho x + c \tag{8.63}$$

where κ is the relative permittivity, $\rho = en_e$ in the n-type semiconductor and $\rho = -en_h$ in the p-type semiconductor. Since E is zero at the edges of the depletion region, $c = -en_e x_n$ in the n-region and $c = -en_h x_p$ in the p-region. Therefore, the potential difference across the boundary is

$$V_0 = -\frac{e}{\kappa\varepsilon_0}\left[n_e \int_0^{x_n} (x - x_n)\, dx - n_h \int_{-x_p}^{0} (x + x_p)\, dx \right]$$

$$= \frac{e}{2\kappa\varepsilon_0}[n_e x_n^2 + n_h x_p^2] \tag{8.64}$$

The condition of overall neutrality gives

$$n_e x_n = n_h x_p \tag{8.65}$$

These two equations lead to

$$V_0 = \left(\frac{e}{2\kappa\varepsilon_0}\right)\left(\frac{n_e n_h}{n_e + n_h}\right)(x_n + x_p)^2 \tag{8.66}$$

For Si with impurity concentration at 300 K of $n_e \approx n_h \approx 10^{22}$ m^{-3}, $\kappa \approx 12$, V_0 is approximately equal to 0.8 V and $x_n \approx x_p \approx 2 \times 10^{-7}$ m.

The capacitance of the double layer can be calculated by noting that the charge Q in each layer is $e\, n_e x_n$ per unit area, so that

$$V = \left(\frac{Q^2}{2\kappa\varepsilon_0 e}\right)\left(\frac{n_e + n_h}{n_e\, n_h}\right) \tag{8.67}$$

From this, the variable capacitance per unit area, is

$$C = \frac{dQ}{dV}$$

$$= \frac{1}{2V^{1/2}}\left[\frac{2\kappa\varepsilon_0\, e\, n_e n_h}{n_e + n_h}\right] \tag{8.68}$$

The variation of C as $V^{-1/2}$ is the basis of the variable capacitance diodes (varactors) which are used in frequency locking and frequency modulation circuits.

8.5 SEMICONDUCTOR DEVICES

The special properties of semiconductors have led to a large number of important applications. Here a few illustrative examples such as diodes, transistors, solar cells and semiconductor lasers are discussed, and also the fabrication of these devices is briefly described.

Semiconductor Diodes

The *pn* junction can be used as a rectifier, a voltage stabilizer and in high frequency circuits.

Consider again the *pn* junction discussed in Sec. 8.4. Though the net current across the junction is zero, the electrons and the holes diffuse across the boundary but the flow in each direction is the same. The densities of electrons and holes in the corresponding states (similarly located with respect to ε_c or ε_y) on the two sides are related by Boltzmann statistics,

$$\frac{n_e(p)}{n_e(n)} = \exp\,(-\,eV_0/kT) \tag{8.69}$$

$$\frac{n_h(n)}{n_h(p)} = \exp\,(-\,eV_0/kT) \tag{8.70}$$

where V_0 is the potential difference across the junction. It follows from these relations that the product $n_e n_h$, of the total number of electrons and holes, has the same value on the two sides.

To be specific, consider the flow of electrons across the boundary. Since the motion of electrons from p to n is 'downhill', the rate of electron flow from p to n is

$$I_e\,(p \rightarrow n) = c_1 n_e\,(p) \tag{8.71}$$

where $n_e\,(p)$ is the total number of electrons in the conduction band on the p-side. On the other hand the electrons flowing from n to p, face an 'up-hill' potential of V_0 and only those which have an energy of eV_0 will be able to cross the boundary. The number of such electrons is proportional to $n_e\,(n)\exp\,(-\,eV_0\,kT)$, so that

$$I_e\,(n \rightarrow p) = c_2\, n_e\,(n)\exp\,(-\,eV_0/kT) \tag{8.72}$$

At equilibrium, there is no net flow, so that

$$I_e (p \to n) = I_e (n \to p) = I_0$$

and hence $$c_1 = c_2 \tag{8.73}$$

where Eq. (8.69) has been used.

If now an external potential $-V$ is applied, the potential difference across the boundary is $V_0 - V$. Since the concentrations are unchanged, the current is

$$I = c_2 n_e(n) \exp [- e (V_0 - V)/kT] - c_1 n_e(p)$$
$$= I_0 [\exp (eV/kT) - 1] \tag{8.74}$$

Thus if $(- eV)$ has the sign opposite to that of eV_0, *i.e.*, eV is positive, what is termed as *forward bias*, the current increases rapidly, whereas if $(- eV)$ has the same sign as that of eV, *i.e.*, eV is negative, known as *reverse bias*, the current is small and quickly tends to the limiting value of $(- I_0)$ (see Fig. 8.10). A similar behaviour is expected for the flow of holes. Effectively, a *pn* junction allows current to flow in only one direction, and hence can be used as a rectifier.

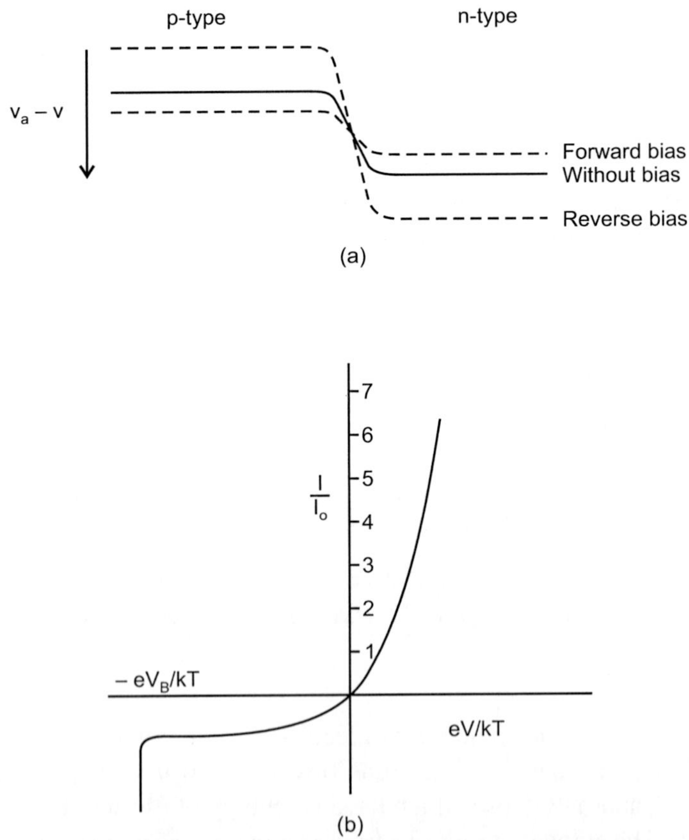

Fig. 8.10. Rectification by a diode, (*a*) potential difference across the boundary, (*b*) current as a function of *eV/kT*, *V* being the applied potential.

If there is reverse bias $|V| \gtrsim |V_B|$ (see Fig. 8.10), Eq. (8.74) for the current is no longer valid. For $|V| > |V_B|$, a very rapid increase in the current is observed (Fig. 8.10). There are two reasons for this increase: (*i*) the large field at the junction speeds up the few electrons in the *p*-region near the junction to such high velocities that they knock out some of the valence electrons into the conduction band. This process continues repeatedly and a large current is quickly built-up. (*ii*) If the potential difference across the boundary is sufficiently large, the conduction band on the *n*-side will overlap the valence band on the *p*-side (see Fig. 8.9). In this case, it was suggested by Zender that the electrons in the valence band on the *p*-side will tunnel across the boundary into the conduction band on the *n*-side. The diodes based on these two effects are known as *avalanche* diodes or *Zener* diodes. They are very useful in voltage stabilization circuits.

If the impurity concentration is very high (of the order of one part in a thousand), the Fermi energy may move into the valence band on the *p*-side and into the conduction band on the *n*-side [Fig. 8.1(*a*)]. In this case, there will be vacant levels above ε_f in the valence band on the *p*-side and electrons below ε_f in the conduction band on *n*-side. When the reverse bias is applied, many electrons will move from the valence band on the *p*-side into the conduction band on the *n*-side, giving rise to a large current [Fig. 8.11(*b*)]. For a small forward bias, electrons from the conduction band on the *n*-side can move not

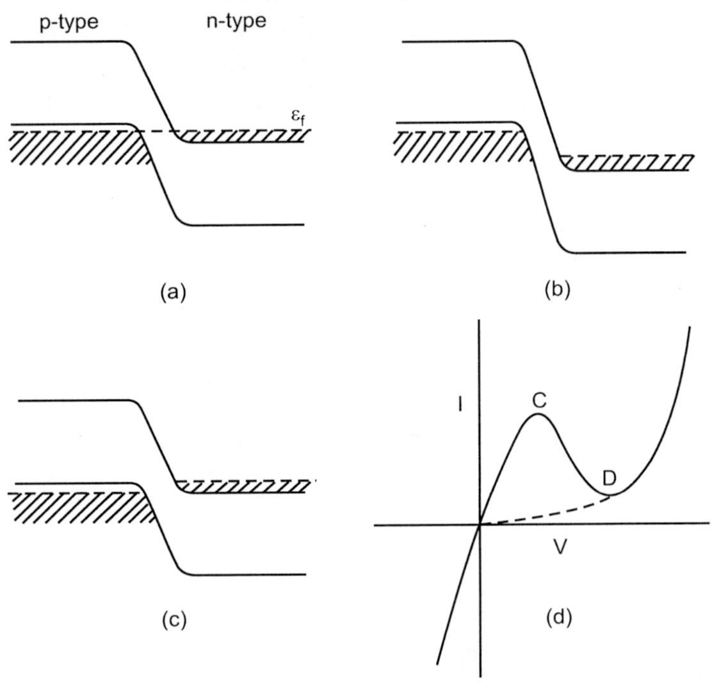

Fig. 8.11 The characteristics of the tunnel or Esaki diode; (*a*) energy bands without bias, (*b*) energy bands with reverse bias, (*c*) energy bands with forward bias, (*d*) current as a function of *V* showing negative resistance between *C* and *D*.

only into the conduction band on the p-side, but also tunnel into some of the vacant levels on the p-side. This again gives rise to a large current. On the other hand, if the forward bias is quite large, there will no longer be any overlap of the valence band on the p-side and the conduction band on the n-side. The resulting current which is due to electrons moving across the potential barrier from the conduction band on the n-side to the conduction band on the p-side, actually shows a decrease [Fig. 8.11(c)]. For still higher potentials, the current will begin to increase again, as in the case of the ordinary diode. The main characteristic of these *tunnel* or *Esaki* diodes is the negative-resistance section, which is used in high-frequency oscillator circuits in the microwave region.

Transistor

An important application of semiconductor junctions is the *transistor*. It consists of two semiconductor junctions close together, which serve as an amplifier of current or voltage.

To be specific, consider a *pnp* transistor (similarly one can have an *npn* transistor) which consists of three regions (Fig. 8.12). The first region is the *emitter*, the small, narrow, middle region is the *base*, and the third region is the *collector*. The equilibrium potential consists of a potential barrier in the n-region. If now the base is connected to a small negative potential V_b (the common emitter may be taken to be at $V_e = 0$), and the collector to a fairly large negative potential V_c, the potential across the two junctions is modified [Fig. 8.12(b)]. There is a forward bias across the emitter-base junction for the holes, and the current across the junction is (Eq. (8.74)]

$$I = I_0 \,[\exp\,(eV_b/kT) - 1] \tag{8.75}$$

Since the base is very narrow (less than 10^{-3} cm in width), most of the holes that enter the base from the emitter roll down into the collectotor, though a few of them will be annihilated by the electrons in the base. Thus, a major part of the emitter current flows through the collector while only a small fraction of it flows out from the base. In a general way, the collector current is controlled by the changes in the barrier introduced by V_b, and the changes in the base current I_b are amplified into the changes in the collector current I_c. The amplification in the current is estimated by

$$\beta = I_c/I_b = \frac{\text{hole lifetime}}{\text{base transit time}} \tag{8.76}$$

(holes annihilated in the base contribute to I_b) which in practice has a value of about 100.

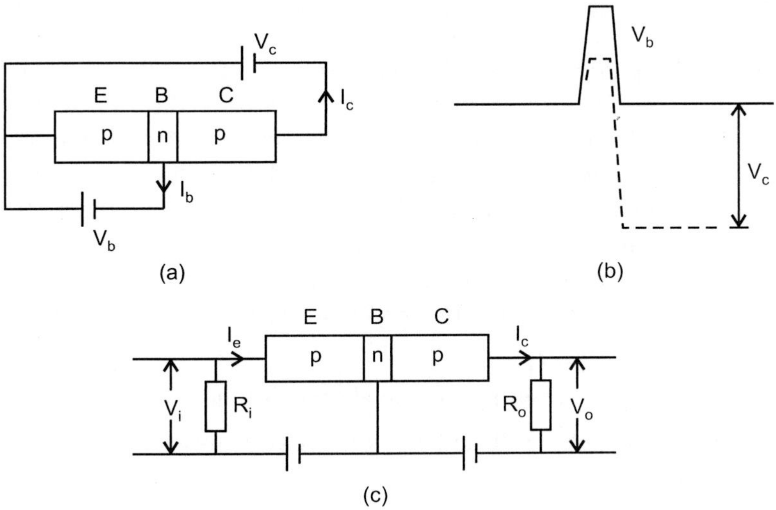

Fig. 8.12. Transistors, (*a*) common emitter circuit, (*b*) the potential distribution, (*c*) common base circuit.

In the analysis so far the motion of only holes has been considered. Regarding the electrons, there is hardly any electron flow from the collector to the base. The flow of electrons from the base to the emitter is minimized by having relatively small amount of doping of donors in the base. Thus the currents in the *pnp* transistor, are mainly due to the motion of the holes. For an *npn* transistor, the analysis is similar except that the signs of all the potentials are opposite and the current in this case is due to the flow of the electrons.

The *pnp* transistor can also be used as a voltage amplifier by having a common base [see Fig. 8.12(*c*)]. In this arrangement, the input potential across R_{in} is amplified into the output potential across R_{out}. Arguments similar to those given above imply that the emitter current I_c is approximately equal to the collector current. Therefore, the voltages across R_{in} and R_{out} are given by

$$V_{in} = I_c R_{in}$$
$$V_{out} = I_c R_{out} \tag{8.77}$$

It therefore follows that

$$\frac{V_{out}}{V_{in}} \approx \frac{R_{out}}{R_{in}} \tag{8.78}$$

Since R_{out}/R_{in} is usually quite large, voltage gains of the order of 500 are quite usual. This arrangement amplifies both voltage and power.

Photodiodes

A photodiode is a *pn* junction used to convert radiation energy into an electric current. Consider a photon of frequency ν

$$\nu \geq (\varepsilon_c - \varepsilon_v)/h \tag{8.79}$$

incident on the depletion region of a *pn* junction (see Fig. 8.9), or near it (within about the diffusion length). This photon may be absorbed by an electron in the valence band as a result of which it may move into the conduction band, thus creating an electron and a hole. The electron and the hole are separated by the electric field in the depletion region, the electron moving into the *n*-region and the hole moving into the *p*-region. The direction of the resulting current I_ν, is that of the current produced by a reverse bias. If therefore an external potential V is applied to the junction ($V > 0$ corresponds to forward bias), the current is

$$I = I_0 \left[e^{eV/kT} - 1\right] - I_\nu \tag{8.80}$$

where the first term is the current in the absence of radiation [see Eq. (8.743)].

If the circuit is open, $I = 0$, and an effective forward bias voltage V appears at the terminals, given by

$$V = \frac{kT}{e} \ln\left(1 + \frac{I_\nu}{I_0}\right) \tag{8.81}$$

This voltage is essentially due to the accumulation of the excess photo-electrons in the *n*-region and photoholes in the *p*-region (this reduces the potential difference across the junction). The expression for the photovoltage in Eq. (8.81) is valid for $V \leq \varepsilon_c - \varepsilon_v$ (for $V = \varepsilon_c - \varepsilon_v$, there is no longer a potential difference across the junction to separate the electrons and the holes). If the terminals are connected to an external load resistance R, a voltage V' somewhat less than V in Eq. (8.81), appears across the junction, and the net current is

$$I_L = I_0 \left(\exp\left(eV'/kT\right) - 1\right) - I_\nu \tag{8.82}$$

The voltage across R is

$$V_L = V' - |I_L| R_c \tag{8.83}$$

where R_c is the resistance of the solar cell. These relations together with $V_L = I_L R$, alow us to determine V', V_L and I_L for given I_ν, R_c and R. Thus a *pn* junction can be used to convert radiation energy into electrical energy. This is the principle behind the use of a *pn* junction in photometers, detectors and in solar cells.

For using a *pn* junction as a photometer, the terminals are usually short circuited *i.e.*, $V = 0$. The resulting current [Eq. (8.80)] is $- I_\nu$. Its magnitude is proportional to the intensity of the incident radiation, and hence it is used in photometers for estimating the intensity of radiation. For the use of a photodiode as a γ-ray or particle detector, it is noted that the energy of a photon in *x*-ray γ-rays is much greater than the energy gap. Such a photon creates a highly energetic electron and a hole. These produce other pairs and the process continues till their energies are comparable to the energy gap. The number of carriers indicated by the current gives a measure of the initial photon energy. It may be noted that in order to collect the carriers quickly (collection time required is about 10^{-8} s) and efficiently, the junction is subjected to a reverse bias so that the carriers at the junction are subjected to a large potential difference. The *pn* junction is also used as solid-state detector for measuring the energy of other particles such as protons and electrons, the only difference in this case being that these particles are not absorbed and come to rest after creating several electron-hole pairs.

In a silicon solar cell [Fig. 8.13(*a*)] there is a very thin, large surface of *n*-type silicon, forming a junction with a large volume of *p*-type silicon. The surface is made large so as to collect a substantial amount of radiation, and thin, about 10^{-6} m, so that the majority of the carriers generated by the photons can diffuse to the junction before recombining. The surface is coated with an anti-reflection coating to increase the efficiency which can reach a value of about 16% in silicon solar cells (efficiency is the ratio of the electrical energy output to the radiation energy input). When the surface is exposed, the holes move across the junction to the *p*-side and the photodiode becomes an energy cell with the *n*-side being at a negative potential and the *p*-side being at a positive potential.

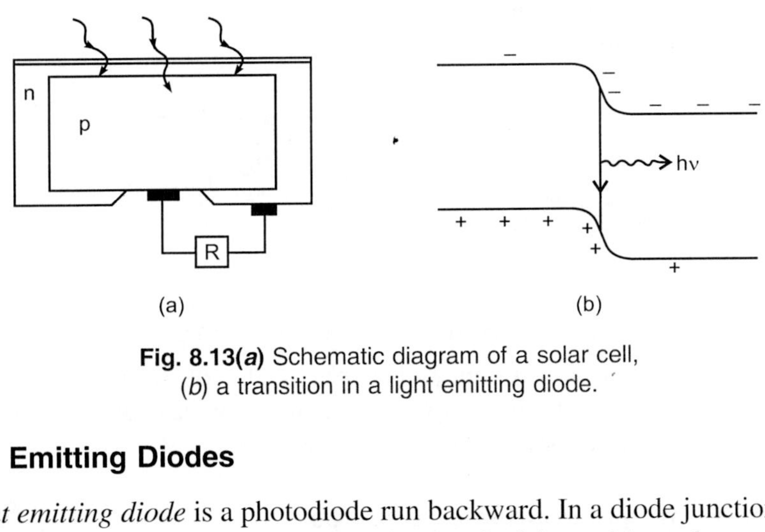

Fig. 8.13(*a*) Schematic diagram of a solar cell, (*b*) a transition in a light emitting diode.

Light Emitting Diodes

A *light emitting diode* is a photodiode run backward. In a diode junction, there is an excess of electrons in the conduction band of the *n*-side, and an excess of holes in the valence band of the *p*-side, with a potential barrier across the junction (Fig. 8.9). When a forward bias is applied, current flows across the junction. As a result, electrons from the *n*-side are injected into the *p*-side and holes are injected from the *p*-side into the *n*-side. Near the junction, the electrons in the conduction band combine with the holes in the valence band with the emission of photons of frequency

$$\nu = (\varepsilon_c - \varepsilon_v)/h \tag{8.84}$$

In practice, these radiative transitions have to compete with nonradiative recombinations due to impurities. While nonradiative recombinations dominate in Ge and Si, radiative recombinations are important in some semi-conductors such as GaAs. In GaAs the energy gap is about 1.4 eV corresponding to a frequency for $\lambda \approx 8900$ Å which is in the infrared region. The gap can be increased by alloying the material with phosphorus, *i.e.*, by using Ga $(As)_{1-x} P_x$ which can produce radiation in the optical region. This radiation can be extracted from openings close to the junction. Light emitting diodes are used in display and warning devices.

Semiconductor Diode Laser

If a *pn* junction is designed properly, a light emitting diode can produce laser action.

In a semiconductor laser, the laser medium is usually a *pn* junction of a semiconductor such as $Ga(As)_{1-x} P_x$ in which radiative recombinations dominate. The opposite surfaces of the crystal, perpendicular to the junction, are taken along the cleavage planes, and are polished so as to form an effective resonance cavity. In this cavity, the photons emitted at the junction and moving in a particular direction parallel to the junction plane, produce stimulated emission giving rise to laser action (Fig. 8.14).

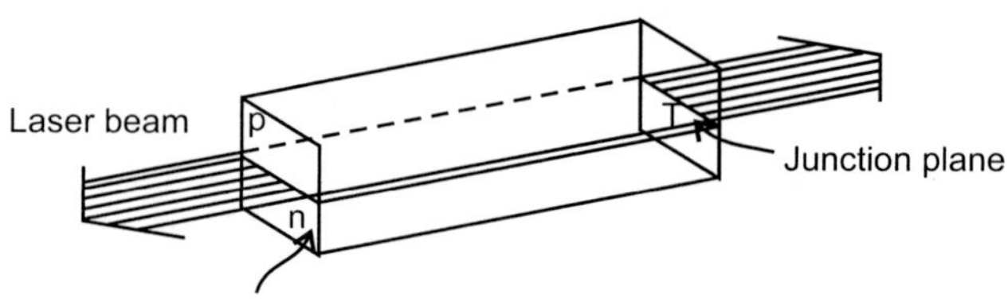

Fig. 8.14 Schematic diagram of a semiconductor laser.

The advantages of semiconductor lasers are that they are compact, efficient and can be fabricated with ease. However, their monochromaticity, coherence and directionality are inferior to those of other lasers.

Fabrication of Semiconductor Devices

The working of semiconductor devices depends on the ordered periodic arrangement of atoms in a lattice and the introduction of impurities in a controlled manner. A highly sophisticated technology is involved in their development. The steps involved in their production are illustrated by considering the specific case of a silicon *npn* junction.

In the first step, a silicon crystal is grown by dipping a seed crystal into the molten silicon at about 1425°C, and slowly pulling it up at a rate of about 100 millimetres per hour. The melt contains a suitable amount of phosphorus to produce new crystal layers of the *n*-type. The crystal is cut into thin slices about 0.25–0.5 mm in thickness and about 10 cm in diameter, by using a diamond saw. The surface of the slice is polished mechanically and chemically, to give what is referred to as the *substrate*.

In the next step, a mixture of silicon tetrachloride, hydrogen and phosphine (PH_3) is passed over the substrate at about 1200°C. The silicon released by the reduction of silicon tetrachloride, along with a suitable amount of phosphorus (from the PH_3), crystallizes on the substrate surface forming what is known as an *n*-type *epitaxial layer* (for producing *p*-type epitaxial layer PH_3 is replaced by diborane, B_2H_6).

The surface of the epitaxial layer is oxidized by heating the substrate to about 1100°C in steam or oxygen so as to produce a thin layer of SiO_2 [Fig. 8.15(*a*)]. The oxide surface is coated with a photosensitive material called the *photoresist*. An area of the surface is covered with a photographic mask and the rest of the surface is exposed to ultraviolet radiation. The photoresist is then developed and the unexposed area is washed off. The oxide in this area is removed by immersing in hydrofluoric acid and then the exposed photoresist is removed. This procedure is known as *window opening* and it effectively removes SiO_2 from specified areas.

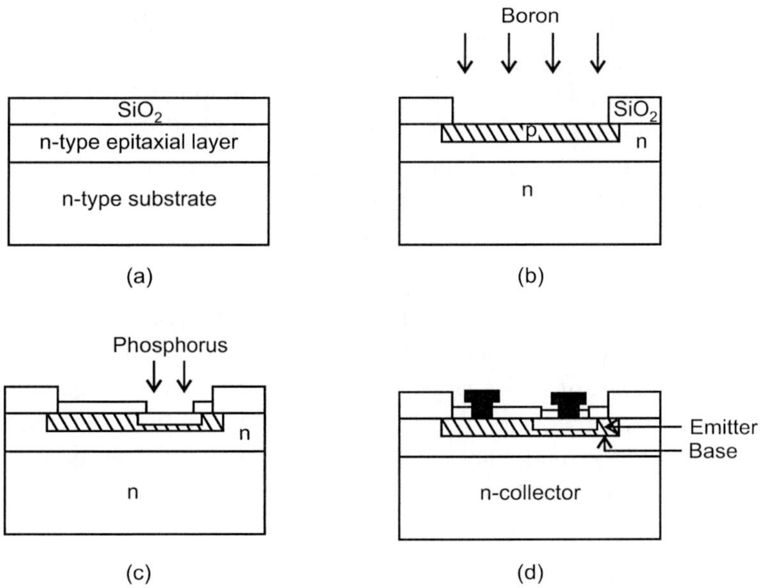

Fig. 8.15 Fabrication of an *npn* transistor: (*a*) oxidization, (*b*) boron diffusion through the windows, (*c*) phosphorus diffusion through the window, (*d*) metal contacts shown by shaded areas.

A *p*-type diffusion is introduced by using boron atoms, producing a *p*-type layer [Fig. 8.15(*b*)] which essentially forms the base of the final *npn* transistor. The surface is re-oxidized and a smaller window is opened (following the same procedure as before) over the *p*-layer. An *n*-type of diffusion is now made to convert part of the *p*-region back to *n*-type to form the emitter [Fig. 8.15(*c*). The surface is again oxidized and two new windows are opened to expose the emitter and base regions and a metal (usually aluminium) is evaporated into those windows forming electrical contact with these regions. The contact with the original epi-layer which forms the collector, can be made through the substrate [Fig. 8.15(*d*)].

Other devices can be produced by the variations of the essential steps involved in the development of the *npn* transistor described.

Amorphous Semiconductors

There are many amorphous substances which have significant electrical conductivity. They are known as *amorphous semiconductors*. In these materials, the conduction is by electrons. They differ from the crystalline semiconductors in that while they have short-range order, long-range order is absent in them. This can be illustrated by amorphous Ge. In this case, though each atom is surrounded by four nearest neighbours, the location of the second-nearest neighbour is not unique. In the amorphous semiconductors, the different possible locations of farther-away neighbours are almost randomly filled leading to disorder at long range. The effect of this long-range disorder is not significant for energy levels deep inside an energy band but is important for those near the edges, *e.g.*, those near the top of the valence band and bottom of the conduction band. It leads to narrowing of the energy gap as compared with the gap in crystalline semiconductors. Some amorphous semiconductors are Ge, Si, Se, As_2Se_3, etc.

Amorphous semiconductors are used in switching and memory components, They are also used in xerographic processes. Here, typically a thin film of amorphous selenium is deposited on a metallic substrate, usually Al. It is charged electrically by means of a discharge. When a pattern of light to be copied falls on this, the lighted areas become photoconductive and discharge their charge whereas the dark areas retain their charge. A finely-powdered pigment is sprayed on the surface. It is retained by the charged areas and then transferred to a sheet of paper.

8.6 MAGNETIC PROPERTIES

In this section, we discuss the magnetic properties of materials. They are related to the spin and orbital angular momentum of the electrons, and are observed in the form of (*i*) diamagnetism (*ii*) free-electron paramagnetism (*iii*) paramagnetism of atoms, ions or molecules, (*iv*) ferromagnetism, and (*v*) antiferromagnetism and ferrimagnetism. Of these the first three are rather weak effects while the last two are strong effects which are also of great technological importance.

The magnetic properties are discussed conveniently in terms of magnetic susceptibility χ defined as

$$\chi = M/H \tag{8.85}$$

where M is the magnetic moment per unit volume, and H is the magnetic intensity. The magnetic intensity H, the magnetic moment M and the magnetic field B are related by

$$B = \mu_0(H + M) \tag{8.86}$$

where μ_0 is the vacuum permeability. Different magnetic properties, in particular the behaviour of the associated magnetic moment and the magnetic susceptibility will be discussed here.

Diamagnetism

When an atom is subjected to a magnetic field, the changing magnetic flux induces currents (via the electron orbits) which, as per Lenz's law, oppose the change in flux. The currents persist, and have a magnetic moment which is opposite in sign to the magnetic field intensity. The associated magnetic susceptibility is negative and the property is known as *diamagnetism*. Diamagnetism is present in all substances but is usually obscured by the larger effects due to permanent magnetic dipole moments of the atoms.

Essentially, diamagnetism is the consequence of the term in which is quadratic in B,

$$\begin{aligned} H_d &= \frac{e^2}{8m}\sum_i (\mathbf{r}_i \times \mathbf{B})^2 \\ &= \frac{e^2}{8m}\sum_i (\mathbf{r}_{i\perp})^2\,\mathbf{B}^2 \end{aligned} \tag{8.87}$$

where the summation is over all the electrons. In perturbation theory, the energy due to this term is the average value

$$E = \frac{e^2}{8m} B^2 \sum_i \left\langle (\mathbf{r}_{i\perp})^2 \right\rangle \tag{8.88}$$

The magnetic moment in this case is defined by

$$\begin{aligned} m_0 &= -\frac{\partial E}{\partial B} \\ &= -\frac{e^2}{4m} B\sum_i \left\langle (\mathbf{r}_{i\perp})^2 \right\rangle \end{aligned} \tag{8.89}$$

The diamagnetic susceptibility, therefore, is

$$\chi = -\frac{e^2 \mu_0 N}{4m} \sum_i \left\langle (\mathbf{r}_{i\perp})^2 \right\rangle \tag{8.90}$$

where N is the number of atoms per unit volume. For the evaluation of this quantity, an approximate value for $\langle (\mathbf{r}_{i\perp})^2 \rangle$ is normally used. For a typical value of $r^2 \sim 10^{-20}$ m^2, the molar susceptibility is

$$\chi_m \sim 5 \times 10^{-8}/\text{kg. mol, in MKS units,} \tag{8.91}$$

(multiply by $10^3/4\pi$ to get the value in Gaussian units of per g mole) which means that diamagnetism is a small effect. It is significant minly in atoms and ions with closed shells, *e.g.*, He, Ne, F^-, Cl^-, etc. which do not have a permanent magnetic moment.

Free-Electron Paramagnetism

Free-electron paramagnetism in metals arises from the intrinsic magnetic moment associated with the spin of the electron. In the absence of any magnetic field, there is no preferred orientation of these magnetic moments. However, in the presence of a magnetic field, the energies of the electron are perturbed by an additional interaction

$$H = -\left(-\frac{e}{m}\mathbf{s}\right) \cdot \mathbf{B} \tag{8.92}$$

and the resulting energy eigenvalues are

$$\varepsilon' = \varepsilon \pm \frac{e\hbar}{2m} B \tag{8.93}$$

ε being the unperturbed energy. The net magnetic moment is obtained by using the Fermi-Dirac distribution:

$$M = \frac{2\pi V (2m)^{3/2})}{h^3} \int \left[\frac{(-e\hbar/2m)}{1 + \exp\left(\varepsilon + \frac{e\hbar}{2m} B - \varepsilon_f\right) \Big/ kT} + \frac{-e\hbar/2m}{1 + \exp\left(\varepsilon - \frac{e\hbar}{2m} B - \varepsilon_f\right) \Big/ kT} \right] \varepsilon^{1/2} d\varepsilon \tag{8.94}$$

For $T \to 0$, this expression reduces to

$$M = \frac{2\pi V(2m)^{3/2}}{h^3}\left(\frac{e\hbar}{2m}\right)\int_{\varepsilon_f - e\hbar B/2m}^{\varepsilon_f + e\hbar B/2m} \varepsilon^{1/2}\, d\varepsilon$$

$$\approx \frac{3}{2}\left(\frac{e\hbar}{2m}\right)^2\left(\frac{NB}{\varepsilon_f(0)}\right) \tag{8.95}$$

where Eq. (7.89) has been used. The susceptibility therefore is positive and given by

$$\chi \approx \frac{3}{2}N\left(\frac{e\hbar}{2m}\right)^2 \frac{\mu_0}{\varepsilon_f(0)} \tag{8.96}$$

This is generally quite small and for sodium [$\varepsilon_f(0) \sim 3.1$ eV] the susceptibility per unit mass 8.3×10^{-9} kg^{-1} in MKS units or 6.6×10^{-7} g^{-1} in Gaussian units. On including the corrections due to exchange correlation and effective mass, the value is 8.8×10^{-7} g^{-1} in Gaussian units, which should be compared with the experimental spin susceptibility of 9.8×10^{-7} g^{-1}. For obtaining bulk susceptibility, the diamagnetic susceptibility due to the free electrons and the ions should also be included.

At finite temperature, there is a slight dependence of χ on T which, for all practical purposes, may be neglected.

Paramagnetism

Atoms, ions and compounds with unpaired electrons (this is the case if the number of electrons is odd and also for some systems with even number of electrons), have a nonzero magnetic moment. In the presence of a magnetic field, they align with the magnetic field and produce a net, macroscopic magnetic moment, giving rise to *paramagnetism*. Since the atoms (most of the subsequent discussion applies to ions and molecules as well) are localized, Boltzmann distribution can be used for the electron states. This gives rise to a temperature-dependent susceptibility.

As was discussed in Sec. 6.2 the energy due to the interaction of an atom with a magnetic field is

$$\Delta E = \frac{e\hbar}{2m} g\, M_J\, B, M_J = -J, -j+1, \cdots, J \tag{8.97}$$

where g is the landé g-factor [Eq. 6.13)] and $\hbar M_J$ is the z-component of the total angular momentum. Using Boltzmann distribution for the populations, the magnetic moment per unit volume is

$$M = -N\frac{e\hbar}{2m} g \frac{\sum_{M_J=-J}^{J} M_J \exp(-aM_J/kT)}{\sum_{M_J=-J}^{J} \exp(-aM_J/kT)} \tag{8.98}$$

where $a = \frac{e\hbar}{2m} gB$ and N is the number of particles per unit volume. The summations can be carried out to yield

$$M = N\left(\frac{e\hbar g}{2m}\right)\left[\frac{d}{dx}\ln f(x)\right]_{x=a/kT}$$

$$f(x) = \frac{\exp\left[\left(J+\frac{1}{2}\right)x\right] - \exp\left[-\left(J+\frac{1}{2}\right)x\right]}{e^{x/2} - e^{-x/2}} \tag{8.99}$$

For $\frac{e\hbar}{2m} B \ll kT$, the expression leads to a susceptibility

$$\chi = M/H$$

$$= N\mu_0\left(\frac{e\hbar}{2m}\right)^2 \frac{J(J+1)}{3kT} \tag{8.100}$$

It is observed that the susceptibility is inversely proportional to temperature. This is stated in the form

$$\chi = C/T \tag{8.101}$$

known as Curie's law, where C is called the Curie constant.

At $T \approx 300$ K, molar susceptibility χ_m is of the order of 5×10^{-7}/kg mol in MKS units (4×10^{-5}/gm mol in Gaussian units), which is rather small, but it becomes much larger at low temperatures. The pedictions of Eq. (8.100) with the J values given by Hund's rule (ground state has the largest S allowed by Pauli principle, the maximum L consistent with this S, and $J = L + S$ when the shell is more than half full and $J = |L - S|$ otherwise, are generally in good agreement with the experimental observations for many paramagnetic crystals, *e.g.*, rare earth ions, where in some cases the effect of the nearby states has to be included.

The predictions of Eq. (8.100) are not in good agreement with experimental observations for the ions of the iron group. The reason for this is that the partially filled 3*d* shell for these ions is the outermost shell and is exposed to the strong field due to the neighbouring ions in the crystal. This field, called the *crystal field*, breaks the rotational symmetry, and the total angular momentum is no longer a 'good' quantum number. Furthermore, the average value of L_z may reduce to zero. This effect is known as the *quenching* of the orbital angular momentum and implies that Eq. (8.97) should be replaced by

$$\Delta E = \frac{e\hbar}{2m} M_S B \; (g = 2 \text{ for spin}) \tag{8.102}$$

This leads to

$$\chi = N\mu_0 \left(\frac{e\hbar}{2m}\right)^2 \frac{S(S+1)}{3kT}, \; \frac{e\hbar B}{m} \ll kT \tag{8.103}$$

for the ions of iron group. As an example, in the case of χ for Mn^{3+} ($^5D_0$), Eq. (8.100) predicts that $\chi = 0$, whereas the prediction of Eq. (8.103) with $S = 2$ is in very good agreement with experimental observations.

It may be noted that adiabatic demagnetization of a paramagnet system can be used for attaining low temperatures, $T < 1$ K. This is done as follows. A magnetic field is applied to a paramagnetic substance in good thermal contact with the surroundings at T_1. The field aligns the magnetic moments along the direction of the field. This increase in order is equivalent to a decrease in the entropy and hence heat flows out of the system. If now the substance is insulated, and the field removed adiabatically, the spins gradually get out of the alignment by absorbing energy from the lattie vibration which leads to a lowering of the temperature of the paramagnetic substance. Temperatures of the order of 10^{-3} K have been reached by this method.

Ferromagnetism

Ferromagnetism is the phenomenon in which some materials like iron, cobalt, nickel, and some of their alloys behave like ordinary paramagnets at high temperatures but which below a critical temperature known as the *Curie temperature* T_c, acquire a nonzero magnetic moment even in the absence of an applied magnetic field. This is due to the interaction between the magnetic ions, which is strong enough to align their magnetic moments against the disorder introduced by thermal effects.

The interaction that aligns the magnetic moments is quantum mechanical in origin and is due to the *exchange properties* of the electron wave functions. When the wave functions of two atoms overlap, the electrons being indistinguishable, belong to both the atoms. In such cases, the symmetry or the antisymmetry of the wave functions will strongly influence the energy of the system (as in the case of covalent bonding, see Chapter 5). In particular, it is the exchange symmetry between the spins and the extent of the overlap of the wave functions that determines the nature and the strength of the exchange interaction. It is reasonable to represent the energy from the exchange interaction by

$$E = -\sum_{i,\,j} J_{ij}\,\mathbf{S}_i \cdot \mathbf{S}_j, \quad i \neq j \tag{8.104}$$

where $\mathbf{S}_i$ is the spin of the i-th atom, and J_{ij} are symmetric constants. If the magnetic moment is assumed to be due to spin alone, $\mathbf{M}_i = b\,\mathbf{S}_i$, as is the case for the iron group, the interaction energy of the i-th atom can be written as

$$E_i = -\,\mathbf{M}_i \cdot \mathbf{B}_{\text{int}} \tag{8.105}$$

where $\mathbf{B}_{\text{int}}$ is given by

$$\begin{aligned}\mathbf{B}_{\text{int}} &= \frac{1}{b^2}\sum_{j} J_{ij}\,\mathbf{M}_j, \quad i \neq j \\ &= \lambda\,\mathbf{M}\end{aligned} \tag{8.106}$$

i.e., the effective internal field is proportional to an average magnetic moment **M**. It is this field, known as the *Weiss field*, which is responsible for the alignment of the spins. In the case of ferromagnetic substances, J_{ij} are quite large and λ is positive, which gives rise to ferromagnetism.

Consider the behaviour of ferromagnets above the curie temperature T_c. Writing

$$B = B_0 + \lambda M \tag{8.107}$$

where B_0 is the applied field, one gets from Eq. (8.99),

$$M = N\left(\frac{e\hbar g}{2m}\right)^2 \frac{J(J+1)}{3kT}(B_0 + \lambda M), \; \frac{e\hbar}{2m} B \ll kT \tag{8.108}$$

This leads to

$$M = \frac{CB_0/\mu_0}{T - T_c} \tag{8.109}$$

$$c = \frac{C}{T - T_c} \tag{8.110}$$

with

$$\begin{aligned}T_c &= N\left(\frac{e\hbar g}{2m}\right)^2 \frac{J(J+1)}{3k}\lambda \\ &= \frac{\mu_0 T_c}{\lambda}\end{aligned} \tag{8.111}$$

The expression in Eq. (8.110) is known as the *Curie-Weiss low* and T_c is known as the Curie temperature. The behaviour of χ given in Eq. (8.110) is valid for $T > T_c$. At $T = T_c$, χ becomes infinite. Since M is finite, this implies that M is nonzero even when $B_0 = 0$, *i.e.*, spontaneous magnetization exists. The Curie temperature is about 1043 K for Fe, 1400 K for Co, and 631 K for Ni.

Below $T = T_c$, the spontaneous magnetization is to be obtained by using the complete expression in Eq. (8.99) for M, with

$$a = \frac{e\hbar g}{2m}(B_0 + \lambda M) \tag{8.112}$$

The solutions are obtained by plotting M in Eq. (8.99) as a function of x, and also

$$M = \left(\frac{2mkT}{e\hbar g\lambda}\right)x, \tag{8.113}$$

obtained from Eq. (8.112) with $a = xkT$, $B_0 = 0$, and looking for the intersection of the two curves [see Fig. 8.16(*a*), it can be shown that the intersection at the origin gives an unstable solution]. At $T = T_c$ the curve given by Eq. (8.113) is tangential to the curve given by Eq. (8.99), at the origin, and there is no spontaneous magnetization for $T > T_c$. When $T < T_c$, there are two equal and opposite solutions for each T, for example corresponding to points A and A' in Fig. 8.16(*a*). One set of these spontaneous magnetizations is plotted as a function of T ($T < T_c$) in Fig. 8.16(*b*).

For $B_0 \neq 0$, the magnetization is obtained from the intersection of the curve given by Eq. (8.99) and

$$\mathrm{M} = \left(\frac{2mkT}{e\hbar g\lambda}\right)x - \frac{1}{\lambda}B_0 \tag{8.114}$$

obtained from Eq. (8.112) with $a = xkT$. There are two solutions for M corresponding to intersections at D and D' in Fig. 8.16(*a*), for each B_0 ($T < T_c$). These solutions trace the boundary of the *hysteresis curve* [Fig. 8.16(*c*)]. It can be shown that the third solution corresponding to intersection F in Fig. 8.16(*a*) is unstable. This solution is the extension of the unstable solution at the origin for $B_0 = 0$.

At $T = 0$, all the spins and the magnetic moments of the atoms are aligned corresponding to the ground state of the system. The direction of the alignment is introduced by the arbitrarily assumed direction of the internal field B_{int}, and obviously the ground state is infinitely degenerate.

For $T > 0$ K, some of the spins go out of alignment. As in the case of lattice vibrations, the disturbances are correlated, and misalignments travel as waves known as *spin waves*. In analogy with photons and phonons, the excitations of spin waves are quantized into quanta known as *magnons*. Magnons, which obey Bose-Einstein statistics, play a significant role in determining the behaviour of $M(T)$ at low temperatures, and contribute to the specific heat and thermal conductivity of ferromagnets.

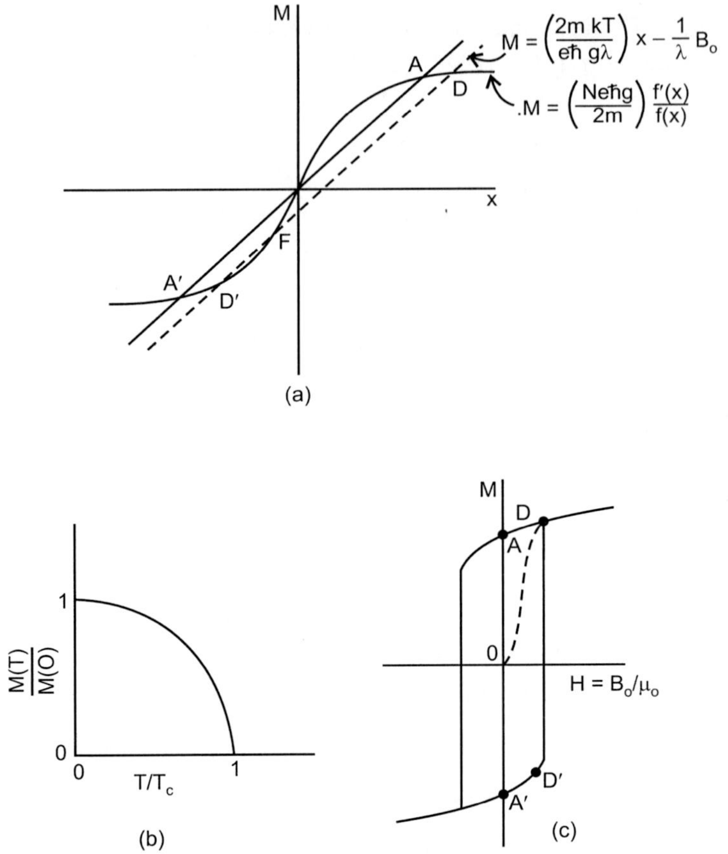

Fig. 8.16 Ferromagnetism, (*a*) determination of M for $B_0 = 0$ and $B_0 \neq 0$, (*b*) spontaneous magnetization as a function of T for $T < T_c$, (c) M as a function of H for $T < T_c$.

An important property of ferromagnets is that for $T < T_c$, they are usually not magnetized. They become magnetized under the influence of an external magnetic field. This behaviour is explained by postulating that a ferromagnet is usually subdivided into what are known as *domains*. Each of these domains is spontaneously magnetized but the direction of magnetization may be different in different domains. This situation is energetically favourable since in addition to the energy of the atoms, energy is stored in the magnetic field also, $E = \frac{1}{2}\mathbf{H} \cdot \mathbf{B}$, and this energy is reduced if the alignment changes from domain to domain. The division into smaller and smaller domains is ultimately restrained by the fact that the formation of domain walls requires additional energy. If the ferromagnet is subjected to an external field, the domain walls move in such a way that the domains with magnetization nearly parallel to the external field grow. For small fields, this movement is reversible. However, if the external field is sufficiently strong, the walls may move irreversibly over potential barriers and finally all the domains will have magnetization along a preferred or an 'easy' direction (determined by the crystal structure) nearly parallel to the direction of the external field. A further increase in the field brings the alignment closer to the direction of the external field. The progress of magnetization as the external field increases is described by the broken line in Fig. 8.16(*c*). If the external field is now removed, some of the magnetization is retained [point *A* in Fig. 8.16(*c*)]. The variation of magnetization with the external magnetic field now produces the well-known *hysteresis curve*, similar to that in Fig. 8.16(*c*) except that usually *B* is plotted against *H*. The magnetization can be destroyed either by heating or by mechanical shocks. The reality of the domains, which are so successful in describing ferromagnets, can be demonstrated by scattering finely divided iron on the surface of a ferromagnet, which collects along the domain boundaries where the field is the strongest.

Antiferromagnetism and Ferrimagnetism

In most substances other than ferromagnets, the exchange energy in Eq. (8.104) is positive when the neighbouring spins are parallel. This leads to a ground state in which the neighbouring spins are antiparallel and hence to a cooperative alternating alignment below temperature T_N known as the *Néel temperature*. The problem can be analysed in terms of alternating lattice points occupied by atoms *A* and *B* with antiparallel spins.

Let the average magnetic moments of *A* and *B* sets of atoms be M_A and M_B respectively. The total magnetic fields at sites *A* and *B* (for antiparallel alignment) are:

$$B_A = B_0 - \lambda_A M_B,$$
$$B_B = B_0 - \lambda_B M_A \qquad (8.115)$$

Substituting these expressions in Eq. (8.99), for temperature above the Néel temperature, the magnetic moments are:

$$M_A = \frac{C_A}{T}(B_0 - \lambda_A M_B), \qquad (8.116)$$

$$M_B = \frac{C_B}{T}(B_0 - \lambda_B M_A)$$

where
$$C_{A,B} = N_{A,B}\left(\frac{e\hbar\, g_{A,B}}{2m}\right)^2 \frac{J_{A,B}(J_{A,B}+1)}{3k} \qquad (8.117)$$

If the atoms *A* and *B* are magnetically equivalent, then

$$\chi = (M_A + M_B)/H$$

$$= \frac{2\mu_0 C_A}{T + T_c}, T > T_c \tag{8.118}$$

with $T_c = \lambda_A C_A$, $\lambda_A = \lambda_B$, $C_A = C_B$, where T_c is the Néel temperature T_N. At $T = T_c$, an additional solution to Eq. (8.116) exists, $M_A = -M_B$ even for $B_0 = 0$. Thus, the sets of atoms A and B are spontaneously magnetized for $T < T_c$ though the net magnetization is zero. Such materials are known as *antiferromagnets*. In antiferromagnets, the crystal field aligns the magnetic moments along a preferred direction below T_c. It can be shown that a field applied perpendicular to this direction is associated with a susceptibility $\chi_\perp$ which is essentially independent of $T(T < T_c)$ whereas a field applied parallel to the preferred direction, is associated with a susceptibility χ_{11} which is equal to $\chi_\perp$ at $T = T_c$ but decreases to zero as $T \to 0$ K.

If the atoms A and B are magnetically inequivalent, Eq. (8.116) can be solved for M_A and M_B and they lead to a susceptibility

$$\chi = \frac{\mu_0[T(C_A + C_B) - C_A C_B(\lambda_A + \lambda_B)]}{(T - T_c)(T + T_c)} \tag{8.119}$$

where $T_c = (C_A C_B \lambda_A \lambda_B)^{1/2}$. Thus, χ tends to infinity as $T \to T_c$, which implies that there is a net spontaneous magnetization for $T < T_c$ (M_A and M_B are opposite in sign but $M_A + M_B \neq 0$). Such meterials are called *ferrimagnets*, Fe_3O_4 being a well known example. An important class of ferrimagnets is the ferrites which have the formula $M^{2+} Fe_2^{3+} O_4^{2-}$ where **M** is a member of the first transition group. They are of great technical importance since they may have large magnetization at room temperature, and high resistivity. They are therefore more suitable than ferromagnets for use at high frequencies when eddy current losses are a serious problem. They are also used for memory storage in computers.

8.7 DIELECTRIC PROPERTIES

In this section, the properties of solids in the presence of an external electric field are discussed. These are important in the propagation of electromagnetic waves in material media, and in the development of many devices such as capacitors, microphones, etc.

An electric field **E** causes a relative displacement of the positive and negative charges of a material. This induces an electric dipole moment which is expressed in terms of the polarization **P** defined as the dipole moment per unit volume. For ordinary electric fields, the polarizability is linear in **E** (it is nonlinear for strong laser fields),

$$\mathbf{P} = \varepsilon_0 \chi_e \mathbf{E} \tag{8.120}$$

where ε_0 is the permittivity of the vacuum and χ_e is the electric susceptibility. It is convenient to define an atomic polarizability α by

$$\mathbf{P} = N \alpha \mathbf{E}_{\text{loc}} \tag{8.121}$$

where N is the number of atoms per unit volume and $\mathbf{E}_{\text{loc}}$ is the effective field at the atom, not including the field due to the atom itself. A displacement vector **D** can also be defined as

$$\begin{aligned}\mathbf{D} &\equiv \varepsilon \varepsilon_0 \mathbf{E} \\ &\equiv \varepsilon_0 \mathbf{E} + \mathbf{P}\end{aligned} \tag{8.122}$$

where ε is the dielectric constant. It can be shown that for a dielectric material with an isotropic or cubic (including simple cubic, bcc, fcc) structure, filling a parallel plate capacitor,

$$\mathbf{E}_{\text{loc}} = \mathbf{E} + \frac{1}{3\varepsilon_0}\mathbf{P} \tag{8.123}$$

This allows us to eliminate $\mathbf{E}_{\text{loc}}$ in Eq. (8.121). Solving for **P** from Eqs. (8.121) and (8.122), and equating the two expressions gives

$$\frac{N\alpha}{3\varepsilon_0} = \frac{\varepsilon - 1}{\varepsilon + 2} \tag{8.124}$$

This is the *Clausius-Mossotti* formula, relating the atomic polarizability α to the macroscopic dielectric constant ε.

For a nonmagnetic material, $\varepsilon = n^2$, n being the refractive index, so that

$$\frac{n^2 - 1}{n^2 + 2} = \frac{1}{3\varepsilon_0} \sum_i N_i \alpha_i \tag{8.125}$$

where a summation over i has been introduced to include the possibility the several mechanisms contribute to the polarizability. For the polarizability of an atom, contributions are electronic, ionic and orientational. It is possible to experimentally separate the different contributions by observing the polarizability as a function of frequency and by noting that the different contributions are generally significant in different ranges. This is illustrated in Fig. 8.17(*a*). The rapid changes in the polarizability are also accompanied by large absorption of radiation [Fig. 8.17(*b*)].

Electronic Polarizability

The electronic contribution to polarizability arises from the displacement of electrons in an atom, relative to the nucleus.

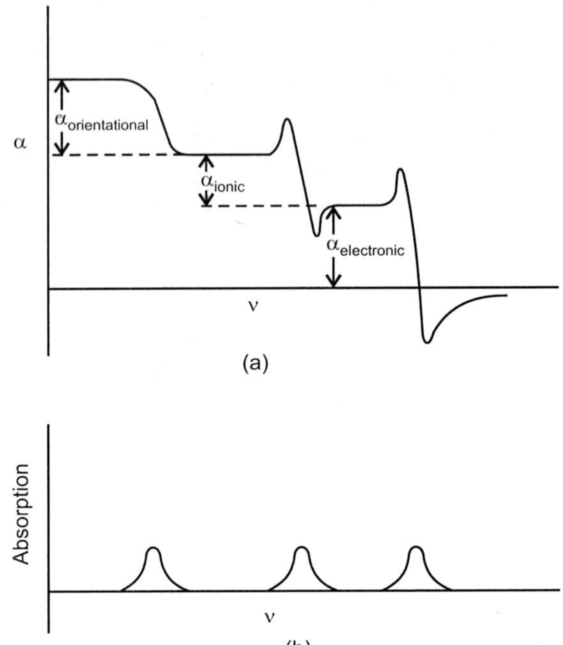

Fig. 8.17 Schematic diagram illustrating the variation of polarizability and absorption as a function of *v*.

Consider the behaviour of a bound electron in the presence of an electromagnetic field **E** cos (ωt) in the z-direction. It was pointed out in Sec. 6.3, that this introduces a potential

$$V = e\,E\,z\cos\omega t \tag{8.126}$$

as a result of which the state changes to the expression in Eq. (6.41) with $a_m(t)$ given by Eq. (6.48). The resulting dipole moment is given by

$$p = -\,e\langle\psi(t)\,|z|\,\psi(t)\rangle \tag{8.127}$$

which on using Eqs. (6.41 and 6.48), after some analysis, gives, to leading order in E

$$p \approx \frac{e^2}{\hbar}E\quad\cos\omega t\sum_j |z_{j0}|^2\left(\frac{1}{\omega_{j0}+\omega}+\frac{1}{\omega_{j0}-\omega}\right) \tag{8.128}$$

where $\omega_{j0} = (E_j - E_0)/\hbar$. From this, the polarizability is

$$\alpha = \frac{e^2}{\hbar}\sum_j |z_{j0}|^2\left(\frac{1}{\omega_{j0}+\omega}+\frac{1}{\omega_{j0}-\omega}\right) \tag{8.129}$$

It is observed that α takes sudden jumps whenever $\omega = (E_j - E_0)/\hbar$. Also at $\omega = (E_j - E_0)/\hbar$, there are transitions to the state j, resulting in an absorption of radiation. Actually the singularity at $\omega = \omega_{j0}$ is displaced by the fact that the state j is an unstable state which essentially requires a replacement of $(\omega_{j0} - \omega)^{-1}$ by the real part of $(\omega_{j0} - \omega - i/2\,\tau_j)^{-1}$, *i.e.*,

$$\frac{1}{\omega_{j0} + \omega} \rightarrow \frac{\omega_{j0} - \omega}{(\omega_{j0} - \omega)^2 + (2\,\tau_j)^{-2}} \tag{8.130}$$

where τ_j is the lifetime of state j [see Eq. (6.76)]. With this modification, Eq. (8.129) provides a qualitative explanation of the polarizability illustrated in Fig. 8.17(*a*). It may also be noted that when $\omega_{j0} \approx \omega$, there is a significant probability for transition to state j, as seen from the expression in Eq. (6.55) for a_j, which leads to absorption of radiation [Fig. 8.17(*b*)]. While Eq. (8.129) is valid for a general charged system, it is practically useful mainly for electronic polarizability for which z_{j0} can be calculated with some reliability. A rough order of magnitude estimation for this electronic polarizability gives

$$\alpha = \frac{(1.6\times10^{-19})^2\,(10^{-10})^2}{(1.6\times10^{-19})} \approx 1.6 \times 10^{-39}\ \text{F.m}^2 \tag{8.131}$$

Ionic Polarizability

The ionic polarizability is due to the displacement of ions with respect to each other. If it is assumed that the forces near equilibrium are simple harmonic, the displacement in the presence of an electric field is given by

$$k\,\Delta x \approx eE \tag{8.132}$$

where k is the force constant. This leads to a polarizability

$$\alpha \approx e^2/k \tag{8.133}$$

Since $k \approx 20$ N/m, $\alpha_{\text{ionic}} \approx 10^{-39}$ F.m^2

The ionic contribution is important at low frequencies (ω_{j0} in Eq. (8.129) is small). This explains the fact that NaCl has $\varepsilon \approx 5.6$ at low frequencies whereas at optical frequencies $\varepsilon \approx 2.25$. The difference may be ascribed to the ionic contribution to polarizability (see Fig. 8.17).

Orientational Polarizability

Molecules with permanent electric dipole moment align themselves in the presence of an external electric field giving rise to an orientational polarizability.

The energy of a dipole $\mathbf{p}$ in an electron field $\mathbf{E}$ is

$$V = -\,p\,E\cos\theta \tag{8.134}$$

where θ is the angle between the dipole and the field. Therefore, the average dipole moment (using Boltzmann distribution) is

$$\overline{p} = p\frac{\int \cos\theta \exp\left(\dfrac{pE\cos\theta}{kT}\right) d\cos\theta}{\int \exp\left(\dfrac{pE\cos\theta}{kT}\right) d\cos\theta} \tag{8.135}$$

$$= p\left[\frac{d}{da}\ln\left(\frac{e^{a}-e^{-a}}{a}\right)\right]_{a=\frac{pE}{kT}}$$

For $pE \ll kT$, $\overline{p} \approx p^2E/3kT$ which leads to

$$\alpha = p^2/3kT \tag{8.136}$$

which at room temperatures is of the order of 10^{-39} F.m^2, comparable to the electronic polarizability. It is distinguished by its temperature dependence, and suggests a relation

$$\alpha_{tot} = \alpha_0 + p^2/3kT \tag{8.137}$$

This expression is quite useful in determining dipole moments of dipolar substances, *e.g.*, HCl (p = 1.1 debyes, 1 debye = 10^{-39} C.m), by looking at the temperature dependence of α_{tot}. It is tempting to substitute Eq. (8.136) in Eq. (8.124) to obtain

$$\varepsilon = 1 + \frac{3T_c}{T - T_c}, T_c = \frac{Np^2}{9\varepsilon_o k} \tag{8.138}$$

which would imply that spontaneous polarization ($E = 0$) sets in at $T = T_c$. It has however been shown by Onsager that Eqs. (8.123) and (8.124) are not valid for permanent electric dipoles. The theory of Onsager for permanent dipoles does not imply the existence of a critical temperature for such dipoles.

Ferroelectric Crystals

The phenomenon of spontaneous polarization ($E = 0$), known as ferroelectricity is observed in (*i*) Rochelle salt and some of the associated salts, (*ii*) some crystals with hydrogen bonds, in which the motion of protons gives rise to ferroelectric behaviour (*e.g.*, KH_2PO_4, RbH_2PO_4, etc.) and (*iii*) ionic crystals with perovskite ($CaTiO_3$) and ilmenite ($FeTiO_3$) structures. The perovskite structure illustrated by $BaTiO_3$ is the simplest structure which exhibits ferroelectricity—it has a cubic structure with Ba at the corners, oxygen at the face centres and Ti at the body centre. Ferroelectricity in barium titanate ($BaTiO_3$) is briefly described here.

Barium titanate becomes ferroelectric at 380 K and exhibits hysteresis curves for $T < T_c$, in the plot of D against E. The ferroelectricity in $BaTiO_3$ is due to induced electronic and ionic dipole moments. From Eq. (8.124)

$$\varepsilon = \frac{1 + \dfrac{2}{3\varepsilon_0}\sum_i N_i\alpha_i}{1 - \dfrac{1}{3\varepsilon_0}\sum_i N_i\alpha_i} \tag{8.139}$$

Now, the contribution of electronic polarizabilities to $\frac{1}{3\varepsilon_0}\sum_i N_i\alpha_i$ is about 0.61. Assuming that the ionic contribution is about 0.39 (estimations show that this is not unreasonable), the dielectric constant tends to infinity. Expanding the denominator as a function of temperature gives

$$\varepsilon = \frac{3/\beta}{T - T_c}, \beta = -\frac{1}{3\varepsilon_0}\left(\frac{\partial}{\partial T}\sum_i N_i\alpha_i\right)_{T=T_c} \tag{8.140}$$

Estimates of β agree well with experimental observations, *i.e.*, $3/\beta \sim 10^5$ K. It may be observed that the (incorrect) expression in Eq. (8.138) for dipolar atoms would have given a value of $3T_c \sim 1140$ K for the residue, considerably smaller than the observed residue, which again rules out the explanation in terms of dipolar atoms.

It may also be noted that (*i*) the description of the hysteresis, etc. in terms of domains is valid for ferroelectricity, and (*ii*) antiferroelectricity is observed *e.g.*, in WO_3, $PbZrO_3$, etc.

Piezo-electricity

Some crystals when deformed by an external stress develop a net dipole moment which produces surface polarization charges. This is known as *piezo-electricity*. Piezo-electric materials exhibit the converse effect as well, *i.e.*, they are distorted when placed in an electric field. The strain produced however is very small. For example in quartz which is the most common piezo-electric substance, an electric field of 10^4 V/m produces a strain of only 1 part in 10^8. Of course, this also means that even a small strain can produce enormous electric fields.

When a crystal is subjected to a strain, there is a displacement of the ions in the crystal. If the charge distribution in the crystal does not have inversion symmetry about a centre, a net polarization of charges may develop giving rise to piezo-electricity. For example, an equilateral triangle with + 3 charge at the centre and – 1 charge each at the vertices will have zero dipole moment. Under strain, the bond lengths may remain the same but make unequal angles with each other giving rise to nonzero dipole moment.

A part from quartz, other examples of piezo-electric materials are Rochelle salt, barium titanate ($BaTiO_3$), etc. In fact, all ferroelectric materials are piezo-electric though the converse is not true. Piezo-electric materials are used to convert electrical energy into mechanical energy and conversely, *i.e.*, as transducers. In particular, they are used in devices such as gramophone pickups, microphones, strain gauges, etc. while the converse effect is used in ultrasonic generators.

8.8 EXAMPLES

Here, some examples that illustrate and extend the ideas about the solid state are considered.

Example 1

The bulk modulus of a crystalline solid can be estimated from Eq. (8.5).

A pressure P produces a decrease in length, Δl,

$$P = C\, l_0\, \Delta l \tag{8.141}$$

where C is a constant, so that the work done is

$$W = 3CV_0 \int_0^{\Delta} x\,dx$$

$$= \frac{3}{2} C\, V_0\, (\Delta l)^2 \tag{8.142}$$

where V_0 is the volume. This causes a change in energy given by

$$\Delta E = N[E(R_0 - \Delta R) - E(R_0)]$$

$$\approx \frac{N}{2} \frac{\partial^2 E}{\partial R^2} (\Delta R)^2 \Big|_{R = R_0} \tag{8.143}$$

N being the number of ion pairs, and where the fact that E is a minimum at R_0 has been used. Equating W and ΔE gives

$$C = \frac{N}{3V_0} \left(\frac{R_0}{l_0} \right)^2 \frac{\partial^2 E}{\partial R^2} \Big|_{R = R_0} \tag{8.144}$$

Therefore, the bulk modulus is

$$K = -\frac{P}{\Delta V / V_0} = \frac{1}{9} \left(\frac{N}{V_0} \right) R_0^2 \frac{\partial^2 E}{\partial R^2} \Big|_{R = R_0} \tag{8.145}$$

where N/V_0 is the number of ion pairs/unit volume, and E is given in Eq. (8.3) with $n \approx 10$. For NaCl, the above expression gives an estimate of about 3.5×10^{10} J/m^3 (experimentally it is about 3×10^{10} J/m^3).

Example 2

It can be shown that when **a**, **b** and **c** are mutually orthogonal, the distance between the planes with Miller indices (h, k, l) is given by

$$d = \left(\frac{h^2}{a^2} + \frac{k^2}{b^2} + \frac{l^2}{c^2} \right)^{-1/2} \tag{8.146}$$

Let one of the planes have intercepts n_1a, n_2b, n_3c along the three axes (where n_1, n_2, n_3 are integers). A translation by an integral multiple of a, or b, or c, along the first, or the second, or the third axis, respectively, gives an equivalent plane. It is found that the number of equivalent planes between the origin and this plane is equal to the l.c.m. of n_1, n_2, and n_3, say N. On the other hand, the Miller indices are

$$h = \frac{N}{n_1}, k = \frac{N}{n_2}, l = \frac{N}{n_3} \tag{8.147}$$

If D is the perpendicular distance of the plane with intercepts n_1a, n_2b, n_3c from the origin, then

$$D = n_1 a \cos \alpha = n_2 b \cos \beta = n_3 c \cos \gamma \tag{8.148}$$

where α, β and γ are the angles made by the perpendicular line with the three axes. Since

$$\cos^2 \alpha + \cos^2 \beta + \cos^2 \gamma = 1, \tag{8.149}$$

$$D = \left(\frac{1}{n_1^2 a^2} + \frac{1}{n_2^2 b^2} + \frac{1}{n_3^2 c^2} \right)^{-1/2} \tag{8.150}$$

Using Eq. (8.147), the separation between two adjacent planes comes out to be

$$d = D/N = \left(\frac{h^2}{a^2} + \frac{k^2}{b^2} + \frac{l^2}{c^2} \right)^{-1/2} \tag{8.151}$$

Example 3

Hall effect provides a convenient method of determining the nature of charge carriers and their nonability.

Consider a current flowing in the x-direction, through a thin sheet in the xy plane. If a magnetic field Bz is applied to the current, the charge carriers are deflected by the $v \times B$ force and build an electric field E_y in the y direction. In the equilibrium condition

$$E_y + (\mathbf{v} \times \mathbf{B})_y = 0 \tag{8.152}$$

or

$$E_y = v_x B_z \tag{8.153}$$

Now the current in the x-direction is nqv_x, n being the carrier density and q their charge, so that the Hall coefficient is

$$R_H \equiv E_y/J_xB_z$$
$$= 1/nq \tag{8.154}$$

Thus, a measurement of E_y for a given J_x and B_z allows us to determine the concentration of the carriers as well as the sign of their charge. From Eq. (8.153) v_x and hence the mobility of the carriers can also be determined as:

$$\mu = v_x/E_x \tag{8.155}$$

Example 4

A silicon crystal contains an arsenic concentration of $1.2 \times 10^{22}/\text{m}^3$ and a boron concentration of $6 \times 10^{21}/\text{m}^3$. What is the density of majority and minority carries at room temperature?

The electrons in the conduction band and in the acceptor levels, are from the donor levels and the valence band:

$$n_c + n_a = h_d + h_v \tag{8.156}$$

This leads to

$$c_0(m_e{}^*T)^{3/2} \exp [(\varepsilon_f - \varepsilon_c)/kT] + \frac{N_a}{\exp[(\varepsilon_a - \varepsilon_f)/kT]+1}$$

$$= c_0\,(m_h{}^*T)^{3/2} \exp [(\varepsilon_v - \varepsilon_f)/kT] + \frac{N_d}{\exp[(\varepsilon_f - \varepsilon_d)/kT]+1} \tag{8.157}$$

where $c_0 = 2(2\pi k)^{3/2}/h^3$, $N_d = 1.2 \times 10^{22}$ m^{-3} and $N_a = 6 \times 10^{21}$ m^{-3}. Assuming the $\varepsilon_f - \varepsilon_a \gg kT$ and $\varepsilon_d - \varepsilon_f \gg kT$, and neglecting the first term on the rhs, gives

$$\varepsilon_f \approx \varepsilon_c + kT \ln\left[\frac{N_d - N_a}{c_0\,(m_e{}^*T)^{3/2}}\right] \tag{8.158}$$

Therefore $n_c \approx N_d - N_a$

$$h_v \approx c_0^2\,(m_e{}^*\,m_h{}^*\,T^2)^{3/2}\,\frac{1}{N_d - N_a} \exp [- (\varepsilon_c - \varepsilon_v/kT] \tag{8.159}$$

where at room temperature, $m_e{}^* \approx 0.25\ m$, $m_h{}^* * 0.3\ m$, and for the given concentrations,

$$\varepsilon_f - \varepsilon_c \approx -0.24 \text{ eV} \tag{8.160}$$

This result shows that the neglect of the first term on the rhs of Eq. (8.157) is justified.

Example 5

The effective masses of the carriers are determined from cyclotron resonance experiments. A resonance is observed in a Si crystal at 3×10^{10} Hz and a field of 0.4 T. What is the value of m^*?

The resonance condition is

$$q\, v\, B = m^*\, \omega\, v \tag{8.161}$$

which gives

$$m^* \approx 0.37\, m \tag{8.162}$$

where m is the electron mass.

Example 6

A germanium *pn* junction has 5×10^{22} phosphorus atoms/m^3 in the n-side and 3×10^{22} gallium atoms/m^3 in the p-side. What is the potential difference across the junction at room temperature? If the current for a large reverse bias is 5×10^{-8} A, what is the current for a forward bias of 0.4 V?

Assuming complete ionization of donor atoms and occupation of acceptor levels, one has the relations

$$c_0\, (m_e^*\, T)^{3/2} \exp\,[(\varepsilon_f - \varepsilon_c)/kT] = N_d \tag{8.163}$$

$$c_0\, (m_h^*\, T)^{3/2} \exp\,[(\varepsilon_v - \varepsilon_f^*)/kT] = N_a \tag{8.164}$$

with $N_d = 5 \times 10^{22}$ m^{-3}, $N_a = 3 \times 10^{22}$ m^{-3}, and $m_e^* \approx m_h^* \approx 0.1m$. This gives $\varepsilon_c - \varepsilon_f \approx 0.072$ eV and $\varepsilon_f' - \varepsilon_v \approx 0.085$ eV. Hence the potential difference across the junction is

$$\begin{aligned} V_0 &\approx 0.72 - 0.072 - 0.085 \\ &= 0.56 \text{ V} \end{aligned} \tag{8.165}$$

Since a large reverse bias gives a current of 5×10^{-8} A, the forward bias current is

$$I \approx 5 \times 10^{-8}\, (\exp\,[e\Delta V/kT] - 1) \tag{8.166}$$

which for $\Delta V = 0.4$ V gives $I \approx 0.24$ A.

Example 7

The diamagnetic susceptibility of helium can be estimated from the approximate helium wave function (see Example 5 of Sec. 5.8)

$$\psi = \left(\frac{1}{\pi a'^3}\right) \exp[-(r_1 + r_2)/a'] \tag{8.167}$$

$a' = 4\pi\varepsilon_0\hbar^2/me^2Z'$, $Z' = 27/16$. The diamagnetic susceptibility is obtained from Eq. (8.89) to be

$$\chi = -\frac{e^2\mu_0 N}{m} a'^2 \tag{8.168}$$

which comes out to be 2.1×10^{-8}/kg mole (1.67×10^{-6}/g mol in Gaussian units, compared to the experimental value of 1.9×10^{-6}/g mol).

Example 8

A ferromagnetic material with $J = 3/2$ and $g = 2$ has a transition temperature $T_c = 120$ K. Calculate the internal field near 0 K. What is the ratio of magnetization at 300 K for $B = 5 \times 10^{-3}$ T compared to that at 0 K?

From the expression for T_c in Eq. (8.111), and

$$B_{\text{int}} = \lambda\left(N\frac{e\hbar g}{2m}J\right) \text{ at 0 K,}$$

one has

$$B_{\text{int}} = \frac{6mkT_c}{e\hbar g(J+1)} \tag{8.169}$$

which is about 108 T. The ratio of magnetization at 300 K to that at 0 K, is

$$R = \frac{B_0(J+1)}{3k(T-T_c)}\left(\frac{e\hbar g}{2m}\right)$$

$$= 3.1 \times 10^{-5} \tag{8.170}$$

which illustrates the fact that paramagnetic effects are, in general, much smaller than ferromagnetic effects.

Example 9

When a photon is incident on a material with an energy gap ΔE, an electron in the valence band may absorb this radiation and go to the conduction band if $h\nu > \Delta E$. The kinetic energy of the electron is given by

$$\frac{1}{2}mv^2 = h\,\nu - \Delta E - (\varepsilon_v - \varepsilon) \tag{8.171}$$

where ε is the initial energy of the valence electron, the maximum energy being observed for $\varepsilon = \varepsilon_v$. Thus, a rapid increase is observed in the absorptivity of radiation as ν increases through the value of $\nu = \Delta E/h$. This property is used to determine the energy gap (the experiments are usually done at low temperatures to reduce thermal effects). Since $\Delta E \sim 1$ eV for semiconductors, they are essentially transparent to infrared radiation but absorb most of the radiation in the optical region.

The excited electrons are de-excited either immediately, in general emitting radiation (*fluorescence*) of a different frequency than that of the original photon, or wander around in the crystal until they are trapped at the luminescent centers (usually produced by adding chemical impurities, *e.g.*, zinc sulphide). The trapped electrons then are de-excited, with the emission of radiation after a time delay (*phosphorescence*). Fluorescence and phosphorescence are together known as *luminescence* and are used in fluorescent lamps, television picture tubes, etc.

PROBLEMS

1. It requires an energy of 5.14 eV to remove the valence electron of Na and an energy of 3.80 eV is released when an electron is added to Cl. Assuming a value of $n = 10$ in Eq. (8.3), and an interatomic spacing of 2.82 Å, obtain the cohesive energy/ion pair, and the repulsive energy.

2. Show that the Madelung constant for an infinite array of alternating positive and negative charges in 1-dimension is $\alpha_1 = 2 \ln 2$. Show that the expressions in 2- and 3-dimensions are:

$$\alpha_2 = 2\alpha_1 - 4 \sum_{n, m=1}^{\infty} \frac{(-1)^{n+m}}{(n^2 + m^2)^{1/2}}$$

$$\alpha_3 = 3\alpha_2 - 3\alpha_1 - 8 \sum_{l, m, n=1}^{\infty} \frac{(-1)^{l+m+n}}{(l^2 + m^2 + n^2)^{1/2}}.$$

3. If the repulsive force is of the form $Ce^{-r/a}$, determine C and a for NaCl if the cohesive energy/ion pair is 6.61 eV, and the interatomic separation is 2.82 Å.

4. It is observed that *x*-rays of wavelength 1.2 Å produce a first order maximum at a Bragg angle of 12.3° when reflected by the (1, 0, 0) planes of NaCl (which has fcc structure as in Fig. 8.1). If the density of NaCl is 2.165 g/cm^3 and its molecular weight is 58.454, obtain the value of Avogadro's number.

5. For a cubic crystal of unit length 10^{-10} m, at what angles will the first order maxima be observed for (1,1, 1), (1, 1, 0) and (1, 0, 0) planes? The incident *x*-ray has a wavelength of 1 Å. Will the second order maxima be observed? Will be (2, 1, 0) planes produce maxima in this case?

6. Hard spheres of radius R are arranged in contact in simple cubic, bcc and fcc structures. Find the radius of the largest sphere that can fit into the largest interstices of these structures.

7. Iron undergoes a phase transition from bcc (at lower temperature) to fcc at 1180 K. If there is no change in the density show that the ratio of the nearest neighbour separation increases by a factor of about 1.029.

8. If an element contains both electron and hole carriers, show that the Hall coefficient is given by

$$R_H = \frac{n_h\mu_h^2 - n_e\mu_e^2}{e(n_h\mu_h + n_e\mu_e)^2}$$

A material has 10^{21} electrons/m^3 and 5×10^{20} holes/m^3. If $\mu_e = 0.05$ and $\mu_h = 0.07$ in MKS units, evaluate the conductivity and the Hall coefficient of the material.

9. Hall coefficient of A1 is $- 0.3 \times 10^{-10}$ MKS units. How many conduction electrons does each atom contribute?

10. The conductivity of germanium is $0.7\Omega^{-1}m^{-1}$ at 0° C and $2\Omega^{-1}m^{-1}$ at 20°C. What is the energy gap for germanium?

11. A measurement of 0.1% change in resistivity is possible in a silicon crystal. What is the sensitivity at room temperature of such a crystal used as a thermistor?

12. What is the conductivity at room temperature of (*i*) pure silicon (*ii*) silicon containing 10^{-5}% of phosphorus? Mobility of electrons is 0.14m^2/Vs, that of holes is 0.05 m^2/Vs and the number of charge carriers in pure silicon is about 2×10^{16} m^{-3}, at room temperature.

13. A current of 10^{-4} A flows when a forward bias of 0.2 V is applied at room temperature. Obtain the currents if (*i*) forward bias of 0.4 V (*ii*) reverse bias of 1 V, are applied.

14. Show that the width of the depletion region at a *pn* junction, when a forward bias of *V* is applied, is given by

$$x = \left[\frac{2\kappa\varepsilon_0(n_e + n_h)(V_0 - V)}{en_en_h}\right]^{1/2}$$

where V_0 is the equilibrium potential difference across the junction. What is the width for a junction with $n_e = 10^{22}$ m^{-3}, $n_h = 10^{22}$ m^{-3}, $\kappa = 14$, $V_0 = 0.7$ V, if a reverse bias of 2 V is applied?

15. Show that the product of electron and hole carriers is

$$n_en_h = c_0^2(m_h{*}m_e{*}T^2)^{3/2} \exp\,[(\varepsilon_v - \varepsilon_c)/kT],$$

independent of the concentration of the impurity. Thus, if an *n*-type of impurity is introduced, the number of holes decreases.

16. A silicon semiconductor is doped with 9×10^{21} donors/m^3 and 4×10^{21} acceptors/m^3. If the electron mobility is 0.15 m^{2}V^{-1}s^{-1}, estimate the resistivity at room temperature.

17. Neglecting the number of holes in the valence band of an n-type semiconductor, show that

$$\varepsilon_f \approx \varepsilon_d + kT \ln \left[\left(c_1 + \frac{1}{4} \right)^{1/2} - \frac{1}{2} \right]$$

where $\boldsymbol{c}_1 = \dfrac{N_d h^3}{2(2\pi k m_e^* T)^{3/2}}$ exp $[(\varepsilon_c - \varepsilon_d)/kT]$. This expression gives the correct expression at $T \to 0$ as well as at room temperature.

18. An electron does not tunnel across the depletion region if the width of the layer is more than 10^{-8} m. What is the minimum doping needed for a silicon tunnel-diode to operate? Take $n_e \approx n_h$ for the two impurities, $\kappa \approx 12$ for Si.

19. A silicon pn junction has 10^{23} gallium atoms/m^3 and 10^{22} arsenic atoms/m^3. What is the approximate potential difference across the junction?

20. The main contribution to parmagnetism of copper sulphate comes from the copper ions which have spin 1/2. Show that its magnetization is given by

$$M = N \left(\frac{e\hbar}{2m} \right) \tanh \left(\frac{e\hbar B}{2mkT} \right).$$

21. For a substance containing paramagnetic ions with $S = 1/2$ and orbital angular momentum quenched, derive an expression for the energy and specific heat of the substance. Discuss its high- and low-temperature limits.

22. A magnetic field is applied to a salt containing Cu^{2+} ions. Given that Cu^{2+} has nine $3d$ electrons, determine the field at which 90% of the ions are in the ground state at 1 K.

23. Show that the magnetization in a ferromagnet tends to a value (use Eq. (8.99) with $a = e\hbar g\lambda\, M/2m$)

$$M \to \frac{Ne\hbar g\, J}{2m} \left\{ 1 - \frac{1}{J} \exp \left[- \left(\frac{e\hbar g}{2m} \right)^2 \frac{Ng\lambda}{kT} \right] \right\}$$

for $T \to 0$ K. However, a more sophisticated calculation in terms of spin waves and magnons shows that the second term vanishes at $T^{3/2}$ rather than an exponential.

9

The Nucleus

Structures of the Chapter

Rutherford's experiment (1911) indicated the existence of a heavy, positively charged nucleus of very small dimensions near the centre of the atoms. It was found that the scattering of α-particles *i.e.* ionized helium atoms, by a thin, metal foil could be described by assigning a charge of Ze to this nucleus. However, it was also noted that when the distance of closest approach was less than 10^{-14} m, the scattering showed deviations from Coulomb scattering, indicating the extension of the nucleus over a distance of 10^{-14} m. Indeed, the descriptions of the details require the existence of a new, shortchange interaction known as the *strong interaction.* The properties of the nuclei and their interactions are quite different from those of an atom, and will be briefly discussed here.

Some of the important properties of the nucleus, such as its mass, size, magnetic moment, etc. will be discussed first. This will be followed by an analysis of nuclear forces and different models of the nucleus. Finally, stability criteria, nuclear reactions and fission and fusion processes will be discussed. At this stage, it may be mentioned that while modern experimental techniques have provided quite detailed information about nuclear properties, an entirely satisfactory framework for the quantitative prediction of these properties has not been formulated. This is primarily because (*i*) nuclear force appear to be structurally much more complicated than electromagnetic forces, and (*ii*) the strength of nuclear forces is quite large which means that perturbative methods cannot be used for calculation.

9.1 PROPERTIES OF THE NUCLEUS

Nuclear properties are most simply described in terms of the nuclear constituents.

Nuclear Constituents

The nucleus is made up of *protons* and *neutrons.* The proton is the nucleus of the simplest atom, the hydrogen atom. It has a rest energy of 938.256 MeV or a mass of 1.0072766 mu (1 atomic mass unit, mu, is equal to 1/12 of the ^{12}C mass and corresponds to 931.478 MeV), a positive charge of e and spin $\hbar/2$. The neutron has a rest energy of 939.550 MeV or a mass of 1.0086654 mu, zero net charge and spin $\hbar/2$. Since protons and neutrons have half-integral spin, they are fermions and satisfy Fermi-Dirac statistics. The near-equality of the neutron and proton masses is an important property and it has a bearing on nuclear interactions.

Both the proton and the neutron have magnetic moments given by

$$\mu_p = g_p \frac{e}{m_p} \mathbf{s} \tag{9.1}$$

$$\mu_n = - g_n \frac{e}{m_p} \mathbf{s}$$

where $g_p = 2.793$, $g_n = 1.913$. A Dirac particle without any structure would be expected to have $g_p = 1$ and $g_n = 0$. The observed values for the magnetic moments suggest a complicated structure for the proton and the neutron. It may be noted that these magnetic moments are smaller than the electronic magnetic moments by a factor of about m_e/m_p.

While the proton is stable (some recent theories however, predict that the proton decays with a lifetime of about $10^{31} - 10^{32}$ years), the neutron decays,

$$n \to p + e + \overline{\nu} \tag{9.2}$$

where $\overline{\nu}$ is the antineutrino (neutrino ν and its antiparticle antineutrino $\overline{\nu}$ are zero-mass, spin 1/2, neutral particles, see Sec. 10.1), with a half-life of about 11 minutes. The decay process is known as the neutron β-decay. Protons and neutrons are together known as *nucleons.*

Binding Energies

Since nuclear forces are strong, nuclear binding energies are a significant fraction of nuclear masses. Thus, nuclear binding energies can be obtained from the masses.

The binding energy E_b of a nucleus of mass m_A, containing Z protons and $(A - Z)$ neutrons, is

$$E_b = c^2 \, [Zm_p + (A - Z)m_n - m_A] \tag{9.3}$$

This is the minimum energy required to separate the nucleus into its constituent nucleons. A nucleus with Z protons and A nucleons is said to have a charge or atomic number Z and mass number A, and is designated by ${}^{A}_{Z}X$ where X stands for the chemical symbol of the nucleus (*e.g.* one has ${}^{1}_{1}\text{H}$, ${}^{2}_{1}\text{H}$, ${}^{3}_{1}\text{H}$ nuclei with 0, 1, 2 neutrons respectively). Nuclei with the same number of protons but different number of neutrons are known as *isotopes e.g.* ${}^{1}_{1}\text{H}$, ${}^{2}_{1}\text{H}$ etc. or ${}^{16}_{8}\text{O}$, ${}^{17}_{8}\text{O}$, etc. Nuclei with the same number of neutrons but different number of protons, are called *isotones, e.g.* ${}^{4}_{2}\text{He}$, ${}^{5}_{3}\text{Li}$, and nuclei with the same A but different Z are called *isobars, e.g.* ${}^{5}_{2}\text{He}$, ${}^{5}_{3}\text{Li}$. One also has some nuclei which are excited states of a stable nucleus, but which have a very long lifetime (say $\tau > 0.1$ s). These are called *isomers*.

A very useful concept in nuclear physics is the binding energy per nucleon, E_b/A. It starts from a very low value of $E_b/A \approx 1.1$ MeV for the deuteron ($E_b \approx 2.225$ MeV), rapidly increases to 7.1 MeV for the α-particle, *i.e.* ${}^{4}_{2}\text{He}$, reaches a peak value of about 8.7 MeV for $A \approx 56$. For nuclei with larger A, it decreases slowly reaching a value of about 7.5 MeV for the heaviest natural element, uranium. The general dependence of the binding energy per nucleon on the mass number is shown in Fig. 9.1, for the stable nuclei. Two important results follow from the general behaviour of E_b/A: Energy can be released (*i*) in the fission of a heavy nucleus into lighter nuclei, and (*ii*) in the fusion of lighter nuclei into a heavier nucleus. For example, a nucleus with $A = 220$ ($E_b/A \approx 7.5$ MeV), breaking into two nuclei with $A = 110$ each ($E_b/A \approx 8.5$ MeV) will liberate an energy of about $220 \times (8.5 - 7.5) = 220$ MeV. Similarly two ${}^2_1H$ nuclei ($E_b/A \approx 1.1$ MeV) can combine into a ${}^4_2H$ nucleus ($E_b/A \approx 7.1$ MeV) to liberate an energy of about $4 \times (7.1 - 1.1) = 24$ MeV. These energies are very large compared to the few electron volts released in chemical reactions which are governed by electromagnetic forces.

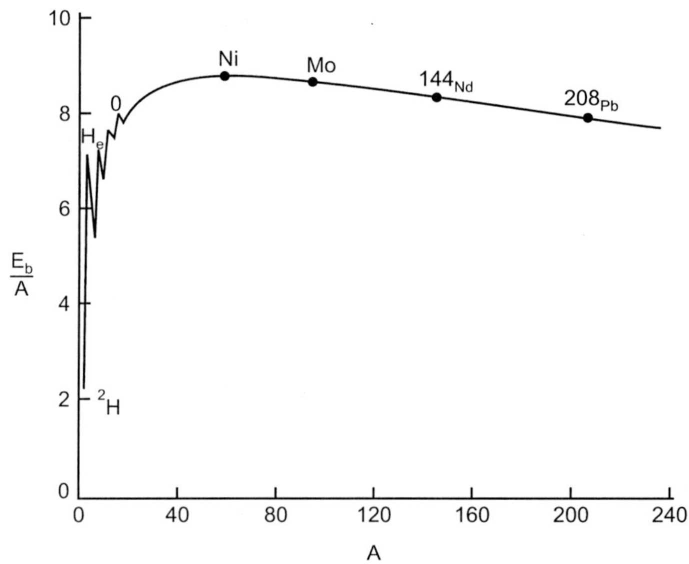

Fig. 9.1 The general behaviour of the binding energy per nucleon as a function of *A*.

Though the fission and fusion processes leading to nuclei with $A \approx 55$ are feasible, it is observed that most of the nuclei are stable. The reason for this is that before a heavy nucleus breaks up, the components must go through an intermediate state with higher energy than the ground state (this can be induced by providing extra energy available in the capture of a neutron). Similarly, lighter nuclei encounter a higher energy intermediate state with large Coulomb repulsion, before they can combine (the fusion can take place at high temperature, *e.g.* in stars).

Unstable Nuclei

If a lower energy state is available to a nucleus, it will, in general, be unstable and decay with the emission of a photon, or an α-particle, or some leptons (*i.e.* e, ν etc. see Sec. 10.1), provided the basic conservation laws, such as conservation of energy, momentum, charge, etc. allow the decay. The neutron β-decay in Eq. (9.2) is one such example. These decays are characterized by the lifetime τ of the nucleus (the lifetime was discussed in Sec. 6.4), such that the number of undecayed nuclei $N(t)$ is given by

$$N(t) = N(0) \exp(-t/\tau) \tag{9.4}$$

where $N(0)$ is the number of nuclei at time $t = 0$. The lifetimes of the nuclei vary from the unmeasurably small values of $\tau < 10^{-6}$ s (they may be indirectly estimated), to the very large values of $\tau \sim 10^{100}$ years.

The important classes of nuclear decays are the following:

1. α-*decay*: It may be described by the process

$$^{A}_{Z}\mathbf{X} \rightarrow {}^{A-4}_{Z-2}\mathbf{Y} + {}^{4}_{2}\mathbf{He} \tag{9.5}$$

an example of which is

$$^{238}_{92}\mathrm{U} \rightarrow {}^{234}_{90}\mathrm{Th} + {}^{4}_{2}\mathrm{He} \tag{9.6}$$

2. γ-*decay*: In a γ-decay, an excited nucleus undergoes transition to a lower energy state by the emission of a photon. This may be represented by

$$\mathrm{X}^* \rightarrow \mathrm{X} + \gamma \tag{9.7}$$

where X* is the excited state. The photon energies in nuclear transitions are of the order of an MeV compared with the few eV in atomic transitions, and the corresponding lifetimes are of the order of 10^{-14} s (compared to $t \sim 10^{-15}/10^{8} = 10^{-23}$ s required for a relativistic particle to traverse a nucleus).

3. β-*decay*: These processes involve electrons and neutrinos, and are exemplified by

$$^{A}_{Z}\mathrm{X} \rightarrow {}_{Z+1}^{\;\;A}\mathrm{Y} + e + \overline{\nu} \tag{9.8}$$

$$^{A}_{Z}\mathrm{X} \rightarrow {}_{Z-1}^{\;\;A}\mathrm{Y} + \overline{e} + \nu \tag{9.9}$$

$$^{A}_{Z}\mathrm{X} + e \rightarrow {}_{Z-1}^{\;\;A}\mathrm{Y} + \nu \tag{9.10}$$

where e and $\overline{e}$ are the electron and the positron, and ν and $\overline{\nu}$ are the neutrino and the antineutrino. In the electron-capture process [Eq. (9.10)], the absorbed electron is usually from the atomic shells.

The activity of the unstable nuclei, known as *radioactivity*, is measured in terms of the *curie* which corresponds to 3.7×10^{10} disintegrations/s.

Nuclear Radius

The wave-function description of a particle does not provide an unambiguous description of the size of a particle. However, since nuclear forces are large only within a distance of a few fermis (1 fermi = 10^{-15} m), it is useful to consider the size of the nucleus. The nuclear radius may be estimated from the scattering of neutrons and electrons by the nucleus, or by analysing the effect of the finite size of the nucleus on nuclear and atomic binding energies.

Fast neutrons of about 100 MeV energy, whose wavelength is small compared to the size of the nucleus, are scattered by nuclear targets. The fraction of neutrons scattered at various angles can be used to deduce the nuclear size. For example, in the scattering of a high energy particle by a hard sphere, $V = \infty$ for $r < R$, $V = 0$ for $r > R$, all the incident particles within a cross-sectional area of $2\pi R^2$ are scattered. The factor of 2 is due to the diffraction of the waves at the edges. The results of these experiments indicate that the radius of a nucleus is given by

$$R \approx r_0 A^{1/3} \tag{9.11}$$

where A is the mass number and $r_0 \approx 1.3 - 1.4$ fm. The scattering can be done with proton beams as well. In this case, however, the effects due to Coulomb interaction have to be separated out. The observations are in agreement with the result in Eq. (9.11) with $r_0 \approx 1.3 - 1.4$ fm.

The scattering of fast electrons of energy as high as 10^4 MeV, with a wavelength of about 0.1 fm, has the advantage that it can directly measure the charge density inside a nucleus. The results of the experiment are in agreement with Eq. (9.11) but with a somewhat smaller value of $r_0 \approx 1.2$ fm. The slight difference in the value of r_0 may be ascribed to the fact that the electron scattering measures the charge density whereas the neutron and proton scattering experiments measure the region of large nuclear potential, which may be expected to be somewhat larger than the size of the nucleus.

The finite size of the nucleus modifies the atomic potential ($-Z/r$) at short distances. This gives rise to a small separation between the spectral lines of atoms with the same Z value but different A values—this is known as *isotope shift*. The shifts can be used to deduce the nuclear ratius. The isotope shift is much larger in muonic atoms (which have a muon in place of an electron) since the radii of the muonic orbits are smaller than the electronic orbits by a factor of about 200 ($m_\mu \approx 200\ m_e$). However, the accuracy of measurements is muonic atoms is lower since the muons have a short lifetime, about 2×100^{-6} s. Finally, the measurement of differences in the binding energies of mirror nuclei can give an estimation of the nuclear radius. The mirror nuclei are nuclei which are identical except that one proton is replaced by a neutron. They may be characterized by ${}^{2Z+1}_{Z+1}\mathrm{X}$, ${}^{2Z+1}_{Z}\mathrm{Y}$. The difference between their binding energies may be ascribed to the two different charges. A model calculation with assumed charge distribution then provides an estimation for the nuclear radius. All these approaches are in essential agreement with Eq. (9.11) with $r_0 \approx 1.2$ fm.

An important consequence of Eq. (9.11) is that the volume per nucleon is the same for all nuclei:

$$V_1 = \frac{4\pi}{3}(r_0 A^{1/3})^3/A$$

$$= \frac{4\pi}{3} r_0^3 \quad (9.12)$$

Thus, the nuclear density is the same for all nuclei. The result is in agreement with what might be expected from the strong, short-range forces in nuclei. Furthermore, it implies that the nuclear forces are independent of the charge of the nucleons. This is known as *charge independence* on nuclear forces.

Angular Momentum and Magnetic Moment

The total angular momentum of nuclei is made up of the spins and orbital angular momenta of the constituent nucleons. Associated with the angular momentum is a magnetic moment.

The angular momentum of the nuclei can be deduced from the hyperfine interaction between the magnetic moments of the nuclei and of the electrons. This interaction is of the form

$$H = A\, \mathbf{I}.\mathbf{J} \quad (9.13)$$

with I and J being the angular momenta of the nucleus and of the electrons respectively. The atomic states are characterized by the total angular momentum

$$\mathbf{F} = \mathbf{I} + \mathbf{J} \quad (9.14)$$

and the corresponding quantum number takes on the values

$$F = J + I, J + I - 1, ..., |J - I| \quad (9.15)$$

The shift in the energy of these states is given by

$$\Delta E = \frac{1}{2} A\hbar^2 [F(F+1) - I(I+1) - J(J+1)] \quad (9.16)$$

which leads to a separation of

$$E_F - E_{F-1} = A\hbar^2 (I + J), A\hbar^2 (I + J - 1), ..., A\hbar^2 (|I - J| + 1) \quad (9.17)$$

between the successive states in Eq. (9.15). The analysis of the spectral lines corresponding to these levels gives the value of I (and also of J).

The spin of the nuclei can also be determined from the spectra of homonuclear molecules. It was observed in Chapter 5 that the transitions in *para* and *ortho* modifications of a homonuclear molecule, have intensities in the ratio of $I/(I+1)$ so that the rotational band will show alternating intensity. A measurement of these intensities allows the determination of the angular momentum I of the nucleus especially for small I values. It is worth noting that this method depends on the exchange symmetries of the wave functions and not on the magnetic moment associated with the nucleus.

The magnetic moment of a nucleus is associated with its angular momentum **I**, and may be expressed (at least for the purpose of calculating the expectation values within multiplets) as

$$\mu = \frac{e}{m_p} g\,\mathbf{I} \tag{9.18}$$

The gyromagnetic ratio g can be obtained from the nuclear magnetic resonance experiments using atomic beams. From the resonance frequency

$$\omega = \frac{e}{m_p} gB \tag{9.19}$$

g and hence the magnetic moment is obtained.

Empirically, it is observed that even A and even Z nuclei have zero angular momentum and zero magnetic moment, even A and odd Z nuclei have integral angular momentum, and odd A nuclei have half-integral angular momentum. The angular momenta of nuclei are generally found to be small. These observations suggest that the angular momenta of protons, as also on neutrons, separately compensate one another. This has a bearing on the validity of nuclear models.

Electric Quadrupole Moment

A nucleus is usually non-spherical (though in a few cases it may be spherical). The distortion which is along the axis of rotation is expressed in terms of the electric quadrupole moment Q,

$$Q = \frac{1}{e}\int (3z^2 - r^2)\,\rho(r)\,dV \tag{9.20}$$

where $\rho(r)$ is the charge density distribution in the nucleus. For a spherically symmetric $\rho(r)$, Q is zero whereas for an ellipsoidal (ellipsoid is obtained by rotating an ellipse about one of its axes) distribution,

$$Q = \frac{2}{5}\left(\frac{q}{e}\right)(a^2 - b^2) \tag{9.21}$$

where a is the semi-axis along the axis of rotation and q is the total charge ($Q > 0$ implies and *elongated* or *prolate* nucleus and $Q < 0$ implies a *flattened* or *oblate nucleus*).

The nuclear quadrupole moments are determined from their effect on the hyperfine structure of the atomic spectra. The observed values of Q range from $Q = -10^{-28}$ m^2 for ^{123}Sb to 8×10^{-28} m^2 for ^{176}Lu, while deuteron has a value of $Q = 2.74 \times 10^{-31}$ m^2.

9.2 NUCLEAR FORCES

The forces that bind nucleons together into a nucleus are very strong forces as indicated by the large binding energies, and have a very complicated structure. These forces are described by what is known as strong interactions. Several important characteristics of these forces follow from a general analysis of the nuclear properties.

1. The nuclear forces are strong, their magnitude being roughly 100 times that of electromagnetic forces. This follows from the large nuclear binding energies.
2. The nuclear forces have a short range. They are dominant over a distance of about 1 fm but vanish rapidly at distances greater than a few fermis. This explains the approximate constancy of nuclear density as well as of binding energy per nucleon. Roughly speaking, the short range of the forces implies that each nucleon interacts with only a small number of nearby nucleons.
3. Nuclear forces are independent of the nuclear charges. It is indeed a striking property that the proton and the neutron have nearly the same mass. It is convenient to ascribe to the nucleons an isotopic spin (or isospin for short) $\tau = 1/2$, the $\tau_2 - 1/2$, $- 1/2$ states corresponding to the proton and the neutron respectively. The properties of the isospin are similar to those of the ordinary spin, and the charge independence of the nuclear forces is equivalently described by the statement that the nuclear forces are independent of the orientation of the isotopic spin.
4. Nuclear forces are not central forces. In particular, they depend on the orientation of the spin. This is forcibly demonstrated by the observation of the deuteron as an $S = 1$ bound state of a proton and a neutron; no such bound state is observed in the $S = 0$ state.

Some important aspects of the nuclear forces are described subsequently.

Yukawa Forces

One of the important modern ideas of forces is that forces between particles arise from the exchange of particles. The form of the resulting potential can be deduced from the following arguments.

The electromagnetic potential (only the scalar potential ϕ is considered here) satisfies the equation

$$\left(\nabla^2 - \frac{1}{c^2}\frac{\partial^2}{\partial t^2}\right)\phi = 0 \tag{9.22}$$

in free space. This equation may be regarded as arising from the relation $\frac{1}{\hbar^2 c^2}(E^2 - c^2\,\mathbf{p}^2) = 0$ for the photon with zero mass, by the quantum mechanical replacements in Eqs. (3.10) and (3.11). For a particle with nonzero mass m, one may instead start with the relation

$$\frac{1}{\hbar^2 c^2}(E^2 - c^2\mathbf{p}^2 - m^2c^4) = 0 \tag{9.23}$$

Implementing the quantum mechanical replacements in Eqs. (3.10) and (3.11) and including a point source, the equation for the potential comes out to be

$$\left(\nabla^2 - \frac{1}{c^2}\frac{\partial^2}{\partial t^2} - \frac{m^2c^2}{\hbar^2}\right)\phi_m = g\,\delta(\mathbf{r}) \tag{9.24}$$

where g is a constant. The static solution to this equation is found to be

$$\phi_m = -\frac{g}{4\pi r}\exp\left[-r(mc/h)\right] \tag{9.25}$$

which is the well known *Yukawa potential*. This potential has an approximate range of r_0 given by

$$r_0 = \hbar/mc \tag{9.26}$$

i.e. essentially the Compton wavelenght of the quantum of the field exchanged. The potential decreases very rapidly for $r \gg r_0$. Yukawa argued that the nuclear forces, which have a range of $r_0 \approx 10^{-15}$ m, arise from the exchange of a particle of mass $m \approx h/r_0\, c$, which for $r_0 \approx 10^{-15}$ m, comes out to be (expressed as rest energy)

$$m \approx 200 \text{ MeV} \tag{9.27}$$

The π-meson with a mass of about 140 MeV, would be a good candidate for the quantum whose exchange gives rise to nuclear forces. Of course, there are additional contribution to the interaction from the exchange of other, heavier particles but with correspondingly shorted ranges [see Eq. (9.26)].

The short-range nature of the nuclear force is due to the rapidly-decreasing exponential function. In contrast, the electromagnetic forces arising from the exchange of zero-mass photons, have a long range.

Nucleon-Nucleon Interaction

The forces between two nucleons from the basis of nuclear interactions. An important part of these forces is from the exchange of π-mesons.

A nucleon is continuously emitting and absorbing virtual π-mesons, and is effectively surrounded by a cloud formed by them. These processes are regarded as virtual since total energy and momentum conservation would forbid them, but the uncertainty principle allows them to take place over short distances and times. The mesons emitted by the nucleon may be absorbed by another nucleon (Fig. 9.2). This gives rise to the forces between the nucleons.

The exchange of a charged meson [Fig. 9.2(*b*)] gives rise to charge-exchange forces. They are observed, for example, when a beam of neutrons passes through hydrogen. The exchange of charged mesons converts some of the neutrons into protons which are then observed in the beam with almost the same energy and momentum as the initial neutrons. Similar to the exchange of charges are processes that lead to an exchange of spin, and exchange of charge and spin. There are additional forces due to relativistic corrections, many-body forces, etc. Clearly, the nuclear forces, even in a two-nucleon system, are very complicated.

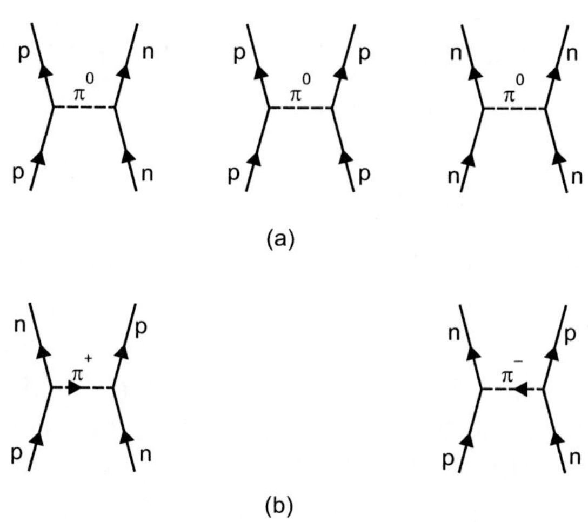

Fig. 9.2 Diagrammatic representation of pion-exchange interaction: (*a*) the exchanged π^0 may travel in either direction, (*b*) exchange of π^+ and π^-.

Though nuclear forces are complicated, their charge independence leads to some simple relations. In particular,

$$V_{pp} \approx V_{nn} \tag{9.28}$$

$$V_{pp} \approx V_{np} \text{ in the same state}$$

The second relation, Eq. (9.28), is valid only for the states which are allowed by Fermi-Dirac statistics for the two protons, *i.e.* even l for $S = 0$ and odd l for $S = 1$. The electromagnetic interaction will violate charge independence and introduce small corrections to the above relations.

The model of a nucleon surrounded by a cloud of virtual π-mesons, provides a qualitative explanation for the magnetic moments of the proton and the neutron. In this picture, a neutron spends part of the time in the virtual $(p + \pi^-)$ state. In this virtual state, the orbital motion of π^- gives rise to a substantial negative, magnetic moment. Similarly a proton spends part of the time in the virtual $(n + \pi^+)$ state. In the $(n + \pi^+)$ state, the orbital motion of π^+ gives rise to a large positive, magnetic moment. These descriptions are in qualitative agreement with the observed gyromagnetic ratios for the neutron and the proton [Eq. (9.1)].

Strength of Nuclear Interaction

The strength of nuclear interaction can be estimated from the binding energy of the deuteron. The energy of deuteron may by written as:

$$E = \frac{1}{2m_p} p_1^2 + \frac{1}{2m_n} p_2^2 + V \tag{9.29}$$

If the deuteron is regarded as a sphere of radius r_0, the uncertainty principle gives:

$$E \approx \frac{h^2}{m_p r_0^2} + V \tag{9.30}$$

where V is the average potential energy. Taking $r_0 \approx 1$ fm and $E \approx -2$ MeV (binding energy of the deuteron)

$$V \approx -40 \text{ MeV} \tag{9.31}$$

This may be compared with an electrostatic energy of about 1 MeV when two protons are separated by a distance of about 1 fm. Nuclear interaction is thus seen to be such stronger than electromagnetic interaction, which explains the term strong interaction used to describe it.

9.3 MODELS OF THE NUCLEUS

An investigation of the properties of a nucleus with several nucleons is immensely difficult both because of the complexity of two-nucleon forces and the absence of a dominant central force. Therefore, model calculations, each of which has a limited aim of investigating only certain aspects of the nuclear properties, have to be done. Here, a few nuclear models which together provide some understanding of the overall structure of the nucleus are considered.

Shell Model

In the shell model, nucleons are assumed to move independently of each other in an average, centrally-symmetric potential. They occupy discrete energy levels in this potential, taking the Pauli exclusion principle into account. The grouping together of some of the energy levels gives rise to a shell structure of the nucleus. The independent motion or equivalently the long, mean free path is partially justified by the Pauli principle which forbids transitions to states that are already occupied.

Experimentally, it is found that nuclei with the number of neutrons or protons equal to,

$$Z \text{ or } (A - Z) = 2, 8, 20, 28, 50, 82, 126 \tag{9.32}$$

are especially stable. These numbers are known as *magic numbers*. The stability is particularly pronounced for nuclei with both the number of protons and neutrons equal to magic numbers, *e.g.* ${}^{4}_{2}\text{He}$, ${}^{16}_{8}\text{O}$, ${}^{40}_{20}\text{Ca}$, ${}^{48}_{20}\text{Ca}$, ${}^{208}_{82}\text{Pb}$. The stability of ${}^{4}_{2}\text{He}$ leads to its being the only composite nucleus emitted in radioactive decays. It is also found that the 3rd, 9th, 21st, 29th, 51st, 83rd and 127th neutron or proton is loosely bound (in fact ${}^{5}_{2}\text{He}$ is unstable).

For describing the shell structure of nuclei, two of the potentials used are the spherical-well potential and the simple harmonic oscillator potential. Since the oscillator levels have already been deduced (Sec. 3.12, example 6), the details for this case are given. The wave functions of the 3-dimensional oscillator are products of the 1-dimensional wave functions and the energies are the sums of the corresponding, 1-dimensional, equispaced energy levels:

$$E = \hbar\omega\,(n + 3/2),\ n = n_x + n_y + n_z \tag{9.33}$$

Now, corresponding to each n level, there are several degenerate states, *e.g.* for $n = 1$, there are three states with n_x or n_y or n_z equal to 1. These states correspond to states with different angular momenta ($n = 0, l = 0$), ($n = 1, l = 1$), ($n = 2, l = 2, 0$) etc. with the number of corresponding states (taking spin states into account), being 2, 6, 12 etc. However, the actual potential cannot be simulated by the oscillator potential at large distances. Since it goes to zero at large distances, the larger angular momentum states are lowered with respect to the smaller angular momentum states. The ordering of the energy levels with the degeneracy removed is shown in Fig. (9.3). It should be noted that while these levels explain the magic numbers 2, 8, 20 corresponding to closed shells, they cannot explain the other magic numbers.

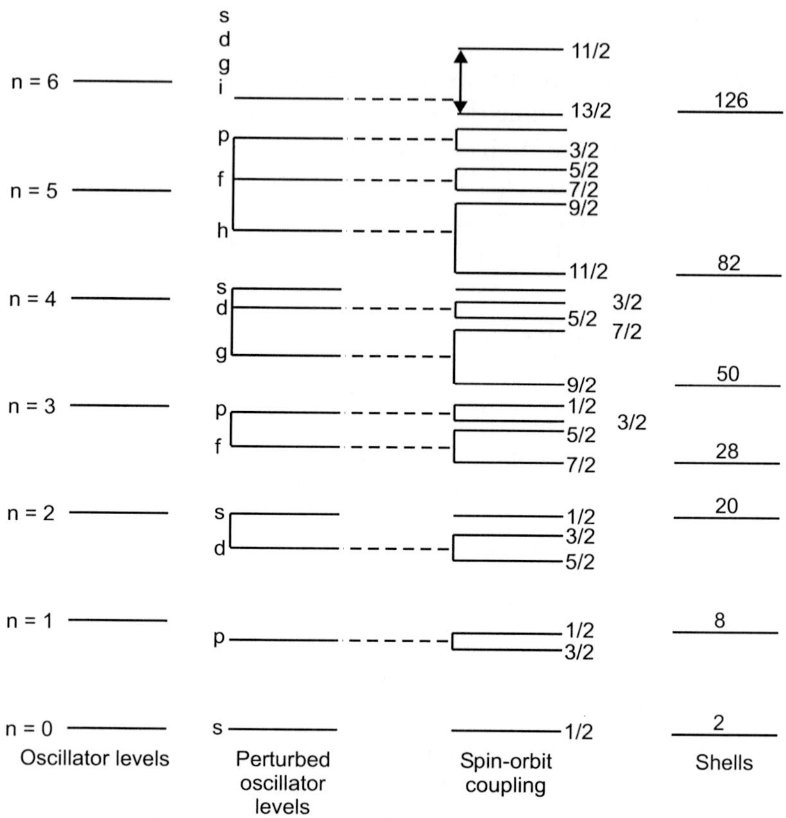

Fig. 9.3 Nuclear energy shells arising from the perturbed oscillator levels with spin-orbit coupling.

To account for the observed magic numbers, Mayer and Jensen (1949) postulated a strong spin-orbit interaction. The postulated interaction is between the spin and the orbital angular momentum of each nucleon, *e.g.* an interaction of the form **l.s**, so that each l level splits into two levels $j = l \pm 1/2$ except for $l = 0$ for which $j = 1/2$, with the $j = l + 1/2$ state being lower. The resulting energy levels clearly reproduce the shells corresponding to magic numbers (Fig. 9.3).

Superheavy nuclei: The energy levels based on the average simple harmonic or spherical-well potential provide only a general description of the shell properties. For the finer properties, more detailed calculations have to be carried out. Such calculations produce reordering of some of the energy levels shown in Fig. 9.3, especially when the number of particles is large. In particular, there are indications that 114 is a magic number for protons. There are also indications that 184 is a magic number for neutrons. This has given rise to speculations regarding the existence of long-lived superheavy nuclei with Z near 114 and $A - Z$ near 184. For example, it has been predicted that a nucleus with $Z = 110$ and $A = 294$ has a half life of about 10^8 years.

Angular momenta: The shell model allows us to predict the angular momenta of the nuclei, for example, the angular momentum of a closed shell is predicted by the Pauli exclusion principle, to be zero. If further, it is postulated that like nucleons in a shell pair off in such a way that their total angular momentum is zero, it follows that (*i*) all even *A*, even *Z* nuclei have zero angular momentum, (*ii*) the angular momentum of odd *A* nuclei is due to the odd nucleon. These results are generally observed to be true with a few exceptions. Some of the exceptions are ${}^{2\,3}_{1\,1}\text{Na}$ with $j = 3/2$ instead of 5/2, ${}^{5\,5}_{2\,5}\text{Mn}$ with $j = 5/2$ instead of 7/2, and ${}^{7\,9}_{3\,4}\text{Se}$ with $j = 7/2$ instead of $j = 9/2$. In all other cases the predictions are consistent with experimental observations, *e.g.* ${}^{209}_{83}\text{B}$ has $j = 9/2$.

Magnetic moments: The magnetic moments of odd-*A* nuclei can be estimated under the assumption that they are due to the odd nucleon (the magnetic moments of even *Z*, even *A* nuclei are zero, while those of odd *Z*, even *A* nuclei are difficult to analyse). The magnetic moment of the nucleon is both due to its spin as well as its orbital angular momentum and is given by

$$\mu = \frac{e}{2m_p}(2g_s\,\mathbf{s} + g_l\,\mathbf{l}) \tag{9.34}$$

where $g_s = 2.793$ for the proton, $g_s = -1.913$ for the neutron, and $g_l = 1$ for the proton, $g_l = 0$ for the neutron. As in the case of atoms (see Chapter 6), μ can be expressed in terms of the total angular momentum **j** as

$$\mu = \frac{e}{2m_p}(2g_s a_s + g_l a_l)\,\mathbf{j} \tag{9.35}$$

where

$$a_s = \frac{\mathbf{j}\cdot\mathbf{s}}{\mathbf{j}\cdot\mathbf{j}} = \frac{j(j+1) + s(s+1) - l(l+1)}{2j(j+1)} \tag{9.36}$$

$$a_l = \frac{\mathbf{j}\cdot\mathbf{l}}{\mathbf{j}\cdot\mathbf{j}} = \frac{j(j+1) - l(l+1) + s(s+1)}{2j(j+1)} \tag{9.37}$$

Now, for a given *j*, the allowed values of *l* are $j \mp \frac{1}{2}$, and the corresponding magnetic moments in units of $e\hbar/2m_p$ called nuclear magneton, are

$$\mu = g_s + (j - 1/2)\,g_l,\; l = j - 1/2,$$

$$\mu = -\frac{j}{j+1}g_s + \frac{j(j+3/2)}{j+1}g_l,\; l = j + 1/2 \tag{9.38}$$

The plots of these moments as functions of *j* give what are known as *Schmidt lines*. The experimental values of the magnetic moments ate not in good agreement with predictions of Eq. (9.38) but do lie between the two values. This suggest the need for a more detailed analysis including a mixing of states, *e.g.* the states may contain components in which the pairs of nucleons do not pair off to give zero angular momentum states.

Quadrupole moments: The predictions of the shell model for electric quadrupole moments are not in good agreement with the experimental values. If the quadrupole moment of an odd Z, odd A nucleus is due to the last proton, it should be approximately of the order

$$Q \approx R^2 \tag{9.39}$$

where R is the radius of the nucleus. While this is the case for small nuclei, some of the nuclei with large A, have Q as large as $10R^2$. Similar, large quadrupole moments are observed for even Z, odd A nuclei as well. Many of these effects are due to collective motions in nuclei, which are considered in the *collective model.*

The shell model can be generalized by taking the average potential to be an asymmetric harmonic oscillator potential. For example, in the *Nilsson model* the force constant in the z-direction is taken to be different from those in the x-and y-directions. This model retains the rotational symmetry in the z-direction while being able to describe the observed large quadrupole moments of nuclei.

Collective Model

For nuclei with a closed shell plus one or a few nucleons, the elementary shell model is quite successful in describing the nuclear properties. However, when there are several nucleons outside the closed shell, the nucleus is significantly deformed. The motion of the deformed nucleus gives rise to collective rotational and vibrational levels of the nucleus.

In the deformed nucleus which is assumed to be ellipsoidal in shape, the rotation can be of two types:

(*i*) it may be *irrotational* as in the case of tidal waves with no part of the nucleus actually going around the nucleus,

(*ii*) the whole nucleus may rotate as a rigid body. Both these motions may contribute to the rotational motion of a nucleus.

In even Z, even A nuclei, the angular momenta of the nucleons pair off to a zero value, so that the total angular momentum is also the angular momentum due to collective rotation. Accordingly, the rotational energy levels are given by

$$E_I = \frac{\hbar^2}{2\mathcal{I}} I(I+1) \tag{9.40}$$

where $\mathcal{I}$ is the moment of inertia and I is the total angular momentum quantum number. However, since the remaining wave function (other than the rotational part) satisfies the required exchange symmetry, the rotational wave function must be even under **r** → – **r** which effects an interchange of particles. Because of the relation $Y_0^m(\pi - \theta, \pi + \phi) = (-1)^l\, Y_l^m(\theta, \phi)$, this implies that only $I = 0$, 2, 4, ... are allowed. The observed energy levels for ^{238}Pu shown in Fig. 9.4(*a*), are in very good agreement with levels predicted by Eq. (9.40) with I even and a moment of inertia

$$\mathcal{I} \approx 1.4 \times 10^{-54} \text{ kg. m}^2 \tag{9.41}$$

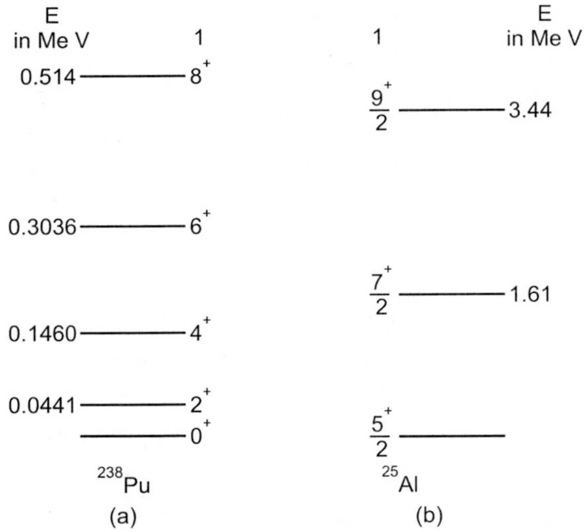

Fig. 9.4 Energy levels for collective rotation for (*a*) ^{258}Pu, even *Z*, even *A* nucleus, (*b*) ^{25}Al, odd *A* nucleus.

This is quite large compared to the value of $m_p R^2 \approx 10^{-55}$ kg. m^2 expected for the motion of a single nucleon, thus justifying the interpretation in terms of collective motion.

For odd *A* nuclei, the angular momentum is due both to the angular momentum of the odd nucleon and to collective rotational motion. Since the nucleon moves in a hemispherical potential, only the component of its angular momentum along the axis of symmetry is a constant of motion. This component, designated by Ω, adds vectorially to the collective angular momentum **R** which is perpendicular to the axis of symmetry, to give the total angular momentum **I**.

When $\Omega \neq \frac{1}{2}$, the rotational levels are given by

$$E_I = \frac{\hbar^2}{2\mathcal{I}}[I(I+1) - I_0(I_0+1)] \tag{9.42}$$

where I_0 corresponds to the ground state angular momentum, and I takes on values I_0, $I_0 + 1$, $I_0 + 2$, ... The observed values of the positive parity $\Omega = \frac{5}{2}$ levels for ^{25}Al, are in good agreement [see Fig. 9.4(*b*) with these predictions with $\mathcal{I} \approx 1.36 \times 10^{-55}$ kg. m^2. The analysis of $\Omega = \frac{1}{2}$ states is more complicated and may be found in specialized books.

In addition to rotational states, collective motion leads to vibrational states also. The vibrational levels may be analysed in terms of quanta of vibrational energy, called phonons. These phonons carry two units of angular momentum and an energy of about 1 MeV. In multi-phonon excitations, the angular momenta of the phonons add vectorially and produce closely-spaced levels with the same number of phonons but different angular momenta. Such levels have been identified experimentally.

The deformation of the core, by the nucleons outside the closed shell, can lead to large quadrupole moments. That this is the correct explanation of the large observed quadrupole moments, is indicated by the fact that large deformations may be expected odd A and odd Z nuclei as well as in odd A and odd $(A - Z)$ nuclei. This would lead to large quadrupole moments for odd Z as well as odd $(A - Z)$ nuclei as indeed it is observed experimentally.

All in all, collective motion is an important component of nuclear structure and energy levels.

Degenerate Gas Model

The degenerate gas model is similar in spirit to the free-electron theory of metals (Sec. 7.5). Here it is assumed that noninteracting nucleons are confined to the volume of the nucleus. The predictions of the model are discussed here briefly so as to emphasize the importance of the Pauli exclusion principle in determining the properties of the nucleus.

As shown in Sec. 7.5, the number of fermions occupying the lowest energy levels in a box of volume V is given by [Eq. (7.89)]

$$n = \frac{8\pi V}{3h^3} p^3 \tag{9.43}$$

where p is the maximum momentum of the occupied states, $p^2 = 2mE$. Substituting $V = \frac{4}{3}\pi r_0^3 A$, the maximum kinetic energy E_m is

$$E_m(n) = C\left(\frac{n}{A}\right)^{2/3} \tag{9.44}$$

where
$$C = \left(\frac{9}{32\pi^2}\right)^{2/3} \frac{h^2}{2mr_0^2}$$

$$\approx 52 \text{ MeV} \tag{9.45}$$

for $r_0 \approx 1.2$ fm. Similarly the total kinetic energy is given by

$$E_r = \int \frac{p^2}{2m} dn$$

$$= \frac{3}{5} n\, E_m \tag{9.46}$$

For a nucleus with Z protons and $(A - Z)$ neutrons, the total kinetic energy is given by

$$E = \frac{3C}{5A^{2/3}}[Z^{5/3} + (A - Z)^{5/3}] \tag{9.47}$$

If $Z = \frac{1}{2} A$, the kinetic energy of the last nucleon is obtained from Eq. (9.44) to be $E \approx 33$ MeV. Since the binding energy of the last nucleon is about 8 MeV, the average depth of the potential is of the order of 41 MeV. This is in agreement with the earlier estimation is Eq. (9.31) based on the uncertainty principle. It is also seen that for a given A,

$$\frac{\partial E}{\partial Z} = 0 \text{ for } Z = \frac{A}{2} \tag{9.48}$$

i.e. the most stable nucleus has equal number of protons and neutrons. Expanding E about $Z = \frac{1}{2} A$,

$$E = \frac{3C}{5(2)^{2/3}} A + \frac{2(2^{1/3})\, C}{3A}\left(Z - \frac{1}{2} A\right)^2 + \ldots \tag{9.49}$$

The second term, which gives the increase in the energy because of the imbalance of the protons and neutrons, is

$$\delta E \approx 43.7 \frac{\left(Z - \frac{1}{2} A\right)^2}{A} \text{ MeV} \tag{9.50}$$

In this analysis, Coulomb interaction which increases the potential for the protons has been ignored. Because of the Coulomb interaction, in stable, heavy nuclei, one finds more neutrons than protons. An important point brought out by the model is that the Pauli principle would push up the energy of a nucleus with Z very different from $\frac{1}{2}A$, and implies that the least-energy state is the one with nearly equal number of protons and neutrons.

9.4 WEIZSACKER'S MASS FORMULA

Many of the nuclear properties are critically controlled by the masses or equivalently the binding energies of the nuclei. For example, a nucleus can decay only into a set of particles with lower masses, with the difference in the masses appearing as kinetic energy of the final particles. Here, a semi-empirical formula for the masses of the nuclei is briefly discussed. This formula is not only useful in the discussion of the stability of the nuclei but also provides a useful insight into the physical properties that determine their masses.

The total energy of a nucleus is largely controlled by the short-range nature of the forces leading to saturation in binding, the Pauli exclusion principle, and Coulombic repulsion between protons with some finer effects due to pairing of like-nucleons.

To start with, the major contribution to the mass of the nucleus comes from the masses of the nucleons.

$$M_1 = Zm_p + (A - Z)m_n \tag{9.51}$$

Since nuclear forces are of short range, the binding energy of the nucleus is proportional to the number of nucleons A, so that the mass is reduced by

$$M_2 = -a_1A \tag{9.52}$$

However, nucleons near the surface are less tightly bound which may be taken into account by a term

$$M_3 = a_2A^{2/3} \tag{9.53}$$

proportional to the surface area. The electrostatic, repulsive interaction may be incorporated by noting that the electrostatic energy of a sphere of charge Ze is proportional to Z^2/R. This brings in a contribution

$$M_4 = a_3Z^2 A^{-1/3} \tag{9.54}$$

The Pauli exclusion principle brings in a term

$$M_5 = a_4 \frac{\left(Z - \frac{1}{2}A\right)^2}{A} \tag{9.55}$$

because of the imbalance of protons and neutrons. Finally, it is noted that the Pauli principle allows pairs of protons and neutrons with spin $\frac{1}{2}$ to occupy the same energy state whereas and odd proton of neutron is forced to go into a higher energy state. This effect is included by a pairing term

$$\delta = \begin{cases} \delta(A) & \text{for odd } Z \text{ and odd } (A-Z) \\ 0 & \text{for odd } A \\ -\delta(A) & \text{for even } Z \text{ and even } (A-Z) \end{cases} \frac{du}{dy} \tag{9.56}$$

The final formula for the mass of a nucleus is

$$M(Z, A) = Zm_p + (A - Z)m_n - a_1A + a_2A^{2/3} + a_3Z^2A^{-1/3} + a_4 \frac{\left(Z - \frac{1}{2}A\right)^2}{A} + \delta \tag{9.57}$$

The constants in Eq. (9.57) are determined empirically, from a fit to the observed masses. The best fits are obtained for

$$a_1 = 15.7,\ a_2 = 17.8,\ a_3 = 0.710 \tag{9.58}$$
$$a_4 = 94.8,\ \delta(A) = 33.6\ A^{-3/4}$$

all in MeV. The expression in Eq. (9.57) is known as the *Weizsacker mass formula* and the values in Eq. (9.58) give the best fit to the binding energy plot in Fig. (9.1). This formula is of considerable use in the analysis of the stability of nuclei.

9.5 NUCLEAR STABILITY

A nucleus can decay by emitting or absorbing electrons (β-decay), emitting α particles (α-decay), emitting protons or neutrons, emitting γ ray (γ-decay) or breaking into smaller nuclei (fission). Each decay mode is characterized by a decay probability λ defined by

$$dN(t) = -\lambda N(t)\, dt \tag{9.59}$$

in terms of which

$$N(t) = N(0)\, e^{-\lambda t} \tag{9.60}$$

λ^{-1} is called the lifetime τ of the nucleus. If a nucleus can decay via several modes, then its lifetime is the inverse of the sum of decay probabilities λ_i,

$$\tau = \left(\sum_i \lambda_i \right)^{-1} \tag{9.61}$$

This quantity τ is the average lifetime of the nucleus (see Sec. 6.4). For example ^{238}U has a lifetime of about 6.5×10^9 years, comparable with the age of the universe.

The allowed decays in general must satisfy certain conservation laws such as charge conservation, energy-momentum conservation, etc. γ-decays involve changes only in the energy levels, the components of the nucleus remaining the same. Here, we concentrate on β-decay, α-decay, and fission, which alter the composition of the nucleus.

Conservation of energy allows only those decays which satisfy the following rules:

1. A nucleus is unstable against the emission of an electron [Eq. (9.8)], *i.e.* β^--decay, if

$$M(Z, A) > M(Z + 1, A) + m_e \tag{9.62}$$

It is unstable against the absorption of an electron [Eq. (9.10)] if

$$M(Z, A) + m_e > M(Z - 1, A) \tag{9.63}$$

It is suitable against the emission of a positron [Eq. (9.9)], β^+-decay, if

$$M(Z, A) > M(Z - 1, A) + m_e \tag{9.64}$$

2. A nucleus is unstable against breakup into two fragments if

$$M(Z, A) > M(Z', A') + M(Z - Z', A - A') \tag{9.65}$$

In particular, it may decay by emitting a portion if

$$M(Z, A) > M(Z - 1, A - 1) + m_p \tag{9.66}$$

or by emitting an α particle if

$$M(Z, A) > M(Z - 2, A - 4) + m_\alpha \tag{9.67}$$

It may be noted that because of the large binding energy per nucleon, the decays via the emission of a heavy particle are important mainly in heavy nuclei. For determining the stability pattern of lighter nuclei, it is necessary to consider only electron emission or absorption.

Beta Decay

Consider first an odd A nucleus. The Z value for the most stable nucleus is given by the condition

$$\left.\frac{\partial M(Z, A)}{\partial Z}\right|_{Z = Z_A} = 0 \tag{9.68}$$

which implies that [using Eq. (9.57)]

$$(m_p - m_n) + 2a_3 Z_A A^{-1/3} + 2a_4 \left(Z_A - \frac{1}{2}A \right) A^{-1} = 0 \tag{9.69}$$

Solving for Z_A,

$$Z_A = \left(\frac{m_n - m_p + a_4}{a_4 + a_3 A^{2/3}}\right)\left(\frac{A}{2}\right) \tag{9.70}$$

For $a_3 = 0.710$, it is seen that Z_A is smaller than $A/2$ for $A \geq 3$, and most of the heavy nuclei will have more neutrons than protons.

Expansion of $M(Z, A)$ about the point $Z = Z_A$ leads to

$$M(Z, A) = M(Z_A, A) + (a_3A^{-1/3} + a_4A^{-1})(Z - Z_A)^2 + \ldots \tag{9.71}$$

This relation is shown schematically in Fig. 9.5(*a*), for a given A. If the mass of the electron m_e is ignored in Eqs. (9.62) and (9.64), it follows that the integral Z value nearest to Z_A corresponds to the stable isobar while the neighbouring isobars will be unstable against either β^--decay or β^+-decay. In some rare cases, it may happen that Z_A is essentially in between the two nearby integral Z values, in which case there may be two stable isobars. Experimentally, it is found that there is one stable isobar for each odd A nucleus, the only exceptions being (^{113}Cd, ^{113}In), (^{115}In, ^{115}Sn) and (^{123}Sb, ^{123}Te).

Going over to the case of even A nuclei, the same procedure can be adopted as in the case of odd A nuclei, *i.e.* expand $M(Z, A)$ about the minimum at $Z = Z_A$. However, in this case, δ is not only nonzero but also takes on two values $\pm$ 33.6 $A^{-3/4}$ corresponding to odd Z and even Z values. This leads to two plots for $M(Z, A)$ as a function of Z. In the typical case shown in Fig. 9.5(*b*), two stable isobars ($Z = 28$, 30) are obtained. There can be a situation in which Z_A may be close to an even integer, say $2n$ and the masses of the nuclei with $Z = 2n \pm 2$ may be higher than those with $Z = 2n \pm 1$ in which case only stable nucleus (with $Z = 2n$) is obtained, *e.g.* ^{194}Pt. It may also happen that the masses of all the three nuclei with $Z = 2n$, $2n \pm 2$ may be lower than those with $Z = 2n \pm 1$. In this case three stable nuclei are obtained, with $Z = 2n$, $2n \pm 2$. Thus, in the case of even A nuclei, one, two or three stable isobars may exist.

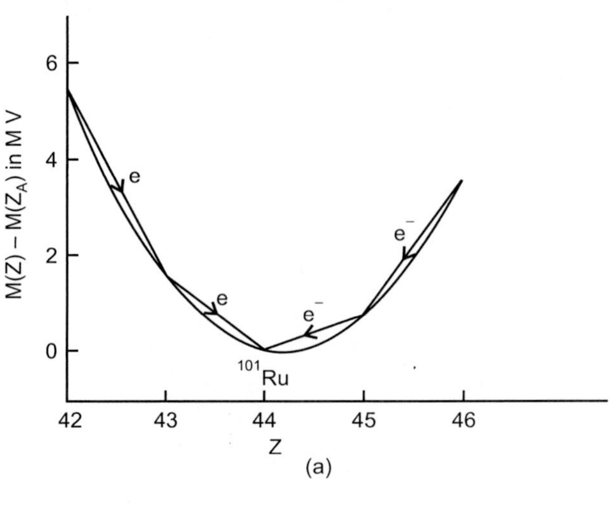

(a)

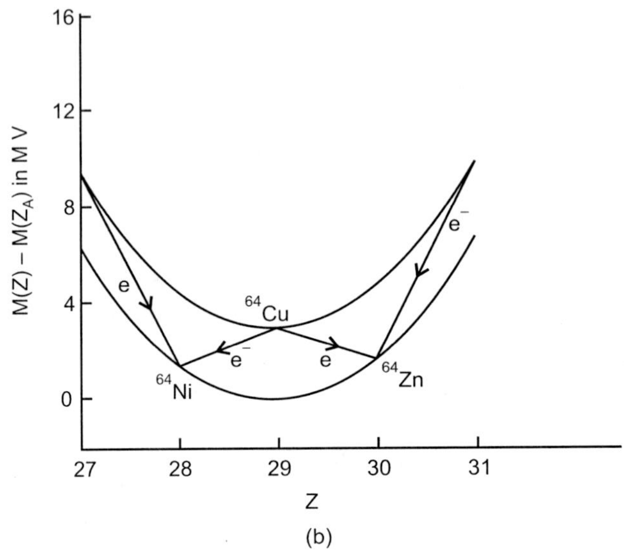

Fig. 9.5 (*a*) Stability of nuclei with $A = 101$, (*b*) stability of nuclei with $A = 64$.

A point of caution: the usual masses of nuclei quoted are the masses of the corresponding neutral atoms and hence include an additional mass of Z electrons. For the sake of clarity, we refer only to the masses of the nuclei.

Alpha Decay

It is observed (Fig. 9.1) that the binding energy per nucleon decreases as nuclear mass increases ($A > 56$). Therefore, a heavy nucleus would, in some circumstances prefer to decay into lighter nuclei. However, a decay by emission of only a proton or neutron is not observed since each nucleon in a nucleus has a binding energy of about 8 MeV whereas the binding energy of a free nucleon is zero. On the other hand, a decay by emission of an α particle (^{4}He), which has a binding energy of about 7.1 MeV per nucleon, is quite likely and is observed in many nuclei.

The properties of α-decay, representing in Eq. (9.5), may be illustrated by taking a specific example. Consider the α-decay of ^{212}Bi,

$$^{212}\text{Bi} \rightarrow {}^{208}\text{Ti} + {}^{4}\text{He} \tag{9.72}$$

If Tl is in its ground state, the initial mass exceeds the final mass by 6.203 MeV. This appears in the form of kinetic energy shared by the final particles. Since momentum conservation implies that Tl and He have equal and opposite momenta, their kinetic energies are inversely proportional to their masses. This means that the α particle is ejected with a kinetic energy

$$E_\alpha \approx 6.203\left(\frac{208}{212}\right) \approx 6.086 \text{ MeV} \tag{9.73}$$

Alternatively, Tl may be in one of its excited states [see Fig. 9.6 (*a*)] in which case the kinetic energy of the α particle is

$$E_\alpha \approx (6.203 - \varepsilon)\left(\frac{208}{212}\right) \tag{9.74}$$

where ε is the excitation energy of the state. Consequently, the α particle is observed with essentially discrete kinetic energies, 27% of the times with 6.086 MeV, 70% of the times with 6.047 MeV, and the remaining 3% of the times with the smaller allowed energies given by Eq. (9.74). The total lifetime corresponding to these transitions is about 87.7 minutes. It is important to note that subsequent to the α-decay, typically in about 10^{-8} to 10^{-15} s, the excited Tl nucleus undergoes transitions to a lower energy state by emitting a γ ray. In the general case, the excited nucleus may also loose its energy by emitting an electron, a proton, a neutron or another α particle. Alternatively, the excess energy may knock out one of the electrons in the atomic orbits. This is known as internal conversion and is usually characterized by the emission of an *x*-ray photon when the vacancy created by the ejection of the electron is filled by an electron from the higher energy levels.

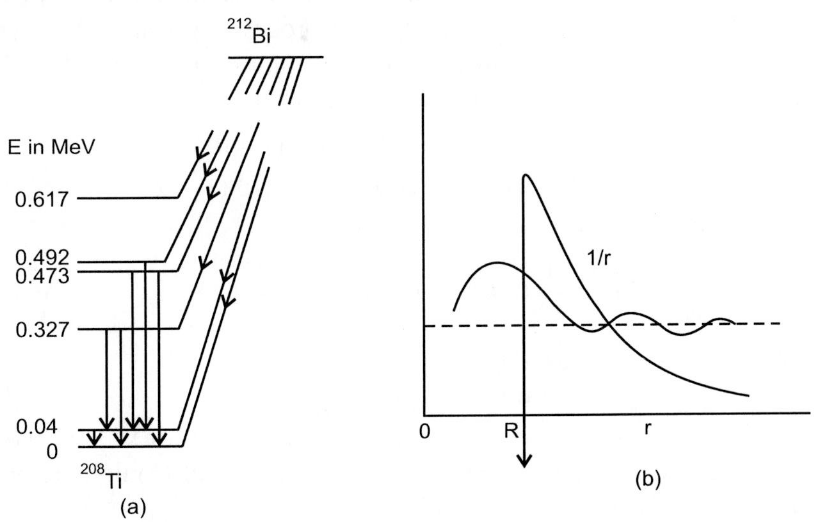

Fig. 9.6 (*a*) Alpha decay of ^{212}Bi and the subsequent gamma decay of ^{208}Tl, (*b*) tunnelling of α-particle wave function leading to alpha decay.

An important question which arises is the following: Why does the nucleus not undergo and instantaneous decay to an energetically allowed state? The reason for this is that when the nucleus decays, it goes through some states which have higher energy and the decay products must overcome the potential barrier. For example, when Tl and He are just touching each other before separating, they have an additional Coulomb energy

$$E_c \approx \frac{e^2\,(81)\,(2)}{4\pi\varepsilon_0\,(r_{\text{Tl}} + r_{\text{He}})} \tag{9.75}$$

Using the values for radii given by Eq. (9.11) with $r_0 \approx 1.2$ fm,

$$Ec \approx 26 \text{ MeV} \tag{9.76}$$

Classically, this is a forbidden domain. Quantum mechanically, the α particle can penetrate and tunnel through a potential [see Fig. 9.6 (*b*)] with some probability. It is this property which permits the decay, with a finite lifetime.

Nuclear Fission

In some cases, it may be energetically favourable for a heavy nucleus to break up into fragments of nearly equal masses, accompanied by the release of a large amount of energy. Since heavy nuclei have an overabundance of neutrons, the fission process is usually followed by the emission of a few neutrons. An example of this is the fission of an excited $^{236}U^*$ nucleus:

$$^{236}\text{U}^* \to {}^{144}\text{Ba} + {}^{89}\text{Kr} + 3n \tag{9.77}$$

An insight into the fission process is obtained by considering a simple model in which a nucleus (A, Z) decays into only two fragments, (αA, βZ) and [$(1 - \alpha)A$, $(1- \beta)Z$]. The energy released in this process is given by the decrease in the masses, which may be estimated from the empirical formula in Eq. (9.57):

$$\begin{aligned}\Delta E = {} & 17.8\,A^{2/3}\,[1 - \alpha^{2/3} - (1 - \alpha)^{2/3}] \\ & + 0.71\,Z^2 A^{-1/3}\,[1 - \beta^2 \alpha^{-1/3}\,(1 - \beta)^2\,(1 - \alpha)^{-1/3}] \\ & + 95\,Z^2 A^{-1}\,[\,1 - \beta^2\alpha^{-1} - (\,1 - \beta)^2\,(1 - \alpha)^{-1}]\end{aligned} \tag{9.78}$$

It is easy to show that

$$\frac{\partial}{\partial\alpha}(\Delta E) = \frac{\partial}{\partial\beta}(\Delta E) = 0 \text{ at } \alpha = \beta = \frac{1}{2} \tag{9.79}$$

so that the maximum energy released is

$$(\Delta E)_{\max} \approx -4.6\,A^{2/3} + 0.26\,Z^2 A^{-1/3} \tag{9.80}$$

For A = 236, Z = 92, this has a value of about 180 MeV. However the fission products, when just in contact, have an additional Coulombic energy given approximately by

$$E_c \approx \frac{e^2\,(Z/2)^2}{4\pi\varepsilon_0\,(2R)} \tag{9.81}$$

where $R \approx 1.2\,(A/2)^{1/3}$ fm. For $A = 236$, $Z = 92$, E_c has a value of about 259 MeV so that the fission products have to escape by tunnelling through this potential barrier [$E_c > (\Delta E)_{\max}$]. The tunnelling is not necessary only if

$$(\Delta E)_{\max} > E_c \tag{9.82}$$

On using Eqs. (9.80) and (9.81) with $R \approx 12(A/2)^{1/3}$ fm, the inequality reduces to

$$Z^2 > 65\,A \tag{9.83}$$

This condition is not satisfied even in the case of the heaviest known nucleus $^{262}_{105}$Ha (hahnium) for which $Z^2/A \approx 42$. Thus all the observed fission processes are expected to produced by tunnelling through the potential barrier.

A fission process is initiated by bombarding a nucleus with neutrons, giving an excited initial state such as $^{236}U^*$, and forms the basis of nuclear fission reactors used for tapping large nuclear energies.

Gamma Decay

A nucleus may be found in an excited state if it is one of the products in an alpha or a beta decay. The excitation energy here is provided by the higher mass of the decaying particle. Alternatively, the excitation may take place as a result of a collision in which the projectile excites the nucleus by transferring to it some of its kinetic energy. In either case, the excited nucleus usually undergoes transition to a lower state by emitting a photon,

$$Z^* \rightarrow Z + \gamma \tag{9.84}$$

An example of γ-decay is the de-excitation of ^{208}Tl discussed earlier. The energy of the photon emitted is

$$h\nu \approx E^* - E \tag{9.85}$$

i.e. the difference in the energies of the number states.

The lifetimes for γ-decay are usually of the order of 10^{-14} s. However, selection rules may inhibit some of these decays leading to mush longer lifetimes of the excited states are sufficiently long so as to be directly measurable, the nuclei are known as *isomers.* An example of extreme isomerism is ^{91}Nb* which has a lifetime of about 87 days.

Radioactive Series

In nature one observes several decays of radioactive elements some of which were created in the early stages of the evolution of the universe, others which are continuously formed by the bombardment of cosmic rays (see Sec. 10.6). Of special interest are the ones which form what are known as the *radioactive series.*

All nuclei with $A > 209$ are unstable and normally undergo either α-decay, β-decay, or γ-decay. Since the changes in the number of nucleons A, are due only to α-decay, the mass numbers A of the series of nuclei produced in a series are related by

$$A = a + 4n \tag{9.86}$$

Thus, four series exist corresponding to a = 1, 2, 3. These are summarized below along with the half-life in years ($\tau^{1/2} \approx 0.69\ \tau_{av}$) of the longest-lived member:

Series	*Mass number*	*Longest lived nucleus*	*Final stable nucleus*
Thorium	$4n$	^{232}Th (1.39×10^{10})	^{208}Pb
Neptunium	4n + 1	^{237}Np(2.25×10^{6})	^{209}Bi
Uranium	$4n + 2$	^{238}U(4.51×10^{9})	^{206}Pb
Actinium	$4n + 3$	^{235}U(7.07×18^{6})	^{207}Pb

Of these, the thorium, uranium and actinium series are observed in nature. The neptunium series is not observed in nature since its longest lived member, ^{237}Np has a half-life of about 2.25×10^6 years and whatever amount was created at the early stages of the universe would have decayed by now. However, ^{237}Np can be artificially produced, *e.g.* from ^{236}U by the capture of neutron followed by β^- decay. All these elements undergo a series of α and β^- decays till they are reduced t.o the final stable nucleus. The details of the ^{238}U series are shown in Fig. (9.7) where the steps with decrease in Z of two correspond to α-decays and steps with unit increase in Z correspond to β-decays.

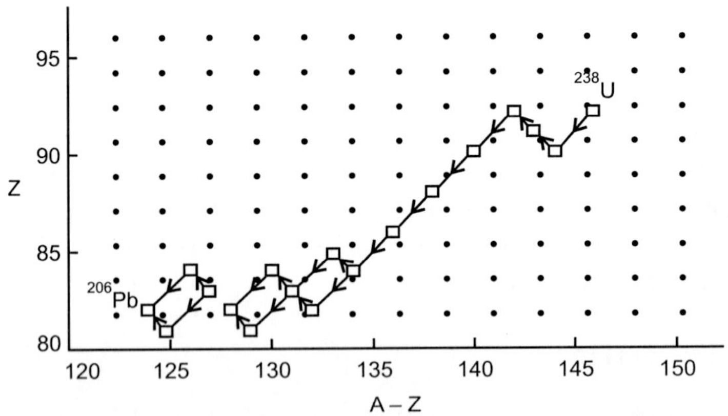

Fig. 9.7 Uranium radioactive series.

The occurrence of radioactive elements in nature provides a means of determining the age of the elements in our planetary system. Assuming that ^{235}U and and ^{238}U were initially created in about same quantities (this is approximately true for many stable nuclei), the ratio of these elements found in nature today is

$$\frac{N_{235}(t)}{N_{238}(t)} \approx \exp\left[-t\,(1/\tau_{235} - 1/\tau_{238})\right] \tag{9.87}$$

Using the observed value of about 1/140 for the relative abundance,

$$t \approx \frac{\tau_{235}\,\tau_{238}}{\tau_{238} - \tau_{235}} \ln 140 \tag{9.88}$$

Since $\tau_{235} \approx 1.02 \times 10^9$ years and $\tau_{238} \approx 6.51 \times 10^9$ years, the age of the elements is

$$t \approx 5.98 \times 10^9 \text{ years} \tag{9.89}$$

This is in reasonable agreement with the age determined from the abundances of other nuclei. It is also of the same order as the age of the universe deduced from the rate of expansion of the universe (see Sec. 11.5), and hence is an important element in the understanding of the universe.

9.6 NUCLEAR REACTIONS

The properties of a nucleus can be studied by bombarding it with energetic particles, and analysing the consequences of the collisions. If the collision leads to a nuclear interaction, it gives what is known as a *nuclear reaction.*

Usually, the projectile is a light particle, it may be a neutron, a proton, a deuteron, an alpha particle or a photon (recently there has been considerable interest in heavy ions as projectiles). A typical collision between a light particle a and a nucleus X, may produce a light particle b and a nucleus Y:

$$X + a \rightarrow Y + b \tag{9.90}$$

Such a two-body process is customarily written in the form $X\,(a, b)\,Y$. If the particle b is the same as particle a, one has a *scattering* process. If the total kinetic energy is unaltered in the collision, the scattering is *elastic,* whereas if the total kinetic energy changes (generally decreases), the scattering is *inelastic.* If b is different from $a,$ we have a special case of nuclear reactions. All these processes must satisfy certain conservation laws, such as charge conservation, energy-momentum conservation, etc.

A nuclear reaction is usually accompanied by either a release or absorption of kinetic energy. This is given by the difference between the total mass of the particles before and after the collision. It is called the reaction energy or Q value and is expressed as

$$Q = \sum_i m_i - \sum_f m_f \tag{9.91}$$

where m_i and m_f are the masses of the initial and final particles respectively, normally expressed as rest energy in MeV. If Q is positive, energy is released in the reaction (*endothermic*) and if Q is negative , energy is absorbed in the reaction (*endothermic*). A reaction cannot proceed unless the energy of the bombarding particle is greater than a minimum value known as the threshold energy, equal to $|Q|$ for negative Q reactions and zero for positive Q reactions.

Two specific examples of historic importance are

$$^{14}N + \alpha \rightarrow {}^{17}O + p,\ {}^{14}N\ (\alpha\ , p)\ {}^{17}O \tag{9.92}$$

$$^{7}Li + p \rightarrow \alpha + \alpha,\ {}^{7}Li\ (p, \alpha\ )\ {}^{4}He \tag{9.93}$$

The first reaction was observed by Rutherford (1919), using α particles of 7.68 MeV energy. from a natural, radioactive source. It is an endothermic reaction with $Q \approx -1.2$ MeV, with a threshold energy of about 1.2 MeV. The process in Eq. (9.93) was the first nuclear reaction produced by artificially accelerated particles. Cockcroft and Walton (1932) obserbed the reaction using protons accelerated through a potential difference of about 0.5 MeV. This is an exothermic reaction with $Q \approx 17.3$ MeV and illustrates the possibility of extracting energy from a nuclear reaction.

Cross-Section

The probability for a nuclear reaction to take place is expressed in terms of the effective cross-section σ.

Consider a beam of projectiles incident on a collection of targets in the form of a thin plate (sufficiently thin so that the nuclei do not overlap). An effective cross-sectional area σ [Fig. 9.8 (*a*)] is associated with each target such that every projectile incident within that area produces a reaction. If there are n targets per unit volume and t is the thickness of the plate, the probability of a single projectile producing a reaction is

$$P = \sigma n t \tag{9.94}$$

(it is the fraction of the effective target area). Therefore, the fraction of projectiles $\Delta N/N$ producing reactions, is

$$\frac{\Delta N}{N} = \sigma n t \tag{9.95}$$

which leads to

$$\sigma = \frac{\Delta N}{N n t} \tag{9.96}$$

This relation allows us to determine the effective cross-section for a nuclear reaction. It is usually expressed in terms of barns, 1 barn = 10^{-28} m^2.

Nuclear reaction cross-sections vary over large ranges. Of particular interest are the cross-sections for neutron projectiles. In this case, the neutrons, being neutral, can enter a nucleus with ease even at low energies. In fact, since the time spent by the neutron inside the nucleus is proportional to $\frac{1}{v}$, v being the neutron velocity, it is expected that the effective cross-section increases as $\frac{1}{v}$ at low energies,

$$\sigma \sim \frac{1}{v}$$

$$\sim E^{-1/2} \text{ for small } E \tag{9.97}$$

This is observed experimentally. However, it is also observed that the reaction cross-section has a sharp maximum for neutrons of some specific energies E. This corresponds to what is known as resonance reaction [Fig. 9.8 (*b*)]. For example, the neutron capture cross-section, with Rh as a target, increases by a factor of about 100 at an energy of about 1.4 eV. The resonance reaction takes place when the energy of the incoming neutron equals the energy required to transfer the combined system, of the target plus the neutron, to an excited energy level. Similar resonant absorptions of photons are observed when the photon energy is equal to the excitation energy of a nuclear level of the target.

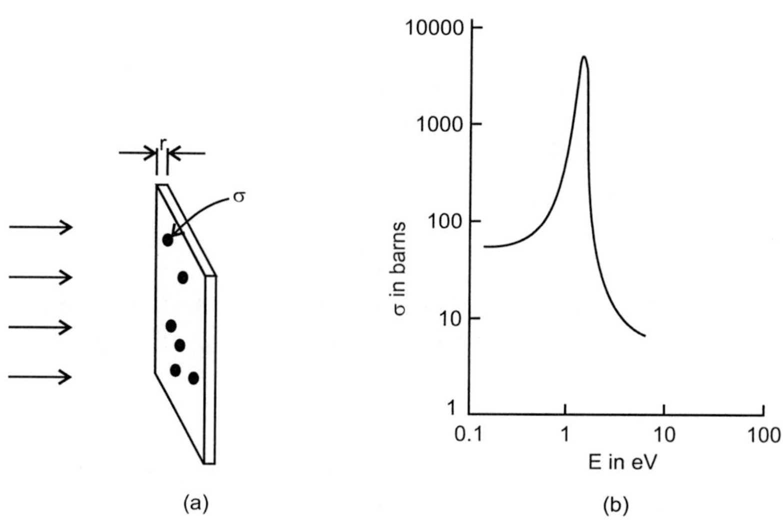

Fig. 9.8 (*a*) Illustration of cross-section, (*b*) resonance scattering of neutrons by Rh, using log-log scale.

Compound Nucleus

A very useful concept in the understanding of nuclear reactions is that of *compound-nucleus* formation proposed by Bohr (1936). It was suggested that for incident energies less than about 50 MeV, a nuclear reaction proceeds in two stages. In the first stage, the projectile *a* is captured by the nucleus, forming a compound nucleus *c** which is usually an excited state of a nucleus. Such a compound nucleus had a lifetime of about 10^{-18} s, considerably longer than the time taken by the projectile to cross the nucleus ($t_{cross} \sim R/v \approx 10^{-21}$ s). During this time, the compound nucleus will have 'forgotten' the way it was formed. In the second stage, this compound nucleus will decay independently of the mode of its formation. The two stage process may be indicated by

$$X + a \rightarrow C^* \rightarrow Y + b \tag{9.98}$$

The corresponding cross-section is factorizable, *i.e.*

$$\sigma\,[X\,(a,b)\,Y] = F\,(X, a)\,G\,(Y, b) \tag{9.99}$$

where F and G depend on the energy, spin, etc. of X, a and Y, b respectively. An important result which follows immediately is that the ratio of the reaction rates at a given energy, is independent of the initial state, *i.e.*

$$\frac{\sigma[X\,(a,b)\,Y]}{\sigma[X\,(a,'b)\,Y']} = \frac{G\,(Y,b)}{G\,(Y',b')} \tag{9.100}$$

An example of the reactions proceeding via a compound nucleus is given by the following processes:

$$\left.\begin{array}{l} p + {}^{26}\text{Mg} \\ d + {}^{25}\text{Mg} \\ \alpha + {}^{23}\text{Na} \end{array}\right\} \rightarrow {}^{27}\text{Al}^* \rightarrow \left\{\begin{array}{l} {}^{27}\text{Al} + \gamma \\ {}^{23}\text{Na} + \alpha \\ {}^{26}\text{Al} + n \\ {}^{26}\text{Mg} + p \\ {}^{25}\text{Mg} + n + p \end{array}\right. \tag{9.101}$$

In these reaction, the relative cross-sections for the five final states are independent of the initial state, but depend on the energy of the system.

Direct Processes

Reactions produced by fast projectiles proceed without the formation of a compound nucleus. They are known as direct processes. As a typical example, consider a deuteron incident on a nucleus, say ^{50}V. When it approaches the target, one of the nucleons may be captured by the nucleus while the other may continue essentially undisturbed along its path (it should be remembered that the deuteron is a loosely bound object). The net reaction is represented by

$${}^{50}\text{V} + d \rightarrow {}^{51}\text{V} + p \tag{9.102}$$

$${}^{50}\text{V} + d \rightarrow {}^{51}\text{Cr} + n \tag{9.103}$$

Such reactions are called *stripping reactions* and are observed for other projectiles, such as α particles, as well. Conversely, a fast-moving neutron (or proton) may collect one or more nucleons and move off as a deuteron or an α particle, *e.g.*

$$^{27}\text{Al} + n \rightarrow {}^{24}\text{Na} + \alpha \tag{9.104}$$

Such processes are known as *pick-up* reactions.

Applications

Nuclear reactions provide a powerful tool for the investigation of nuclear structure and properties. From the conservation of energy, the masses of the nuclei can be deduced, while conservation of angular momentum and the angular distribution of particles help us to determine their angular momenta. Furthermore, the cross-sections give important information on the interaction between nuclei.

Nuclear reactions allow us to produce trans-uranic elements as well as elementary particles. For example, plutonium (Z = 94) is produced by bombarding uranium with 40 MeV α particles

$$^{238}\text{U} + a \rightarrow 241\ \text{Pu} + n \tag{9.105}$$

Similarly, other elements such as berkelium (Z = 95), californium (Z = 98), einsteinium (Z = 99), fermium (Z = 100), nobelium (Z = 102), etc. have been produced from nuclear reactions, the heaviest of them being hahnium $^{262}_{105}\text{Ha}$. Many elementary particles also have been produced and investigated in nuclear reactions. Two examples are:

$$p + p \rightarrow \text{P} + n + \pi^{+} \tag{9.106}$$

$$p + p \rightarrow p + \Sigma^{+} K^{\circ} \tag{9.107}$$

π^{+} and K° being the positively charged pi-meson and the neutral K-meson, respectively, and Σ^{+} is the positively charged sigma baryon (see Sec. 10.1).

From the point of view of practical applications, products of nuclear reactions are of considerable utility in industry, medicine, agriculture, etc.

Their uses may be grouped in the following broad categories:

1. Radioactive isotopes, which have the same chemical properties as a given element but decay with the emission of photons, are of general use as tracers. These isotopes are easily detected by their characteristic radiation and half-life. For example, the phosphorus isotope ^{32}P (decays via beta decay with half-life of 14.2 days) may be used to determine the proper application of fertilizers to plants. Radioactive tracers are also used for analysing blood circulation, flow rates of fluids, etc.
2. Neutron activation analysis is used for detecting small amounts of impurities. The impurities are activated by exposing them to neutron beams. The radioactive isotope formed as a result is estimated by its

characteristic radiation. For example, traces of cobalt may be detected by observing the radiation from cobalt isotope ^{60}Co formed on absorption of neutrons.

3. In place of x-rays, the more energetic gamma rays from radioactive isotopes may be used to detect flaws in metals, in medicine for the treatment of some diseases such as cancer, for preservation of food materials, for crop mutations in agriculture and in various industries. The commonly used γ rays are from the radioactive ^{60}Co.

A particularly interesting application is radioactive ^{14}C dating based on a nuclear reaction taking place in the atmosphere. Neutrons produced in the atmosphere by cosmic rays, collide with nitrogen nuclei, producing ^{14}C and protons

$$^{14}\text{N} + n \rightarrow {}^{14}\text{C} + p \tag{9.108}$$

^{14}C is radioactive an decays by β-emission,

$$^{14}\text{C} \rightarrow {}^{14}\text{N} + e + \overline{\nu} \tag{9.109}$$

with a half-life of 5730 years. This radioactive carbon is assimilated by plants in photosynthesis, so that living organisms contain a small fraction of ^{14}C. When the organism dies, the intake of ^{14}C stops stops and because of the decay of ^{14}C, its concentration relative to ^{12}C begins to decrease. Hence, a measurement of the concentration of ^{14}C in the remains of the organism (*e.g.* wood, bones, etc.), gives the data when it died. This is known as *radioactive dating* and permits us to determine the age of organic relics which may be thousands of years old.

Two important applications of fission and fusion reaction, which lead to the extraction of energy from nuclear processes, are considered in the next two sections.

9.7 FISSION REACTORS

It was noted in Sec. 9.5, that it is energetically favourable [see Eq. (9.80)] for a heavy nucleus to break into two nearly equal parts. Qualitatively, this is due to the fast that the binding energy per nucleon is about 7.6 MeV for nuclei with $A \sim 230$, whereas it is about 8.5 MeV for nuclei with $A \sim 115$. This means that the fission of a nucleus with $A \sim 230$ into two equal parts would release an energy of about $0.9 \times 230 = 2078$ MeV. However, the fission products must pass through intermediate states of high energy provided by the Coulomb barrier, similar to the situation in Fig. 9.6(b). The fission can take place by tunnelling through the barrier, but with a rather long lifetime (for ^{238}U it is about 10^{16} years).

The fission of a nucleus can be induced by bombarding the nucleus with neutrons. Consider, for example, the capture of a slow neutron by a nucleus of ^{235}U. This leads to the formation of the compound nucleus ^{236}U in which the small kinetic energy and the binding energy of the neutron are released. The nucleus therefore is in an excited state and can decay by fission, generally into parts with mass numbers of the order of 95 and 140. The neutron numbers in these fragments are close to the magic numbers 50 and 82 which lead to stable nuclear structures.

Since the heavy nuclei have an excess of neutrons, the fission process is usually accompanied by the emission of neutrons. There is a further reduction of the neutrons in the nuclei, due to either β-decay or emission of delayed neutrons. A typical process would be

$$^{235}\text{U} + n \rightarrow {}^{236}\text{U}^* \rightarrow {}^{137}\text{I} + {}^{97}\text{Y} + 2n \tag{9.110}$$

followed by

$$\begin{aligned} &^{97}\text{Y} \rightarrow {}^{97}\text{Zr} + e + \overline{\nu} \\ &\qquad\quad \downarrow 17\text{h} \\ &^{97}\text{Nb} + e + \overline{\nu} \\ &\qquad\quad \downarrow 74\ \text{min} \\ &^{97}\text{Mo} + e + \overline{\nu} \end{aligned} \tag{9.111}$$

$$\begin{aligned} &^{137}\text{I} \rightarrow {}^{137}\text{Xe} + e + \overline{\nu} \\ &\qquad\quad \downarrow 22\ \text{s} \\ &\qquad ^{136}\text{Xe} + n \end{aligned} \tag{9.112}$$

^{137}I emits a neutron after its β-decay into ^{137}Xe, and since β decays proceed slowly, there is a large item delay between the fission and the emission of this neutron. Such neutrons are called *delayed neutrons.*

The most important feature of neutron-induced fission is that the fission itself provides additional neutrons which can produce additional reactions. Thus, the process can build into a self-sustaining, possibly a growing, chain reaction which is the basis of fission reactors. In the following, the main elements of a fission reactor are discussed briefly.

Reactor Fuel

The essential process in a reactor is the fission of nuclei, accompanied by neutrons and a release of energy. From a practical point of view, it would be required that sufficient quantities of these nuclei, which form the reactor fuel, occur naturally or can be produced. Five such nuclei are ^{235}U, ^{239}Pu, ^{233}U which contain odd number of neutrons, and ^{238}U and ^{232}Th which contain even number of neutrons. The first three of these can undergo fission with the capture of a thermal neutron, whereas that last two undergo fission mainly through the capture of fast neutrons (the captured neutron in these two cases would be an odd neutron and hence would release less energy). It is, however, important to note that the capture of neutrons by ^{238}U and ^{232}Th leads to fissionable fuel:

$$
\begin{aligned}
{}^{238}\text{U} + n \rightarrow {}^{239}&\text{U} + \gamma \\
&\downarrow 23 \text{ min} \\
{}^{239}\text{Np} &+ e + \bar{\nu} \\
&\downarrow 2.3 \text{ days} \\
{}^{239}\text{Pu} &+ e + \bar{\nu},
\end{aligned}
\tag{9.113}
$$

$$
\begin{aligned}
{}^{232}\text{Th} + n \rightarrow {}^{233}&\text{Th} + \gamma \\
&\downarrow 24 \text{ min} \\
{}^{233}\text{Pa} &+ e + \bar{\nu} \\
&\downarrow 27 \text{ days} \\
{}^{233}\text{U} &+ e + \bar{\nu}
\end{aligned}
\tag{9.114}
$$

In most cases, the fuel is in the form of rods or plates which are placed in a regular array within a moderator which serves the purpose of slowing down the neutrons to thermal energies, *i.e.* energies of the order of 0.1 eV. The fission reaction is triggered either by secondary cosmic ray neutrons i.e. neutrons produced by the cosmic rays, or neutrons from a small neutron source (usually containing a source of α particles which react with beryllium to produce neutrons). The neutrons emitted, if suitably controlled, can then produce chain reactions.

As a specific example, the source may be ^{235}U which occurs in nature (0.7% ^{235}U along with 99.3%^{238}U). It may be used in the natural form or after concentration. One of the many known reactions produced was indicated in Eq. (9.110). The fission of ^{235}U produces, on the average 2.5 neutrons per nucleus of which about 0.7% are delayed neutrons which play an important role in the control of reactor rates. The energy released in each fission is about 200 MeV which is distributed among the main fission fragments (about 165 MeV), neutrons (about 5 MeV), electrons and photons (about 20 MeV), and neutrinos (about 10 MeV).

Neutron Economy

In order that the fission process be self-sustaining, the neutrons produced in the fission reactions should not all be lost.

In neutrons may be lost by being captured by ^{238}U. The resulting ^{238}U does not lead to fission, but decays by emitting a photon. The capture cross-section for ^{238}U decreases to small values, about 3 barns, for thermal neutrons (note that the cross-section goes through a large resonant value, 2.3×10^4 barns at 7 eV). The capture cross-section for ^{235}U, on the other hand, increases as $1/E^{1/2}$ for small energies [see Eq. (9.97)], and has a value of about 580 barns for thermal neutrons. Thus, the fission-effectiveness of neutrons is increased by the thermalization of neutrons, achieved by the moderators surrounding the fuel. Successive scattering of neutrons by the moderator transfers the neutron energy to the moderator and hence slows down the neutrons.

Some neutrons may be lost from the surface of the active zone. In this connection, it is noted that the fission-effectiveness of the neutrons is proportional to the volume (*i.e.* l^3) whereas the surface losses are proportional to the surface area (*i.e.* l^2). Hence, the relative surface losses can be decreased by increasing the active volume, and the size at which the chain reaction is just self-sustaining is known as the *critical volume,* and the corresponding mass of the active material is known as the *critical mass.*

In addition, there are other sources of neutron losses, such as absorption by the impurity in the moderator. These losses must be carefully controlled, for example, by sing moderators which are nearly free of impurities. The essential requirement of continued chain reactions is that the number of fission neutrons remaining after taking into account all the losses, must be greater than the initial neutrons which induced the fission.

Moderators

The role of a moderator is to slow down the neutrons without absorbing them. Elementary considerations show that maximum energy is transferred to the target if the target mass is equal to the projectile mass.

The ideal moderator would have been hydrogen. Unfortunately, hydrogen can capture a neutron according to the reaction $p\ (n, \gamma)\ d$. More suitable moderators are the deuteron d (nucleus of deuterium), graphite (C) or beryllium (Be). About 25 collisions are adequate to thermalize 2 MeV neutrons in heavy water (D_2O), and about 100 collisions in C or Be.

Control Rods

The number of fission neutrons available for the controlled chain reaction, after taking into account the various losses, must be slightly greater than the neutrons which caused the initial fission. To produce sustained, stable chain reactions, the excess neutrons must be removed and controlled. This is usually done by inserting what are known as *control rods* into the core of the reactor. These rods are made of an element with a large cross-section for neutron capture.

The element often used in control rods is cadmium which has a very large capture cross-section for thermal neutrons, $^{113}Cd\ (n, \gamma)\ ^{114}Cd$. The insertion of these rods decreases the reactivity of the reactor whereas withdrawal increases the reactivity. It is important to observe that the response of the chain reaction to fluctuations in neutrons, is slow, because of the delyed neutrons produced in fission (see Example 6, Sec. 9.9). This permits the use of the control mechanism with a time delay of about 1 min.

Coolant

The heat generated in the active region of the reactor is carried away by a heat-carrying agent, usually water or an alkali metal such as sodium (the agent should have a large thermal capacity). This agent or the *coolant* gives this energy to water (see Fig. 9.9) transforming it into steam which operates the steam turbines.

Breeder Reactors

So far mainly moderator-operated reactor bases on thermal neutrons has been discussed. It is possible to run a reactor on fast neutrons by using a fuel with enriched ^{235}U or ^{239}Pu. The fast neutrons released in the fission are used to transform ^{238}U into ^{239}Pu (or ^{232}Th into ^{233}U) which can undergo fission by capturing thermal neutrons. The core of such a reactor will contain two materials, say, fissionable fuel ^{239}Pu and the potential fuel ^{238}U. If the conditions are such as to produce more fuel than the amount burnt, the reactor is known as breeder reactor (it breeds fissionable fuel).

Uncontrolled Chain Reactions

If two or more pieces of almost pure ^{235}U (or ^{239}Pu), each of sub-critical mass, are brought together, the total mass may be over critical. In such a case, the neutrons rapidly multiply and the energy release will be explosive, as in the case of an atomic bomb.

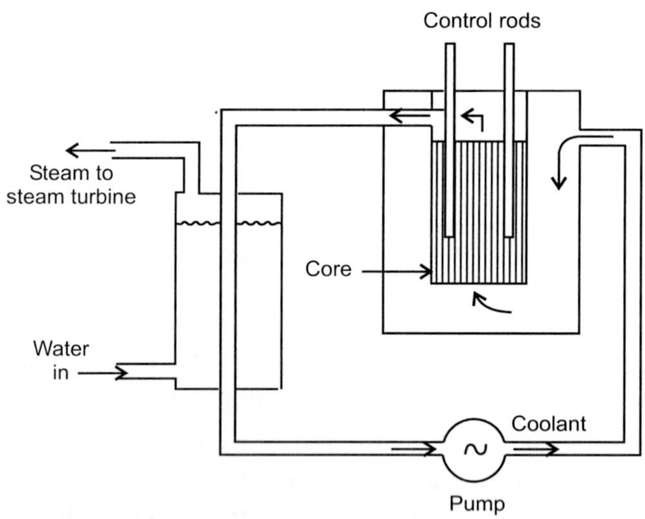

Fig. 9.9 Schematic diagram of a power reactor.

In an atomic bomb, the chain reaction can be triggered by secondary cosmic ray neutrons. Here, the pieces must be kept together under pressure so that the chain reaction can build up to explosive proportions. This is done by using ordinary explosives to shoot one piece into another. Nevertheless, only a part of the fuel has time to react before the final explosion.

Nuclear reactors are used for producing power, producing fissionable material, and to obtain strong neutron sources. The neutron sources may be used for conducting scientific experiments and for producing radioactive isotopes which are of enormous use in medicine, and in industry.

9.8 THERMONUCLEAR FUSION

It may be observed (Fig. 9.1) that the binding energy per nucleon is small for light nuclei, and increases to maximum value for $A \approx 60$. It is therefore energetically preferable for lighter nuclei to fuse into larger nuclei. Such a process would be accompanied by a release of energy, *e.g.* in the fusion of deuterium and tritium,

$$^{2}\text{H} + {}^{2}\text{H} \rightarrow p + {}^{3}\text{H} \tag{9.115}$$

$$^{2}\text{H} + {}^{3}\text{H} \rightarrow n + {}^{4}\text{He} \tag{9.116}$$

energies of 4.0 MeV and 17.6 MeV respectively, are released.

For a fusion reaction to take place, the lighter nuclei must overcome Coulomb repulsion between them [see Fig. 9.6(*b*)], *i.e.* they must have an energy

$$E_c \approx \frac{Z_1 Z_2\, e^2}{4\pi\varepsilon_0\,(r_1 + r_2)} \tag{9.117}$$

which for $Z_1 = Z_2 = 1$ and $r_1 + r_2 = 2$ fm has a value of about 0.7 MeV. Thus, each of the two nuclei must have an energy of about 0.35 MeV. A temperature of about 2×10^9 K would provide an average thermal energy of about 0.3 MeV, and hence promote the fusion reaction. However, the fusion reaction can proceed even at lower temperatures. This is due to the fact that (*i*) the energies of the nuclei at a given temperature have Maxwell-Boltzmann distribution so that there are always some nuclei with enough energy to overcome the Coulomb barrier, (*ii*) nuclei can tunnel through the potential barrier. Therefore, and appreciable amount of fussion takes place at temperatures of about 10^7 K. Since the reaction is induced by high temperatures, it is known as *thermonuclear fusion*.

Controlled Fusion

To have controlled fusion reactions, it is necessary to maintain nuclei at a temperature of about $10^7 – 10^8$ K in a confined region, so that nuclear reactions can take place. At such high temperature, the atoms are ionized into positively charged ions and electrons, forming what is known as a *plasma state*. The two main problems in achieving controlled fusion are the containment of the plasma within a suitable volume, and the heating of the plasma to the required high temperatures.

For the confinement of the plasma, one cannot use the walls of any vessel. Any contact with the wall will not only quickly cool the plasma but also cause the wall to evaporate. What is usually done is to confine the plasma in a suitable magnetic field. The nuclei spiral along the magnetic field lines. By a suitable arrangement of the field, the nuclei are reflected back and for the between bottle necks provided by the converging lines (the lines tend to converge in regions where the magnetic field is stronger). Such an arrangement is called a *mirror machine*. Alternatively, the plasma may be confined in toroidal region formed by a solenoid bent in the form of a torus. In this case, the nuclei spiral along the closed field lines inside the torus. However, there are as yet serious difficulties in controlling the instabilities of confinement over appreciable time periods.

There are two important methods of heating a plasma. In one method, fast neutral atoms are injected into the magnetically-confined system and are ionized by collisions with the plasma. The energetic ions are now trapped by the magnetic field for long enough to transfer their energy to the plasma by collisions. For example, a plasma of H^+ may be heated by a beam of energetic H, or a plasma of D^+ (nucleus of deuterium) and T^+ (nucleus of tritium) by a beam of energetic D (Deuterium). The beam energies are generally of the order of a few tens of keV to several hundreds of keV. The energetic beams are usually produced by accelerating low energy ions in an electrostatic field and then passing the ions through a target gas where the ions capture electrons and are neutralized. The other method of heating a plasma is by radio-frequency electromagnetic waves. When the waves are incident of a plasma, under suitable conditions, their energy is converted into ordered particle energy which is then thermalized by collisions.

An alternative approach to controlled fusion is through what is known as *inertial confinement*. Here, the fusion fuel, *e.g.* mixture of deuterium and tritium, in the form of a pellet, is imploded from all sides by energy sources such as laser beams, high energy electron or ion beams. The intense compression pressures and the high temperatures produced in the pellet may produce conditions conductive to fusion (it is the particle interial which provides the basis for confinement over the required period and hence the term inertial confinement). The difficulties in this approach are the low efficiencies of laser or other sources, and the need to produce stable symmetrical implosion.

For controlled fusion to be a meaningful source of energy, the output energy must be more than the input energy. There are several technical difficulties which remain in achieving the break-even point, such as instabilities in confinement, inefficient heating, etc. As such, controlled fusion has not yet been realized. When realized, it will provided a virtually inexhaustible source of energy. Deuterium, which is suitable for a fusion reaction (ordinary hydrogen has a very small cross-section for fusion, and hence is not suitable), is readily available, 0.03% by mass of hydrogen in water being in the form of deuterium. Furthermore, the fusion reactions have important advantages over other sources of energy, in that they have hardly any radioactivity problem, produce negligible pollution, and their sources are widely distributed.

Uncontrolled Fusion

Uncontrolled fusion can be achieved by using an atom bomb whose explosion produces temperatures of the order of 10^7 K. For example, such an atom bomb can ignite a fuel of deuterium and tritium, leading to the fusion reaction in Eq. (9.116). This is the source of energy in what is known as the *hydrogen* or *thermonuclear bomb.*

Fusion reactions are the source of energy in the sun and the stars, inside which temperatures are of the order of 10^7-10^8 K. The energy there is produced in two ways. In the proton-proton cycle which is dominant at lower temperatures ($T \sim 10^7$ K), the fusion of hydrogen takes place in the following steps:

$$
\begin{aligned}
{}^1\text{H} + {}^1\text{H} &\rightarrow {}^2\text{H} + \overline{e} + \nu \\
{}^2\text{H} + {}^1\text{H} &\rightarrow {}^3\text{He} + \gamma \\
{}^3\text{He} + {}^3\text{He} &\rightarrow {}^4\text{He} + 2\,{}^1\text{H}
\end{aligned}
\tag{9.118}
$$

The net release of energy in this sequence is about 25 MeV. Alternatively the fusion may take place through the carbon cycle which becomes dominant at higher temperature:

$$
\begin{aligned}
{}^1\text{H} + {}^{12}\text{C} &\rightarrow {}^{13}\text{N} + \gamma \\
&\qquad\downarrow \\
&\qquad {}^{13}\text{C} + \overline{e} + \nu \\
{}^1\text{H} + {}^{13}\text{C} &\rightarrow {}^{14}\text{N} + \gamma \\
{}^1\text{H} + {}^{14}\text{N} &\rightarrow {}^{15}\text{O} + \gamma \\
&\qquad\downarrow \\
&\qquad {}^{15}\text{N} + \overline{e} + \nu \\
{}^1\text{H} + {}^{15}\text{N} &\rightarrow {}^{12}\text{C} + {}^4\text{He}
\end{aligned}
\tag{9.119}
$$

The net result is the fusion of four hydrogen nuclei fusion into one helium atom with ^{12}C serving only as a catalyst.

It may be noted that the Coulomb potential barrier (Eq. 9.117) is larger for nuclei with higher Z values so that it is more difficult for the fusion of heavier nuclei to take place. However, when the temperatures of stellar interiors rise, fusion of heavier nuclei begins to take place. In particular, there is helium burning,

$$
\begin{aligned}
{}^4\text{He} + {}^4\text{He} &\rightarrow {}^8\text{Be} \\
{}^4\text{He} + {}^8\text{Be} &\rightarrow {}^{12}\text{C} + \gamma
\end{aligned}
\tag{9.120}
$$

producing carbon. At higher densities and temperatures, fusion of heavier elements also takes place, ultimately leading to elements in the iron mass region

($A = 56$) where the binding energy per nucleon has a maximum. Here, exothermic fusion reactions cease. Elements heavier than ^{56}Fe may be produced by capture of neutrons produced in some reactions, and subsequent β-decays. Thus, elements upto and just beyond uranium are produced. Still heavier elements have short lifetimes and if created would quickly decay either by α-emission or fission.

9.9 EXAMPLES

Some examples to illustrate the properties and interactions of the nucleus are considered here.

Example 1

In the early stages of the development of nuclear physics before the neutron was discovered, one of the models of the nucleus considered was that it was made up of A protons and $(A - Z)$ electrons. There are several arguments against this model.

An electron confined to a volume of nuclear dimensions would be highly relativistic, and its energy would be estimated by the uncertainty principle to be

$$T \approx pc \approx \frac{\hbar c}{\Delta x} \tag{9.121}$$

$$\approx 100 \text{ MeV for } \Delta x \approx 2 \text{ fm}$$

The confinement of such energetic electrons would require the existence of very strong forces for which there is no evidence (such potentials would also create many electron-positron pairs which are not observed). A conclusive evidence against the proton-electron model of the nucleus is that even A, odd Z nuclei have integral spin. In the proton-electron picture, such a nucleus would have A protons and A-Z electrons, *i.e.* the nucleus has an odd number of fermions, and hence would be expected to have a half-integral spin. This is contrary to the experimental observations, *e.g.* ^{14}N has $I = 1$. Finally, the proton-electron picture would imply the existence of nuclear magnetic moments of the order of $\frac{e\hbar}{2m_e}$ whereas the observed moments are much smaller, of the order of $\frac{e\hbar}{2m_p}$.

Example 2

One can estimate the strength of the deuteron potential, by assuming the potential to be a square well of depth V_0 and radius a.

The radial Schrödinger equation for the ground state with $l = 0$, is

$$-\frac{\hbar^2}{2m}\left(\frac{1}{r^2}\frac{\partial}{\partial r}r^2\frac{\partial\psi}{\partial r}\right) = (E - V)\psi \tag{9.122}$$

where m is the reduced mass $\left(m \approx \frac{1}{2}m_p\right)$, $V = 0$ for $r > a$ and $V = -V_0$ for $r \le a$.

It is easy to show that

$$\psi = \frac{1}{2}\begin{cases} A_1 e^{-kr} & \text{for } r > a \\ A_2 \sin \alpha r & \text{for } r < a \end{cases} \tag{9.123}$$

where

$$k = \left(\frac{2m|E|}{\hbar^2}\right)^{1/2}, \alpha = \left(\frac{2m(V_0 + E)}{\hbar^2}\right)^{1/2}$$

Continuity of the wave function and the derivative of the wave function, at $r = a$, give the condition

$$\alpha \cot \alpha a = -k \tag{9.124}$$

Since $|E|$ is known to be small compared to V_0, one has an approximate relation

$$\alpha a \approx \pi/2 \tag{9.125}$$

which implies $V_0 \approx 23$ MeV for $a \approx 2$ fm. A better approximation would be

$$\alpha a \approx \frac{\pi}{2} + \frac{k}{\alpha} \tag{9.126}$$

which yields a value of V_0 » 33 MeV for $|E| \sim 2$ MeV. In any case, it is seen that the interaction is quite strong but the deuteron is a shallow bound state, *i.e.* $(|E|/V_0) \ll 1$.

Example 3

As an application of the shell model, the spin and magnetic moments of ^{17}O and ^{127}I are considered.

The odd nucleon in ^{17}O is the ninth neutron which is in the $d_{5/2}$ state. Therefore, ^{17}O has $j = 5/2$ and its magnetic moment [see Eq. (9.38)] is expected to be about $-\frac{e\hbar}{2m_p}(1.91)$. Experimentally j is found to be 5/2 and the magnetic moment to be $-\frac{e\hbar}{2m_p}(1.89)$.

For ^{127}I, the odd nucleon is the 53rd proton. Shell model would predict that it is in the $g_{7/2}$ state. However, one finds $j = 5/2$ for the nucleus. Assuming that is in the $d_{5/2}$ state (see Fig. 9.3), shell model predicts $\mu \approx \frac{e\hbar}{2m_p}$ (4.79) whereas the experimental value is $\frac{e\hbar}{2m_p}$ (2.81). The two example, illustrate the usefulness and limitations of the shell model.

Example 4

A mass spectrometer is used for determining the masses of nuclei. It is based on the principle that a moving particle subjected to mutually perpendicular electric and magnetic fields, which are also perpendicular to the velocity of the particle, is undeviated if

$$q\mathbf{E} + q\mathbf{v} \times \mathbf{B} = 0 \tag{9.127}$$

or $v = |E|/|B|$. If such a particle is now subjected to a magnetic field, it moves in a circle of radius

$$r = \frac{p}{qB} \tag{9.128}$$

where p is its momentum. Thus, from the knowledge of v and p, the mass of the particle can be determined (provided q is known).

Example 5

It is after the case that only a small amount of the target is exposed to a beam of particles. The reaction produces an unstable isotope which decays. It is of interest to know the number of unstable nuclei remaining after an exposure to the beam for time t.

If the target contains N nuclei, the number of reactions per second is

$$n = N\sigma F \tag{9.129}$$

where F is the flux of the beam, *i.e.* particles/m^2/s. The net increase dP in the isotope population over a period dt is

$$dP = N\sigma\, F\, dt - \lambda\, P\, dt \tag{9.130}$$

where λ is the probability for decay. The solution to this equation is

$$P(t) = N\sigma\, F(1 - e^{-\lambda t})/\lambda \tag{9.131}$$

For example, consider 1 mg of ^{23}Na exposed to a neutron beam of flux 10^{14}/cm^2 s. The cross section for the reaction ^{23}Na(n, γ) ^{24}Na is about 0.56 barns. Since $1/\lambda \approx 21.7$ h, and $N \approx 2.6 \times 10^{19}$, the number of isotope nuclei is

$$P(t) \approx 1.15 \times 10^{14} (1 - e^{-t/21.7}) \tag{9.132}$$

where t is in hours.

Example 6

The determination of the age of a sample by ^{14}C dating is illustrated here.

Let M grams of a sample of organic carbon decay at the rate of $r(t)$ per hour. Then the number $n(t)$ of ^{14}C atoms is given by

$$r(t) = \frac{1}{\tau} n(t) \tag{9.133}$$

where the lifetime τ is about 7.242×10^7 hs (half-life is 0.6931 times τ). The fraction of ^{14}C in atmospheric carbon is 1.3×10^{-12}, which implies that at the beginning there are

$$n(0) = \left(\frac{6.03 \times 10^{23}}{12}\right) M \times 1.3 \times 10^{-12} \tag{9.134}$$

Hence,

$$\frac{n(t)}{n(0)} = 1.109 \times 10^{-3} \left[\frac{r(t)}{M}\right] \tag{9.135}$$

which, with the help of the decay law in Eq. (9.60), leads to

$$t = \tau \ln \left[902 \frac{M}{r(t)}\right] \tag{9.136}$$

For example, if a sample of 1 g yields 300 decays/h, its age is $t \approx 9100$ years.

Example 7

The delayed neutrons through small in number, play an important role in reactor control.

Let there by $N(t)$ neutrons at time t, and let $\tau_0 \approx 10^{-2}$ s be the period of the cycle between two fissions in the chain reaction. In the fission, 99.3% of $N(t)$ are multiplied by a factor of about 2.5 and the remaining 0.7%, though multiplied by the same factor, are emitted later, say after time $\tau_1 \approx 9$ s. Under the equilibrium condition 1.5 $N(t)$ would be lost so that once again $N(t)$ is got back after time τ_0.

Suppose now that the equilibrium condition is disturbed and that $(1.5 - \delta)$ $N(t)$ are lost ($\delta > 0$). Then the number of neutrons at time $t + \tau_0$ is

$$N(t + \tau_0) = 2.5\,(0.993)\,N(t) + 2.5(0.007)\,N(t - \tau_1) - (1.5 - \delta)\,N(t) \tag{9.137}$$

Using $N(t + \tau) \approx N(t) + \tau N'(t)$,

$$(\tau_0 + 0.0175\,\tau_1)\,N'(t) = (\delta)N(t) \tag{9.138}$$

which on integration yields

$$N(t) = N(t_0)\exp\left[\frac{\delta(t - t_0)}{\tau_0 + 0.0175\,\tau_1}\right] \tag{9.139}$$

Since δ is usually of the order of 1% or less and $\tau_1 \gg \tau_0$, the delayed neutrons (characterized by τ_1) dominate the time variation and the time-scales of change are of the order of $0.0175\,\tau_1/\delta \approx 10$ s – 1 min.

Example 8

An interesting mechanism of producing fusion is by screening the Coulomb repulsion between the nuclei while they are being brought together. This can be done by using a muonic hydrogen atom (bound state of a proton and a muon). Since the muon is about 200 times heavier than the electron, its Bohr radius is correspondingly smaller, about 0.25×10^{-12} m. Thus, the muon effectively screens the proton charge, allowing it to approach another nucleus with greater ease. For example, it may fuse with a deuteron to give ^{3}He,

$$(\mu^- p) + {}^2\text{H} \rightarrow \mu^- + {}^3\text{He} \tag{9.140}$$

with a release of about 5.5 MeV. Since μ^- has a short lifetime ($\tau \approx 2.2 \times 10^{-6}$ s) and is not easily produced, this does not appear to be an economical way of producing fusion. However, it is being considered as a triggering mechanism to start controlled fusion.

PROBLEMS

1. For a charge q distributed uniformly in a sphere of radius r, show that the electrostatic energy is $\frac{3}{5}\left(\frac{q^2}{4\pi\varepsilon_0 r}\right)$. For a proton this has a value of about 0.86 MeV if $r \approx 1$ fm. It is believed that the small proton-neutron mass difference is of electromagnetic origin, which depends crucially on the magnetic properties to give a heavier neutron.

2. Nuclei ${}^{2Z+1}_{Z}X$, ${}^{2Z+1}_{Z+1}Y$ are examples of mirror nuclei (which are obtained by $n \leftrightarrow p$). Charge independence of the nuclear forces implies that the mass difference between these nuclei is electromagnetic in origin. Using the result of problem 1, obtain an expression for the mass difference of mirror nuclei. Using $r = 1.2\,A^{1/3}$ fm, determine the mass difference between ^{11}B and ^{11}C, ^{13}C and ^{13}N, ^{35}Cl and ^{35}Ar. Compare with the experimental values of 2.8, 3.0 and 6.7 MeV respectively.

3. Show that for an ellipsoid with semi-major axis a along the axis of rotation and semi-minor axis b perpendicular to it, the quadrupole moment is $\frac{2}{5}Z(a^2 - b^2)$. Estimate $(a - b)/R$ for ^{176}Lu from the information that its quadrupole moment is about 8×10^{-28} m^2.
4. Show that the density of nuclear matter is about 2.3×10^{17} kg/m^3. This is the type of density expected in a neutron star which may be regarded as a giant nucleus of mass comparable to that of the sun.
5. Assuming that the deuteron is a bound state in the Yukawa potential given in Eq. (9.25), and the energy of about 30 MeV is approximately the value of the potential at $r \approx 1$ fm, show that $| g\varepsilon_0/e^2 | \approx 60$. This gives an idea of the strength of nuclear forces compared to the electromagnetic forces.
6. The nucleus ^{121}Sb has spin 5/2. What is its expected magnetic moment? Compare the result with the observed value of $3.36 \dfrac{e\hbar}{2m_p}$.
7. Deduce the spin and magnetic moment of ^{3}He, ^{15}N, ^{39}K and ^{209}Bi from the simple shell model and compare with the experimental values of $j = 1/2, 1/2, 3/2, 9/2$ and $\mu = -2.13, -0.28, 0.39, 4.1$ in units of nuclear magnetons, respectively.
8. Determine the moment of inertia of ^{234}Th given that its lowest rotational energy levels are at 0, 0.048, 0.16 MeV. Compare it with the moment of inertia of the whole nucleus regarded as a rigid sphere. What can you deduce? What is the next expected rotational level?
9. The rotational ground state of ^{237}Np has $I = 5/2$. The observed excited levels have energies 0.033, 0.060, 0.076, 0.103 and 0.159 MeV. Which of these may be expected to belong to the rotational band?
10. Obtain the masses of ^{106}Ru, ^{106}Rh, ^{106}Pd, ^{106}Ag, and ^{106}Cd, from the semiempirical formula and discuss the stability of these nuclei against $\beta^{\pm}$ decays and electron capture.
11. Obtain the masses of ^{65}Ni, ^{65}Cu and ^{65}Zn from the semi-empirical formula and discuss the stability against $\beta^{\pm}$-decay and electron capture.
12. Estimate the Coulomb barrier for α emission by ^{238}U. What is the energy of the α particle and of ^{234}Th in the process $^{238}\text{U} \rightarrow {}^{234}\text{Th} + {}^4\text{He}$? $(m_{\text{U}} - m_{\text{Th}} - m_{\text{He}} \approx 4.3$ MeV).
13. The Q value for the α-decay of ^{213}Po into ^{209}Pb is 8.52 MeV. What is the energy of the α particle in the transition between these states? If some α particle come out with 7.60 MeV, what is the energy of the corresponding excited state of Pb?

14. The element ^{32}P decays into ^{32}S ($m_s \approx 31.972072$) mu including the mass of electrons) by β^--emission. If the maximum kinetic energy of the electron is 1.7 MeV, what is the mass of ^{32}P?
15. The element ^{7}Be decays by electron capture. If the masses of ^{7}Be and ^{7} Li are 7.016930 and 7.016004 mu respectively, what is the energy and momentum of the recoil nucleus?
16. In the thorium series, the initial nucleus is ^{238}U and the final nucleus is ^{206}Pb. How many α particles are emitted by each uranium nucleus? How many electrons are emitted by each nucleus? If the lifetime is 6.5×10^9 years, how much helium is released from 1 g of ^{238}U in 1 year? A mineral sample contains ^{206}Pb and ^{238}U in the ratio of 1 to 4. Assuming that the Pb is from the decay of U, estimate the age of the sample.
17. For the reaction $^{6}Li + n \rightarrow {}^{3}H + {}^{4}He$ with thermal neutrons, determine the kinetic energy of ^{4}He. It is given that the Q value for the reaction is 4.78 MeV.
18. A beam of neutrons is incident on a piece of gold. Show that the intensity of the beam as a function of the depth t of penetration is given by $I(t) = I(0) \exp(-\sigma n t)$ where σ is the capture cross-section and n is the number of target nuclei per volume. If the emerging intersity is 74% of the original intensity for $t = 0.05$ cm, what is the capture cross-section of gold?
19. If an average energy of 200 MeV is released in the fission of each ^{235}U nucleus, how mush ^{235}U is used in one day in a reactor operating at a power of 50 MW?
20. There is evidence to believe the sum has been in the present stable condition for the last 5×10^9 years. Assuming that the stable condition implies that the change in the composition is less than 10%, argue that nuclear fusion is the only feasible source of energy and that the sun is likely to remain in the present condition for another 5×10^9 years. The sun radiates an energy of about 4×10^{26} J/s, and its mass is about 2×10^{30} kg.

10

Elementary Particles

Structures of the Chapter

While considering the microscopic structure of matter, it is pertinent to ask whether the different forms of matter have a common basis. Is it possible to understand the various properties in terms of a few elementary particles with prescribed rules for their interactions? If these constituent particles and their interactions are analysed, then in principle, one can construct all the forms of matter and explain their properties.

In this chapter, the present status of out understanding of the elementary particles, their various interactions and the unification of these interactions that is emerging are briefly discussed.

10.1 ELEMENTARY PARTICLES

What are the elementary particles? This is not always an easy question to answer. The answer will in general depend upon the existing knowledge and the calculational tools provided by the underlying theory, which allow us to explain the properties of the composite objects. Therefore, it might happen that entities that are regarded as elementary particles at some time can later be described as composite particles as out knowledge and calculational techniques improve. For example, atoms which were regarded as the building blocks in the nineteenth century are now regarded as composites of electrons and a nucleus, and the nucleus itself is regarded a composite of protons and neutrons (the nucleons). The generally accepted ideas of elementary particles are presented here and their implications discussed. The interesting feature of these ideas is that many particles that are thought to be elementary, have not yet been observed, and indeed may not be observable in principle.

To start with, there are *leptons* which appear in pairs:

$$\begin{bmatrix} \nu_e \\ e \end{bmatrix}, \begin{bmatrix} \nu_\mu \\ \mu \end{bmatrix}, \begin{bmatrix} \nu_\tau \\ \tau \end{bmatrix} \tag{10.1}$$

In each pair, the first particle is a *neutrino*, which carries zero electric charge, and which is associated with the corresponding negatively-charged lepton. So far, there is no evidence that neutrinos ave nonzero mass (experimentally $m\upsilon_{\nu e} < 2$ eV, $m_{\nu\mu} < 0.170$ MeV, $m_{\nu\tau} < 15.5$ MeV and $m_j \approx 1777$ MeV where mass is expressed in terms of rest energy). The leptons are fermions with spin 1/2, and the charged leptons have masses.

$$m_e = 0.511 \text{ MeV},\ m_\mu = 106 \text{ MeV},\ m_\tau \approx 1777 \text{ MeV} \tag{10.2}$$

and carry a negative charge – *e*. These doublets in (10.1) are called the leptons of the first, second and third generation , in the order of increasing mass. Every particle in nature is accompanied by an antiparticle which has the same mass but with all its other properties (such as the charge) being opposite to those of the particle. Therefore, along with the leptons we have *antileptons* with the same mass as the leptons but with other properties such as the charge, being opposite to those of the leptons (antiparticles for fermions are the holes in the negative-energy sea of Dirac discussed in Sec. 4.8).

Similar to the lepton doublets, there are doublets of *quarks* which have not so far been observed, but which have proved extremely useful in describing the properties of their composites, such as the proton, the neutron, etc. They are

$$\begin{bmatrix} u \\ d \end{bmatrix}, \begin{bmatrix} c \\ s \end{bmatrix}, \begin{bmatrix} t \\ b \end{bmatrix} \tag{10.3}$$

They have half-integral spin and have the properties shown in Table 10.1. Since they have not been observed directly, their masses are inferred from the masses of their composites, and are known as first, second and third generation quark doublets, in the order of increasing masses. For reasons mentioned later, each quark is supposed to come in three varieties, known as *red, white* and *blue* quarks, equal mixtures of which are said to be *colourless*. As in the case of leptons, one has *antiquarks* which have the same mass as the quarks, but all the other properties such as charge, baryon number, etc. are opposite to those of the quarks.

Table 10.1 Properties of quarks, charge in units of *e*, and masses in MeV

Quarks	*Charge*	*Baryon number*	*Isospin I* (I_z)	*Strange-ness*	*Inferred mass (MeV)*
Up (u)	2/3	1/3	1/2 (1/2)	0	1.7–3.3 MeV
Down (d)	–1/3	1/3	1/2 (–1/2)	0	4.1–5.8 MeV
Charm (c)	2/3	1/3	0(0)	0	1.27+0.07–0.09 GeV
Strange (s)	–1/3	1/3	0(0)	–1	101+29–21 MeV
Top (t)	2/3	1/3	0(0)		172.0±0.9±1.3 GeV
Bottom (b)	–1/3	1/3	0(0)		4.61+0.18–0.06 GeV

Leptons and quarks interact with each other via the exchange of some bosons. These bosons are *gluons* (not yet observed directly) which give rise to strong forces between quarks binding them into observed strongly interacting particles such as the proton, the pi-meson, etc., the photons which give rise to electromagnetic forces, the W-bosons which give rise to weak-interaction forces, and the gravitations which given rise to gravitational interaction forces. These bosons are necessary for the description of the interaction between different particles.

The properties of the interactions and the classification of particles are now considered.

10.2 STRONG INTERACTION

Quarks interact with each other strongly by exchanging gluons which are supposed to be eight in number. The strength of this interaction called the *strong interaction* is given by

$$\alpha_s \approx 0.4 - 0.2 \tag{10.4}$$

which may be compared with the corresponding strength of electromagnetic interaction given by the fine structure constant $\alpha \approx 1/137$. As a result of this interaction (one also postulates an additional confining potential), bound states are obtained, which are the observed strongly interacting particles. They are known as the *hadrons,* and two of the hadron sets are listed in Table 10.2, namely the *baryons* with spin 1/2 (*positive parity*), and *pseudoscalar mesons* with spin 0 (*negative parity*). Baryons are the bound states of three quarks (baryon number and its conservation are introduced to explain the stability of matter which is primarily made up of baryons) while mesons are the bound states of a quark and an antiquark. The quark composition of these bound states is shown in Table 10.2, where it should be understood that these states are formed by taking appropriate combinations of the spin-states of the quarks. Hadrons interact with each other through strong interaction of essentially unit strength and with a short range of about 10^{-15} m.

Table 10.2 List of baryons $(1/2)^+$, and mesons $(0)^-$, along with their masses (in MeV), charge, isospin, strangeness and their quark composition

	Hadron	*Mass (MeV)*	*Charge*	$I(I_z)$	*Strange-ness*	*Constituent quarks*
	P	938.2	1	1/2 (1/2)	0	*uud*
	N	939.5	0	1/2 (–1/2)	0	*udd*
	Λ	1116	0	0(0)	–1	*uds*
Baryons	Σ^+	1189	1	1(1)	–1	*uus*
	Σ^0	1192	0	1(0)	–1	*uds*
	Σ^-	1197	–1	1(– 1)	–1	*dds*
	Σ^0	1315	0	1/2(1/2)	–2	*uss*
	Σ^-	1321	–1	1/2(– 1/2)	–2	*dss*

	Hadron	*Mass (MeV)*	*Charge*	$I(I_z)$	*Strange-ness*	*Constituent quarks*
	π^+	139.6	1	1(1)	0	$u\bar{d}$
	π^0	135.1	0	1(0)	0	$u, \bar{u}\ d\bar{d}$
	π^-	139.6	–1	1(–1)	0	$d\bar{u}$
Mesons	K^+	494	1	1/2(1/2)	1	us
	K^0	498	0	1/2(–1/2)	1	$d\bar{s}$
	$\bar{K}^0$	498	0	1/2(1/2)	–1	$s\bar{d}$
	K^-	494	–1	1/2(–1/2)	–1	$s\bar{u}$
	η	549	0	0	0	$u\bar{u}$, $\overline{dd}$, $s\bar{s}$

Isospin Symmetry

The most striking feature to be noted in the properties of the hadrons is that they come in multiplets with components of very nearly the same mass, *e.g.* 938.2 MeV for the proton and 939.5 MeV for the neutron. Such a property has been noted for bound states in central potentials, where the states with different m values but the same l value, have the same energy. It is therefore suggested that we postulate and abstract space in which there is an abstract spin called *isospin* I, and the different components of a multiplet are states with the same I but different I_z. For example, P and N have $I = 1/2$ and $I_z = 1/2, -1/2$ respectively. The equality of the masses of the different components, would then follow from the invariance of the interaction under rotations in the abstract isospin space.

It may be observed in Table 10.2, that the different components of an isospin multiplet differ only in their u, d components. Therefore, an equality of the masses of the u and d quarks would imply an equality of the masses of the components of each multiplet. Thus, if the interactions do not distinguish between u and d quarks, it is suggested that the interactions are invariant under the transformations

$$\begin{aligned} | u' \rangle &= x_1 | u \rangle + x_2 | d \rangle \\ | d' \rangle &= y_1 | u \rangle + y_2 | d \rangle \end{aligned} \tag{10.5}$$

with the condition that $| u' \rangle$ and $| d' \rangle$ are orthonormal (the notation $| u \rangle$, etc. is used to designate the states), which implies

$$\begin{aligned} | x_1 |^2 + | x_2 |^2 = | y_1 |^2 + | y_2 |^2 = 1 \\ x_1 y_1^* + x_2 y_2^* = 0 \end{aligned} \tag{10.6}$$

One also imposes a phase condition for the determine of the matrix formed by the coefficients x_i and y_i

$$x_1y_2 - x_2y_1 = 1 \tag{10.7}$$

The linear transformations in Eq. (10.5), with the conditions in Eqs. (10.6) and (10.7) define the group SU(2) (group of *special unitary* transformations in 2-dimensions) which is closely related to the usual 3-dimensional rotations. Invariance under these transformations gives rise to SU(2) or *isospin symmetry.* It allows the characterization of states by isospin I and its z-component. I_z (similar to l and m in the case of ordinary rotations). Thus, (u,d) have $I = 1/2$ and $I_z = \pm\, 1/2$, (s) has $I = 0$ and $I_z = 0$, (P, N) have $I = 1/2$ and $I_z = \pm\, 1/2$, $(\Sigma^+, \Sigma^0, \Sigma^-)$ have $I = 1$ and $I_z = 1, 0, -1$, etc. Furthermore, these quantum numbers are conserved in processes which involve only strong interaction.

For specific applications, the (P, N) system is considered which can have $I = 1$ or 0. Designating the isospin states by $|\, I, I_z\rangle$,

$$|\,1,1\rangle = |\,PP\rangle \tag{10.8}$$

$$|\,1,0\rangle = \frac{1}{2^{1/2}}(|\,PN\,\rangle + |\,NP\,\rangle) \tag{10.9}$$

$$|\,1,-1\rangle = |\,NN\rangle \tag{10.10}$$

$$|\,0,0\rangle = \frac{1}{2^{1/2}}(|\,PN\rangle - |\,NP\rangle) \tag{10.11}$$

These relations follow from the usual quantum-mechanical rules for combining two angular momenta (also see Problem 1). Isospin symmetry then implies that the probability amplitudes T, which are essentially the probability amplitudes for the processes, satisfy the relations

$$\begin{aligned}\langle PP\,|\,T\,|\,PP\rangle &= \langle\, NN|\,T\,|NN\rangle \\ &= \frac{1}{2}\langle PN + NP\,|\,T\,|\,PN + NP\rangle\end{aligned} \tag{10.12}$$

Another useful application is obtained by noting that the deuteron D appears in only one charge state and hence is assigned $I = 0$. Since the π^- meson multiplet has $I = 1$, the $D\pi$ state is an $I = 1$ state. Conservation of isospin then gives the result

$$\langle\, D\pi^0\,|\,T\,|\,PN\rangle = \frac{1}{2^{1/2}}\langle D\pi^+\,|\,T\,|\,PP\rangle \tag{10.13}$$

Experimentally, this relation was verified at an energy of 340 MeV, to with in a few per cent by Hildebrand (1953), which supports the general ideas of isospin invariance in strong interaction.

SU(3) and Higher Symmetries

If the masses of the baryons in Table 10.2 are examined, it is observed that even baryons with different I have approximately equal masses (the difference are small compared to the baryonic masses). If these differences in masses are ignored, an extra degeneracy exists between the states with different *I*, which is equivalent to having the same mass for *u*, *d* and *s* quarks. This is reminiscent of the accidental degeneracy of the hydrogenic levels where the states with different *l* have the same energy. The interaction between the quarks is approximately invariant under transformations which mix *u, d* and *s*. As in Eq. (10.5), these transformations retain orthonormality and satisfy the phase condition for the determinant. This invariance gives rise to what is called the *SU* (3) *symmetry.* The multiplets of this symmetry which transform among themselves, have dimensions 1, 8, 8, 10 for the three-quark system, 1 and 8 for the quark-antiquark system etc. The observed multiplets are the *baryon octet* [shown in Fig. (10.1)], the *pseudoscalar meson octet,* the *baryon decuplet* with spin 3/2 and positive parity, etc. It should be mentioned that the symmetries not only imply the equality of the masses of the components of each multiplet, but also relate many of their other properties such as the magnetic dipole moment, the interaction strengths. etc. With the addition of charmed, top and bottom quarks, the symmetry can be extended to *SU* (6) *symmetry,* which implies symmetry under transformations which involve all the six quarks.

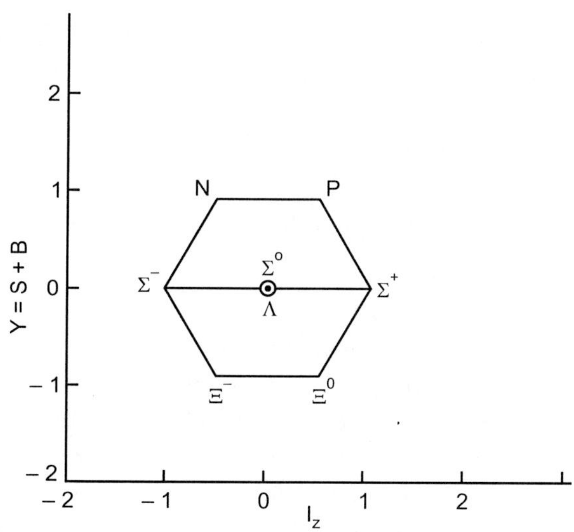

Fig. 10.1 The baryon octet shown in terms of I_z and $Y = S + B$ (Y is called the hypercharge).

It may be noted that processes which involve only strong interaction conserve charge, I_z, I, strangeness, charm, etc. Indeed, it appears that even the constituently quarks retain these properties. An interesting example of this was the discovery of a group of spin 1 particles called Ψ, Ψ', etc. which have masses greater than 3×10^3 MeV but very long lifetimes (about 10^4 times the lifetime of other vector mesons such as the ρ, etc.). They have a net charm of zero, but are made up of $c\bar{c}$ (c has charm 1 while $\bar{c}$ has charm −1). Since each charmed quark retains its charm these Ψ-mesons cannot decay into particles which do not contain c or $\bar{c}$ as constituents which explains their long lifetimes (other particles which contain charm are too heavy to provide decay channels).

10.3 ELECTROMAGNETIC INTERACTION

All charged particles (leptons, quarks and hadrons) interact electromagnetically. Even neutral particles have this interaction because of their charge distribution, for example, the neutron has zero charge but a fairly large magnetic dipole moment. The electromagnetic interaction between particles is propagated by the exchange of photons. An emission by a particle, of a photon of energy E introduces an uncertainty in the energy of the particle. The uncertainty principle implies that the system can remain in this state for a period of time

$$\Delta t \sim \hbar/E \tag{10.14}$$

During this time the photon can travel a distance

$$x \sim \hbar c/E \tag{10.15}$$

which essentially defines the *range* of the interaction. Since photons have zero mass, they can have an indefinitely small energy which means that the electromagnetic forces are long-range forces. The strength of these forces is characterized by the fine structure constant α

$$\alpha \approx 1/37 \tag{10.16}$$

which is much smaller than the strength of strong interactions characterised by α_s in Eq. (10.4).

The interaction between hadrons is dominated by the strong interaction, with electromagnetic interaction providing small corrections. For example, the proton an the neutron (as also π^+ and π^0, and other multiplets of a given isospin multiplet) have slightly different masses which is generally attributed to the difference in their electromagnetic interactions. Small differences due to electromagnetic interaction are observed in the so-called *mirror nuclei,* which have the same strong interactions but different charges. However, strong interactions are of short range $r_0 \sim 10^{-15}$ m, so that the interaction between hadrons at large distances is dominated by the electromagnetic forces, *e.g.* Rutherford scattering.

The electromagnetic interaction comes into its own domain in the description of the properties of charged leptons. The dominant interaction of the charged leptons (which do not have strong interaction) is the electromagnetic interaction. Fortunately, electromagnetic interactions of charged particles are well-defined through a generalization of the minimal electromagnetic interaction introduced by Eqs. (4.89) and (4.90). Furthermore, since the strength of the electromagnetic interactions, given by α in Eq. (10.16), is quite small, results can be obtained in powers of α using perturbation theory.

The predictions of the theory of electromagnetic interaction of leptons (*quantum electrodynamics*), are quite impressive. For example, including the effects of vacuum polarization, self-interaction and vertex correction, it is found that the magnetic dipole moment of the electron comes out to be

$$\mu = \frac{e\hbar}{2m_e}\left(1+\frac{\alpha}{2\pi}-0.328\frac{\alpha^2}{\pi^2}\right) \quad (10.17)$$

$$\approx 1.0011596\ \frac{e\hbar}{2m_e}$$

which is in excellent agreement with the experimental observation of

$$\mu = (1.001156 \pm 0.000012)\ \frac{e\hbar}{2m_e} \quad (10.18)$$

Detailed calculations have also been made for the Lamp shift, that is, the separation between the $2\ ^2S_{1/2}$ and the $2\ ^2P_{1/2}$ levels of the hydrogen atom. The theoretical calculations yield

$$\nu = \frac{\Delta E}{h} = (1.05720 \pm 0.0002) \times 10^9\ \text{s}^{-1} \quad (10.19)$$

which may be compared with the experimental observation of

$$\nu = (1.05777 \pm 0.00010) \times 10^9\ \text{s}^{-1} \quad (10.20)$$

Other processes which are accurately described by quantum electrodynamics are:

(*i*) electron-electron scattering called the *Moller scattering*

(*ii*) electron-positron scattering called the *Bhabha scattering*

(*iii*) electron-positron going into muon and antimuon, etc. It is appropriate to say that attempts to describe the electromagnetic interactions of hadrons have been, at best, only partially successful.

10.4 WEAK INTERACTION

There are some processes observed in nature that cannot be described either by strong interaction or by electromagnetic interaction of particles. An striking example of such processes is the transmutation of a radioactive nucleus by the emission of an electron. In this process, a nucleus of charge Z undergoes a transition

$$n(Z) \rightarrow n'(Z+1) + e + \bar{\nu}_e \quad (10.21)$$

with the emission of an electron. The lifetime for many of these decays is of the order of minutes, compared with the lifetime of order 10^{-22} s for decays involving only hadrons and 10^{-8} s for decays of excited atoms emitting electromagnetic radiation. The interaction which causes these decays is called the *weak interaction.* It was noticed in these decays that the electron does not carry away all the energy ($\Delta E \approx m_n c^2 - m_n' c^2$) but has a continuous energy distribution with a cut-off in the energy equal to $(m_n - m_n')c^2$. It was also found experimentally that no photons were emitted in the process.

In order to save the law of conservation of energy, Pauli made a bold suggestion (1930) that an electrically neutral particle with spin 1/2, accompanies the emission of the electron. This particle is the *neutrino.* It has a very small mass, $m_v < 60$ eV, possibly zero (theories prefer a zero mass for the neutrino), and as in the case of other particles, there is an *antineutrinio* as well. Indeed, the emission of an electron in Eq. (10.21) is accompanied by the emission of an antineutron (a neutrino would accompany the emission of a positron). The basic β-decay process of radioactive decay is

$$N \rightarrow P + e + \overline{v}_e \tag{10.22}$$

The neutrinos do not have direct strong or electromagnetic interaction. They interact very weakly with matter (a neutrino of 1 MeV energy has a path length of 10^{18} m in lead) and hence are very difficult to detect. However, nuclear reactors provide intense beams of neutrinos (about 10^{17} $m^{-2}s^{-1}$), and were detected by Reines and Cowan (1956) in the reaction

$$\overline{v}_e + P \rightarrow N + \overline{e} \tag{10.23}$$

which is essentially the inverse β-decay process ($\overline{e}$ is the positron). There are other examples of reactions due to weak interaction in which neutrinos accompany other leptons, *e.g.*

$$\pi^+ \rightarrow \overline{\mu} + v_m \tag{10.24}$$

The beam of neutrinos (of energy about 500 MeV) from the decay of π^+ produced in accelerators, was allowed to interact with neutrons (Lederman and Schwartz, 1962) and produced reactions

$$v_\mu + N \rightarrow P + \mu \tag{10.25}$$

but not v_μ, $+ N \rightarrow P + e$. Thus, the neutrinos produced in reactions of the type given in Eq. (10.24), accompanying muons, are different from those produced in the β-decay. There are, therefore, two types of neutrinos, v_e and v_μ which are associated with the electrons and the muons respectively. With the recent discovery of τ leptons, there should also be v_τ associated with τ leptons, six different neutrinos along with the antineutrinos, there would be all together six different neutrinos and antineutrinos.

Strangeness

The weak interaction plays an important role in the behaviour of what are known as *strange particles.*

The K-mesons and Λ, Σ, Ξ baryons were discovered in the cosmic rays (which are rays of generally high energy particles originating from the outer space) in the early fifties. After the construction of high energy accelerators, they could be produced and studied in a controlled manner. They are produced in reactions of the type

$$\pi^- + P \rightarrow K^0 + \Lambda \tag{10.26}$$

The rate of their production is typical of strongly interacting particles (*e.g* comparable to the production of π^0 N). However, the decay of Λ,

$$\Lambda \rightarrow \pi^- + P \text{ or } \pi^0 + N \tag{10.27}$$

is very slow. The lifetime of strange particles is generally of the order of 10^{-8} to 10^{-10} s (except for Σ^0 which decays into $\Lambda + \gamma$ in less than 10^{-14} s) whereas the typical lifetimes of decays of strongly interacting particles are of the order of 10^{-22} s. The unusual behaviour of these particles, as strongly interacting particles in production and as weakly interacting particles in decays, brought them the name of strange particles.

It was observed that the strange particles are produced in pairs [K^0 and Λ in Eq. (10.26)], called *associated production,* whereas the decay processes involve individual strange particles. This is reminiscent of a neutral system (*e.g.* radiation) producing a pair of oppositely charged particles (*e.g.* $\overline{e}$ and e^+) but a charged particle being forbidden to decay into a neutral system by charge conservation. Using this analogy, Gell-Mann and Nishijima introduced a new quantum number S called the strangeness which is conserved in strong interaction. Thus K^0 is assigned strangeness 1 while Λ is assigned strangeness –1, and π^- and p are assigned strangeness zero. Thus, the total strangeness is conserved in the reaction given in Eq. (10.26) (being zero both before and after the reaction). However, it is not conserved in the decay process given in Eq. (10.27) and hence the decay would be forbidden by strong interaction. Strangeness is conserved in electomagnetic interactions as well, so that the decay in Eq. (10.27) proceeds via the weak interaction which does not conserve strangeness. This would explain the long lifetime of Λ. Indeed the strength of the interaction for the decay in Eq. (10.27) is of the same order as the strength of the interaction which leads to the β-decay of the neutron in Eq. (10.22), ones the dependence of the decay on the masses is separated out. Thus, it is the weak interaction which governs the strangeness-changing processes, *e.g.* decay of Λ.

The strangeness of a particle is given by the relation

$$Q = \frac{1}{2}(S + B) + I_z \tag{10.28}$$

where Q is the charge, B is the baryon number, and I_z is the z-component of the isotopic spin. The combination $S + B$ is called the hypercharge Y and is often more convenient to use than strangeness. This relation implies that the conservation of strangeness is equivalent to the conservation of charge and I_z. It needs a slight modification once particles with nonzero charm are included:

$$Q = \frac{1}{2}(S + B + C) + I_z \tag{10.29}$$

where C is the charm of the particle. The relation will require further modification if top and bottom quarks are included.

Parity Violation

Weak interaction violates another important conservation low, namely the conservation of parity.

It had been observed that the laws of nature generally do not appear to distinguish between a coordinate frame and an inverted coordinate frame, *i.e.* the equations of motion are the same whether coordinates (x, y, z) or $(-x, -y, -z)$ are used. This is termed as invariance under space inversion or parity transformation. Let us define an operator P called the parity operator, which takes the wave function $\psi(x, y, z)$ in a coordinate frame, into a wave function $\psi'(x, y, z)$ observed for the same state but in a coordinate frame with inverted axes:

$$\psi'(x, y, z) = P\psi(x, y, z) \tag{10.29a}$$

However, $\psi'(x, y, z)$ is essentially the same as $\psi(-x, -y, -z)$ except for a possible phase factor A, so that

$$\psi'(x, y, z) = p_i\psi(-x, -y, -z) \tag{10.30}$$

Now a second operation by P leads us back to the original wave function so that

$$\begin{aligned}\psi(x, y, z) &= P\psi'(x, y, z) \\ &= p_i^2\,\psi(x, y, z)\end{aligned} \tag{10.31}$$

which implies that

$$p_i = \pm 1 \tag{10.32}$$

The states with $p_i = 1$ are said to be *even intrinsic parity* states and the states with $p_i = -1$ are said to be *odd intensity parity* states. Furthermore, let us assume that

$$\psi(-x, -y, -z) = p_e\,\psi(x, y, z) \tag{10.33}$$

where p_e is called the spatial parity of the state. This relation together to p_i, p_e Eqs. (10.30), (10.29) implies

$$P\psi(x, y, z) = p_i p_e\,\psi(x, y, z) \tag{10.34}$$

so that there exist eigenstates of parity with eigenvalues equal to $p_i p_e$ which are the product of intrinsic parity and spatial parity eigenvalues.

Now, if nature does not distinguish between the coordinate frame and the inverted coordinate frame, *i.e.* space inversion symmetry exists, then $\psi'(x, y, z)$ also must satisfy the Schrödinger equation

$$i\hbar \frac{\partial}{\partial t} P\psi(x, y, z) = HP\psi(x, y, z) \tag{10.35}$$

Since $\quad P^2 = 1,$

$$PHP = H \tag{10.36}$$

or

$$HP = PH \tag{10.37}$$

Thus, P commutes with the Hamiltonian. Therefore, parity is conserved and states which are simultaneous eigenstates of H and P can be obtained (See Eq. (3.42) and the dissuasion which follows it). This is valid provided nature exhibits space-inversion symmetry (which has been shown to be equivalent to having $HP = PH$).

In the fifties, two mesons called the τ-meson and the θ-meson were discovered in cosmic rays. They have the same mass, around 498 MeV, the same lifetime, around 1.2×10^{-8} s, and the same production rates in nuclear reactions of the type $\pi^+ N \to \Lambda + (\tau^+ \text{ or } \theta^+)$. However, they had different decay modes: the τ-meson decayed into only two π-mesons ($\tau^+ \to 2\pi^+ + \pi^-$) while the θ-meson decayed into only two π-mesons ($\theta^+ \to \pi^+ + \pi^0$). Now, the intrinsic parity for a π-meson is -1 (as deduced from the strong interaction of the π-mesons) so that $p_i = 1$ for the two π–meson states. It was also shown by Dalitz (1953) from the energy distribution of the pions, that $p_e = 1$ for both two π-meson final states. Thus, the decay products of the τ-meson decay are in a negative parity state $p_i p_e = -1$, while the deacy products of the θ-meson decay are in a positive parity state $p_i p_e = 1$. If parity is conserved, an unusual situation occurs, viz. that there are two particles with almost the same mass but opposite party. The other possibility is that τ and θ are one and the same particle but the weak interaction which is responsible for the decay (lifetimes indicate that the interaction is weak), violates parity invariance, *i.e.* parity is not conserved in weak interaction. Lee and Yang (1956) suggested this possibility after a critical examination of processes involving weak interaction, and proposed an experiment to test parity noncompensation in weak interaction.

Before describing the experiment, it is noted that space inversion is equivalent to a reflection and a rotation, *e.g.* reflection in the xz plane (change $y \to -y$), and a rotation about the y-axis by 180° (changes $x \to -x$ and $z \to -z$). Since rotational invariance is a universal symmetry, it gives the result that in addition to $\psi'(x, y, z) = P\psi(x, y, z)$ satisfying the Schrödinger equation, the space-reflected wave function also satisfies the Schrödinger equation. *e.g.* $R\psi(x, y, z)$ where R denotes the operator which changes $y \to -y$. This is known as *right-left symmetry.* Consider. The nuclei of Co^{60} whose spins are aligned along the z-direction with the aid of a magnetic field. It was then found that the β-decay electrons are preferentially emitted in the direction opposite to that of the nuclear spins. The mirror reflection of this (Fig. 10.2) would show that the electrons come out parallel to the nuclear spin. Since this reflected process is different from the physically observed process, the experiment implies the violation of parity invariance. The observation of parity violation in weak decays resolves the puzzle of τ-θ decays, as being due to parity-violation weak decays of a single particle, the *K*-meson.

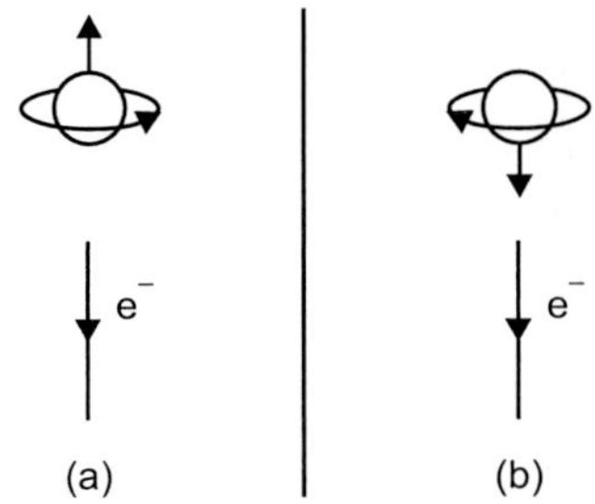

Fig. 10.2 Illustration of parity transformation in Co^{60} decay. If the electrons come out antiparallel to spin as shown in (*a*), they come out parallel to spin in the mirror reflection shown in (*b*).

V-A Theory of Weak Interaction

According to the quantum theory of electromagnetic radiation, the decay of an excited atom accompanied by the emission of a photon is a spontaneous process, *i.e.* the photon is produced at the moment of emission. In analogy, Fermi postulated that in the β-decay, an electron antineutrino pair is spontaneously produced at the moment of the emission. Furthermore, it was assumed that the electron-anti-eutrino pair is produced in a state of unit angular momentum and negative parity (some angular momentum and parity as that of the photon state in atomic decays). Since negative parity objects with unit angular momentum are called *vectors* (number of components is three in both the cases), Fermi's theory is a *vector-theory*.

The vector theory of Fermi conserves parity and had to be modified to incorporate the observed parity violation. This was done in an elegant formulation by Sudarshan and Marshak (1957) and also by Fenynman and Gell-Mann (1958). This theory is called the *V-A* theory and it contains only those neutrinos which have spin opposite to the direction of their motion (called *left-handed neutrinos*) and only those antineutrinos which have spin parallel to the direction of their motion (called *right-handed antineutrinos*). It is clear that parity invariance is violated in this theory since under mirror reflection, a left-handed neutrino goes into a right-handed neutrino (this can be seen from a diagram similar to Fig. 10.2, with the neutrinos moving downwards) which is excluded from the theory.

The *V-A* theory brings in the possibility that the weak interaction is invariant under the combined operation of spatial reflection and charge conjugation which changes a particle into an antiparticle. As observed before, mirror reflection takes a left-handed neutrino into a right-handed neutrino which under charge conjugation goes into a right-handed antineutrino which is included in the theory. However, even this invariance under *CP*, *i.e.* the combined operations of parity and charge conjugation, is violated to a small extent. This was observed by Christenson, Cronin, Fitch and Turaly (1964) in the decay of the long-lived Component, Cronin, Fitch and Turlay (1964) in the decays into three π-meson states with $CP = -1$, also decays to a small extent into two π-meson states with $CP = 1$, indicating a small violation of *CP* invariance.

10.5 UNIFIED APPROACH

There have been efforts to unify the various interactions and deduce them as different manifestations of the same underlying theory. Such a unification was achieved for electric and magnetic forces by Maxwell (1864). Physicists have now succeeded in unifying electromagnetic and weak forces and further efforts are being made to bring in strong and gravitational forces as well. Here the ideas that have led to the unification of electromagnetic and weak forces are briefly described.

As was noted in the lest section, the electron-antineutrino pair in the β-decay is produced in the same angular momentum state as the photon. This similarity between weak and electromagnetic processes can be further extended by postulating that the neutron changes into a proton by emitting a massive W^- (which carries spin $\hbar$) and the massive W^- then decays into an electron-antineutrino pair. The basic weak process here is a vector boson W^-interacting with a fermion pair similar to a photon interaction with a fermion pair. The important differences are (*i*) the W^- is charged and is accompanied by its antiparticle the W^+ whereas the photon is neutral and is its own antiparticle, (*ii*) the W^- is massive since weak interaction is a short-range force. From the analysis of the weak processes, it is estimated that

$$m_W \gtrsim 7.5 \times 10^4 \text{ MeV} \tag{10.38}$$

(*iii*) The interaction of the $W^\pm$ violates parity invariance whereas the interaction of the photon conserves parity.

In spite of the difference mentioned above, the interaction of the *W* bosons and the photon with matter can be combined. This is done by allowing a triplet and a singlet of vector bosons to interact with the lepton and quark doublets in Eqs. (10.1) and (10.3) (the left-handed parts of charged leptons are a part of the doublet and their right-handed parts form singlets). Demanding that the photon does not interect directly with the neutrinos and that its interaction conserves parity identifies the photon and relates the strengths of the weak and the electromagnetic interactions. This model developed by Weinberg (1967) and Salam (1968) has the following interaction features:

1. There are three massive bosons W^+, W^-, Z (which is neutral), with masses

$$m_w = 80.4 \text{ GeV}$$
$$m_z = 91.2 \text{ GeV} \tag{10.39}$$

and have parity violating interaction. It is large mass of these bosons which suppresses the usual weak interaction. The photon has the usual electromagnetic interaction. In high energy processes (energies comparable with $m_W c^2$) the two interactions would become comparable. It may be noted that these bosons with $m_W \approx 8.04 \times 10^4$ MeV and $m_Z \approx 9.12 \times 10^4$ MeV have been observed in a recent experiment performed at CERN.

2. The neutrinos and leptons have another interaction with mater, in addition to the electromagnetic and β-decay type of interactions. This interaction has been observed (1973) in processes of the type

$$\nu_\mu + N \rightarrow \nu_\mu + \text{harrons}$$
$$\overline{\nu}_\mu + N \rightarrow \overline{\nu}_\mu + \text{harrons} \tag{10.40}$$

where N is a nucleus. It has also been observed (1978) in the scattering of polarized electrons by a deuteron target. It was found that there was a difference in the scattering of left-handed and right-handed electrons, indicating violation of the right-left symmetry or the violation of parity invariance in the interaction of electrons. The amount of parity violation is in agreement with the prediction of the Weinberg-Salam model.

With the successful unification of weak and electromagnetic interactions, efforts have been directed towards unifying strong, weak and electromagnetic interactions. In most of these theories, leptons and quarks are put in the same multiplet. Therefore, there is the possibility of quarks transforming into leptons which means that the protons would be unstable against decay into leptons. Search for such decays (the lifetime of protons expected in these theories is greater than 10^{30} years) is being pursued vigorously by different groups of experimentalists.

Finally, there is the gravitational interaction which is generally insignificant for interactions between elementary particles but becomes important in astrophysics and cosmology. It is discussed in Chapter 11.

10.6 PRODUCTION AND DETECTION OF PARTICLES

The progress in our understanding of the properties of elementary particles and their interactions, has been made possible by important advances in the techniques of production and detection of particles. A few of them are briefly discussed here.

Cosmic Rays

Before the development of powerful accelerators, the cosmis rays were the only source of particles with sufficient energy to produce mesons and strange baryons. Many particles such as the positrons, the μ-mesons, the π-mesons, and several strange particles were first observed in the cosmic rays.

Primary cosmic rays are a flux of energetic charged particles, mainly protons (about 89% protons, 9% helium nuclei, 1% remaining heavier elements and about 1% electrons) that are incident on the earth. The energy of these particles varies from about 10^3 MeV to 10^{-14} MeV(average energy is about 10^4 MeV). When these energetic particles encounter the earth's atmosphere, they undergo inelastic collisions, producing what are called *secondary* cosmic rays which consist of mesons, protons, neutrons, strange particles etc. These secondary cosmic rays will themselves undergo additional elastic collisions, producing nucleonic cascades. Ultimately they reach the ground with a composition of about 80% μ-mesons, the remaining being protons, neutrons and some strange particles.

The origin of the rays is thought to be supernova explosions, with additional contributions from the sun (low energy cosmic rays), the centra of the galaxy, etc. Some high energy particles may be from outside our galaxy.

While the cosmic rays have proved to be important for the discovery of many particles, they have the disadvantage that neither their energy, nor their intensity, can be monitored to our convenience.

Van de Graaff Generator

This was one of the earliest generators (Fig. 10.3). In this generator, charges are sprayed onto a cloth belt which then transports the charges to a large metal sphere. The charges leave the belt by way of fine metal points and move to the outside surface of the sphere. The sphere then forms one electrode of an accelerating tube in which charged particles

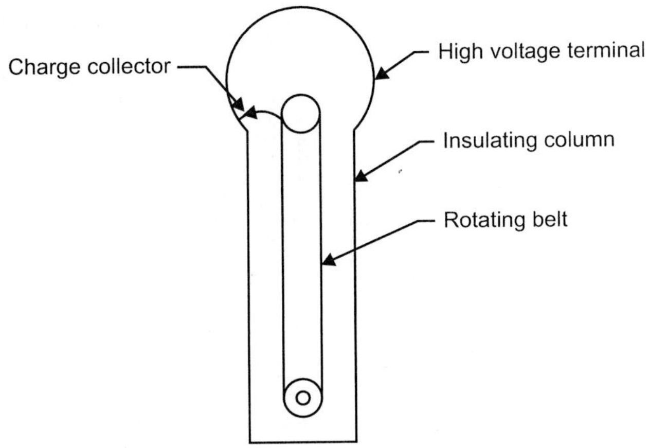

Fig. 10.3 Schematic diagram of a van de Graaff generator.

(such as the protons) can be accelerated. This accelerator can be used for accelerating protons to energies of about 15 MeV and is very useful in low-energy nuclear physics. The limitations of linear accelerators are in general due to the length, instability and loss of voltage.

The Cyclotron

The cyclotron is based on the principle that charged particles (nonrelativistic) in a constant magnetic field B perform circular motion whose frequency is independent of the magnitude of the velocity. The frequency is obtained by the force-acceleration relation

$$evB = mv^2/r \tag{10.41}$$

or

$$\omega = \frac{eB}{m} \tag{10.42}$$

In a cyclotron, protons in spiral orbits (Fig. 10.4) between the poles of two magnets and are accelerated by pulses across the hollow D-shaped electrodes which enclose the particle chamber. The radio frequency pulse across the electrodes has the frequency given by Eq. (10.42) and gives an extra energy of eV for every traversal across the gap (V being the voltage across the electrodes). The cyclotron can be used for obtaining protons of energy about 20 MeV (this would require about 1000 pulses of V = 20 000 V). The limitations of the cyclotron are due to the fact that the relativistic effects reduce the frequency in Eq. (10.42) as the particle speeds up so that it is no longer independent of the velocity of the particle.

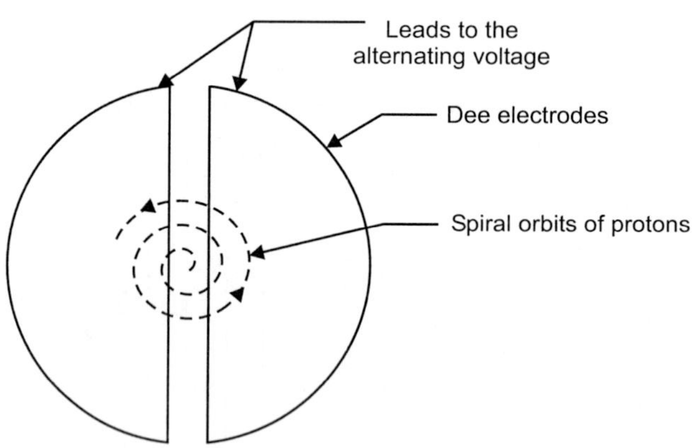

Fig. 10.4 Schematic diagram of a cyclotron.

Synchrotron

To overcome the voltage pulses getting out of phase with the rotational frequency in Eq. (10.42), the pulse frequency can be gradually changed so as to keep it in step with the circulating particles. Machines based on this idea are called *synchro-cyclotrons.* A further modification was to change the magnetic field as well as the pulse frequency so as to keep the protons in circular orbits of approximately the same radius (magnetic field must increase as the velocity of the protons increases) and to keep the pulse frequency in step with the particles. Such an accelerator is called the *synchrotron, and* is capable of providing proton beams of an energy of a few GeV (GeV = 10^9 eV).

One of the problems of synchrotrons is the focussing of the particles. If there is no focussing, particles with velocities slightly different from the average velocity will spread out and only a few particles with the final energy will be obtained. In velocity focussing the particles are kept together by adjusting the timing of the pulses. In spatial focussing, the particles are kept together (though they perform small oscillations) by controlling spatial variation of the magnetic field. An important advance in the accelerators was introduced by what is known as *strong focussing*. This was achieved by using magnet sections with alternating magnetic field gradients—that is the magnetic field increases radially in one section and decreases in the next sections. This allows one to obtain focussing in both radial and axial directions and to reduce the radial and axial oscillations. Synchrotrons using strong focussing are called *alternating-gradient synchrotrons* (AGS) (see Fig. 10.5) and have been used to obtain protons of energies of about 30 GeV.

High-energy beams of electrons have been obtained by linear accelerators (where the radiation loss in the energy due to radial acceleration, synchrotron radiation, is avoided), and beams of photons have been obtained from the synchrotron radiation of electrons, $e^- - e^+$ annihilation, etc.

Colliding Beams

For the production of heavy particles and the observation of interactions at high energies, it is advantageous to work with colliding beams of energetic particles. To see this, consider a collision between two particles of mass m each. The effective energy for a process may be standardized in terms of the total energy in the centre of mass (cm) system ($\mathbf{P}_{\text{tot}} = 0$ in the cm system). If one of the particles is at rest and the other is moving with an energy E (E includes kinetic energy and rest energy) and momentum $\mathbf{p}$, the total cm energy E_t is given by

$$\begin{aligned}\frac{1}{c^2}E_t^2 &= \frac{1}{c^2}(E+mc^2)^2 - \mathbf{p}.\mathbf{p} \\ &= 2m^2c^2 + 2Em \qquad (10.43)\end{aligned}$$

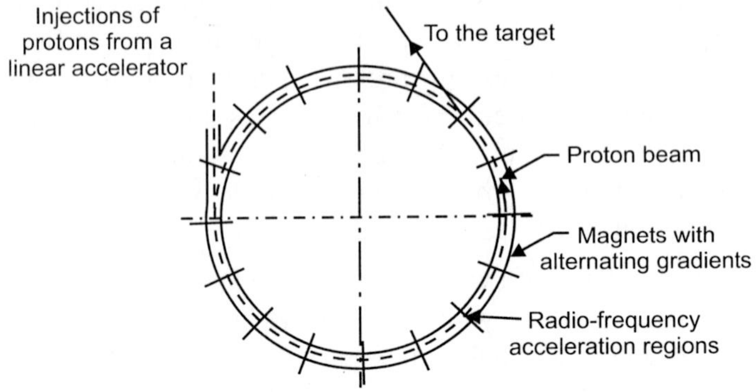

Fig 10.5 Schematic diagram of an alternating gradient synchrotron.

In obtaining this relation, we have equated the invariant scalar product **p.p** [see Eq. (1.56)] for the total energy momentum four-vector, evaluated in the cm system and the frame in which the target is at rest. If, on the other hand, the two particles are moving in opposite direction with energy E each and momenta **p, – p,** the total cm energy is given by

$$\frac{1}{c^2} E_t^2 = \frac{1}{c^2}(2E)^2$$

$$= \frac{4}{c^2} E^2 \tag{10.44}$$

In the CERN super proton synchrotron, the beam energy available for protons and antiprotons is about 270 GeV. Since mc^2 for the proton is about 1 GeV Eqs. (10.43) and (10.44) give

$$E_t \approx 23 \text{ GeV, for target at rest,}$$

$$E_t \approx 540 \text{ GeV, for colliding beams} \tag{10.45}$$

Thus, the colliding-beam facilities allow us to produce particles of mass up to 540 GeV/c^2 whereas with the target as rest, only practices mass up to 23 GeV\\c^2 can be produced.

The development of 270 GeV proton and antiproton beams at CERN has generated considerable interest. Here, antiprotons produced in the collision of protons of energy 26 GeV, with a target, are gathered in a ring shaped accumulator. It must be appreciated that producing antiproton beams of sufficient intensity (technically described in term of luminositi) is quite a difficult task since every million collisions of the protons at this energy, produce only about two antiprotons. In fact, the accumulation proceeds for a period of about 40 hours before realizing a sufficient number of antiprotons. The antiprotons in the accumulator are subjected to suitable electric fields so that most of them (about 60%) have an energy close to 3.5 GeV. After a beam of adequate luminosity has been formed, the antiprotons are first speeded up in a proton synchrotron to an energy of 26 GeV and then to an energy of about 270 GeV in a super proton synchrotron. The same synchrotrons are used for speeding-up protons as well, and providing a beam of 240 GeV protons. Since the protons and the antiprotons are oppositely charged, they move in opposite directions in the synchrotrons.

It was in the collision of 270 collision of 2w70 GeV proton antiproton beams at CERN that the *W* and *Z* bosons were produced recently; and identified by their characteristic decays. These bosons, which are essential for the propagation of unified weak-electromagnetic interaction, were found to have a mass of $m_W \approx 81$ GeV and $m_Z \approx 95$ GeV. This has been an important step in the confirmation of the theory of unified weak-electromagnetic interaction.

Finally, it may be noted that colliding-beam experiments have also been performed with electron and positron beams, which have provided important information about the properties of weak-electromagnetic interaction and of elementary particles.

Scintillation Counters and Semiconductor Counters

Elementary particles are observed by the traces left by their electromagnetic interaction with matter. In scintillation counters, charged particles passing through the substance (known as a *scintillator*) of the counter, excite the atoms. These atoms emit visible light on returning to their normal state. This light is supplied to photomultipliers which then give information about the charged particles.

In semiconductor counters, a charged particle passing through a junction layer with an applied potential, gives rise to electrons and holes. The electric pulse generated by them gives information about the number of charge carriers and hence about the particle which created them.

Wilson Cloud Chamber and Diffusion Chamber

In the Wilson cloud chamber (1912), the track of a charged particle becomes visible due to the condensation of the supersaturated vapour of a liquid, on the ions formed in the track. The supersaturated condition is obtained by the sudden expansion of a mixture consisting of a noncondensing gas (such as helium, argon, nitrogen) and water vapour, ethyl alcohol, etc. The track produced by the condensation of the vapour can be photo-graphed from different angles to reproduce a three dimensional picture. It may be noted that the sensitivity of the state of supersaturation lasts only for a short period so that the instrument is operated in cycles.

The diffusion chamber works along the same lines as the cloud chamber except that the state of super saturation is produced as a result of diffusions of alcohol vapour, from the top (kept at about 10°C) to the bottom (kept at about – 70° C by using solid carbon dioxide). This gives a layer of supersaturated vapour (a few centimeters thick) near the bottom. The diffusion chamber has the advantage that it can work continuously.

Bubble Chamber and Spark Chamber

In the bubble chamber (Glaser, 1952), the supersaturated vapour is replaced by a superheated liquid and the particle track is observed by the boiling of the liquid along the path of the particle. The superheated condition is obtained by first heating a liquid (such as hydrogen, xenon, propane, etc.) under pressure so that it is near the boiling point at that pressure. A sudden lowering of pressure produces a superheated state which survives for a short time. If high-energy, ionizing particles pass through this liquid, bubbles are formed along the track, on account of the electrons knocked out of the atoms by the charged particles, and the track can be photographed. It is a very important feature of bubble chambers that the working liquid itself can serve as a target for the charged particles. A bubble chamber, like the cloud chamber, works in cycles, since the superheated state lasts only for a short time.

In a spark chamber (Fig. 10.6), there is a series of plane, parallel, metal electrodes, alternatively grounded or connected to a source of periodic, short, high-voltage pulse (10-15 kV, lasting for about 10^{-7} s). A high-energy, ionizing particle passing through the chamber will produce a chain of sparks between the electrodes which can be analysed (an associated counter triggers the voltage pulse when the particle passes through).

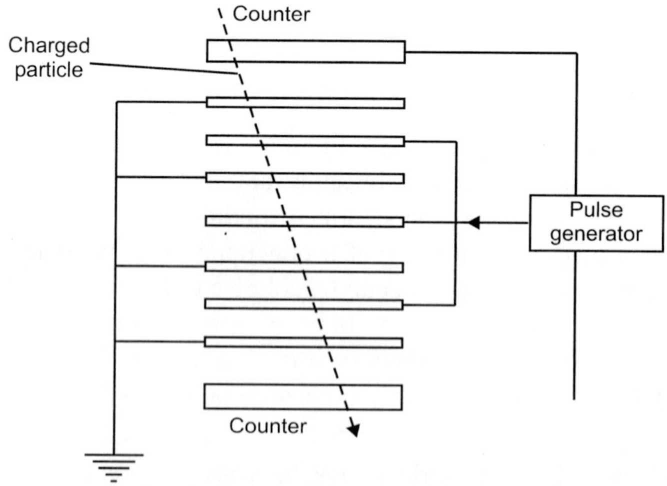

Fig. 10.6 Schematic diagram of a spark chamber.

Emulsion Chamber

Charged particles interact with photographic emulsions in the same way as photons, and hence the emulsions can be used for recording the tracks of charged particles. Photographic emulsions such as silver bromide (which has a density of about 4 g/cc) is an efficient stopper of high-energy ionizing particles. Several hundreds of layers of these emulsion sheets, each about 1/2 mm thick, may be exposed to cosmic rays or to high energy particles from accelerators. After the development of the sheets, the path of the particles can be traced from layer to layer.

Geiger Counter

This is a very compact and sturdy instrument used for detecting energetic particles. It consists of a metal tube with a wire along its axis. It is filled with a suitable gas mixture at low pressure. The tube is insulated from the wire and a high potential difference is maintained between the tube and the wire. When an energetic particle enters the tube, it produces some ions in the gas. This causes a small discharge current to flow between the wire and the tube. Recording of these signals allows us to count the number of incident particles. Since the ions and the electrons take a time of about 10^{-3} s to recombine, the counter can record only a few hundred counts per second. Its main advantage is that it is very inexpensive and easy to operate.

10.7 EXAMPLES

A few examples are discussed in this section.

Example 1

It is interesting to note that several particles in the SU (3) multiplet scheme, such as the η-meson, were discovered only after the scheme predicted the existence of these particles. Particularly noteworthy is the Ω^- which is a member of the baryon decuplet. Gell-Mann predicted it to have spin of 3/2, strangeness –3, and a mass of about 1676 MeV. Since there is no baryonic system with strangeness –3, and which is lighter than 1676 MeV, the Ω^- is stable against decay through strong interaction (which conserves strangeness). The Ω^- can decay via weak interaction (which does not conserve strangeness) but would have a lifetime of about 10^{-10} s characteristic of weak decays. A particle with precisely these characteristics, with a mass of 1675 MeV was observed in 1964, through its weak decay

$$\Omega^- \rightarrow \Xi^- + \pi^0 \qquad (10.46)$$

which does not conserve strangeness.

The Ω^- is made of three strange quarks (*sss*). The three quarks are in the ground state with spatial angular momentum equal to zero and in the totally symmetric spin 3/2 state. This would violate Fermi statistics. In order to get out of this difficulty, an additional property called *colour* was assigned to the quarks. The three quarks in the baryons are supposed to be of different colours so that the exchange statistics is not applicable.

Example 2

The β-decay in Eq. (10.21) is to be regarded as a spontaneous emission of *e* and $\overline{\nu}_e$ and not an escape of an electron bound in the nucleus. This is indicated by the fact that a bound electron would have a momentum $p \sim \hbar/r_0$ (follows from the uncertainty principle) and a kinetic energy

$$\text{KE} \sim (m^2c^4 + c^2\ \hbar^2/r_0{}^2)^{1/2} - mc^2 \tag{10.47}$$

which for $r_0 \sim 10^{-15}$ m comes out to be approximately 200 MeV. Such a highly-energetic electron would quickly escape in about 10^{-23} s, and the slow rate of the β-decay cannot be explained.

Example 3

The parity of π^- is determined from the reaction

$$\pi^- + d \rightarrow N + N \tag{10.48}$$

The low-energy π^- is captured is a Bohr orbit (with a radius about 2×10^{-13} m) in the deuterium. The reaction in Eq. (10.48) proceeds with $l = 0$ (nuclear forces become effective only at a short range of about 10^{-15} m) so that the initial total parity for the reaction is that of the intrinsic parity of the pion.

Now if the orbital angular momentum of the *NN* state had $l = 0$, the spin of the *NN* system would be 1 (equal to the spin of the deuteron, by the conservation of the total angular momentum). This implies that the spins of the two neutrons would have to be parallel, and hence, this state is forbidden by Fermi statistics. Therefore, the orbital angular momentum of the *NN* state has $l = 1$. This means that the parity of the final state is -1 and therefore the intrinsic parity of the π-meson is -1 (parity being conserved in strong interaction).

Example 4

The behaviour of neutral *K*-mesons provides a very interesting application of the superposition principle. Neutral *K*-mesons have the decay modes

$$\begin{aligned} K^0 &\rightarrow \pi^+ + \pi^-,\ \pi^0 + \pi^0 \\ \overline{K}^0 &\rightarrow \pi^+ + \pi^-,\ \pi^0 + \pi^0 \end{aligned} \tag{10.49}$$

where a lifetime of about 10^{-10} s, and

$$K^0 \to \pi^+ + \pi^- + \pi^0,\ 3\pi^0$$

$$\bar{K}^0 \to \pi^+ + \pi^- + \pi^0,\ 3\pi^0 \tag{10.50}$$

with a lifetime of about 10^{-7} s. Now, weak interaction does not conserve strangeness so that the states with well-defined energy (and also lifetime) are not expected to be states with well-defined strangeness.

Considering the possibility of *CP* being conserved in weak interaction, one may define

$$CP\,|\,K^0\rangle = |\,\bar{K}^0\rangle$$

$$CP\,|\,\bar{K}^0\rangle = |\,\bar{K}^0\rangle \tag{10.51}$$

where *P* is the parity operator and *C* is the charge-conjugation operator (which changes a particle into an antiparticle with opposite charge, strangeness, etc.). If the Hamiltonian commutes with *CP*, the eigenstates of energy can also be eigenstates of *CP*. Such eigenstates are the superpositions

$$|\,K_1^0\rangle = \frac{1}{\sqrt{2}}(|\,K^0\rangle + |\,\bar{K}^0\rangle),\ CP = 1 \tag{10.52}$$

$$|\,K_2^0\rangle = \frac{1}{\sqrt{2}}(|\,K^0\rangle - |\,\bar{K}^0\rangle),\ CP = -1 \tag{10.53}$$

and it is these states which are expected to have well-defined masses and lifetimes.

In the 2π-meson decays given in Eq. (10.49) 2π-mesons have $l = 0$ and hence are in $CP = 1$ state. Hence while the $|\,K_1^0\rangle$ can decay into two π-mesons, the decay of $|\,K_2^0\rangle$ into two π-mesons is forbidden. Thus $|\,K_1^0\rangle$ has a short lifetime of about 0.86×10^{-10} s whereas $|\,K_2^0\rangle$ has a much longer lifetime of about 5.4×10^{-8} s (the masses of $|\,K_1^0\rangle$ and $|\,K_2^0\rangle$, also are slightly different). This leads to some interesting observations for the $|K^0\rangle$ created in a process such as Eq. (10.26) . The $|K^0\rangle$ may be regarded as

$$|K^0\rangle = \frac{1}{\sqrt{2}}(|\,K_1^0\,\rangle + |\,K_2^0\,\rangle) \tag{10.54}$$

of which $|\,K_1^0\,\rangle$ decays quickly into two π-mesons [giving the decays in Eq. (10.49)] so that after a few lifetimes of $|\,K_1^0\,\rangle$, only the $|\,K_2^0\,\rangle$ is left. This long-lived component decays slowly into other modes such as the ones in Eq. (10.50). It may be observed that though only $|\,K^0\,\rangle$ existed at first, the component $|\,K_2^0\,\rangle$ which remains at the end contains equal mixtures of $|K^0\,\rangle$ and $|\,\bar{K}^0\,\rangle$.

Finally, it is noted that the long-lived component $|\overline{K}_2^0\rangle$ was found (Christenson, Cronin, Fitch and Turlay, 1964) to decay to a small extent into two π-mesons which means that CP is not conserved and that $|\overline{K}_1^0\rangle$ and $|\overline{K}_2^0\rangle$ are only approximate eigenstates of the Hamiltonian.

PROBLEMS

1. Using the operators in Eq. (4.127), except the factor of $\hbar$, for the isospin operators for $I ==1/2$, and the representation in Eq. (4.126) for the P and N states respectively, obtain the isospin values of the states for the proton-neutron system in Eqs. (10.8) to (10.11)
2. If a positron of energy E annihilates an electron at rest, giving out two photons, $e^+ + e \rightarrow \gamma + \gamma$, obtain the angular distribution of energy.
3. Determine the maximum energy of π^- when the K^+ at rest decays into $\pi^+ \pi^+ \pi^-$.
4. What is the lifetime of Σ^0 so much smaller than that of Σ^+ or Σ^-?
5. What is the range of weak-interaction forces originating from the exchange of W-bosons of mass about 80 GeV?
6. Obtain from dimensional arguments, the lifetime of β-decay, given that it is proportional to $\left(\frac{m_W{}^2}{\alpha}\right)^2$ and that the remaining factor is a function of $\hbar$, c and m_e.
7. For a particle moving with high velocity, the force-acceleration relation in Eq. (10.41) changes to $evB = mv^2/r\ (1 - v^2/c^2)^{1/2}$. What os the field needed to keep a proton of 5 GeV energy, in a circular orbit of radius 20 m?

11

General Relativity and Cosmology

Structures of the Chapter

Our discussion so far has been about modern ideas in the domain of high velocity (special theory of relativity), and small distances (quantum theory and its applications). There have been important developments in our understanding of large-distance, large-body phenomena as well. These refer primarily to the *general theory of relativity* as applied to large stars, galaxies and cosmology, *i.e.* the science of the universe as a whole.

In cosmology, the events taking place in distant objects such as galaxies, which not only may be moving with high velocity with respect to us but may also be accelerating, have to be interpreted. It is to be expected that observation and interpretation are simpler in the galaxy where the events are taking place. Therefore, a relativistic theory, which can relate observations in frames which are in arbitrary motion with respect to each other is needed. Einstein's theory of general relativity provides the frame-work for relating observations of space-time events in arbitrary frames. Indeed, the theory accomplishes more than that. It incorporates gravitational effects as well. This is based on the observation that the motion of a particle in a gravitational field, with a given initial position and velocity, is independent of its mass. This implies, as will be discussed later, that the effect of gravity can be locally simulated by an accelerating frame but without gravity, so that gravitational effects can be described by the theory of general relativity. Furthermore, in the discussion of cosmological dynamics, it is mainly the long-range gravitational forces which are important. Therefore general relativity is also an appropriate theory for the analysis of the development of the universe.

Here, an elementary and brief consideration of the main ideas of the theory of general relativity is presented and the ideas are applied to discuss some predictions of the dynamical properties of the universe. It is quite appropriate to end with the description of general relativity, a theory which is grand in its concepts and structure and awe-inspiring in its predictions. An exposition of modern ideas in physics cannot be said to be complete without a discussion of general relativity, and any thing to follow would only be an anticlimax.

11.1 FRAMES OF REFERENCE

An event is a space-time occurrence. To specify an event, we need a frame of reference, which consists of three spatial coordinate axes and a time coordinate. It is known that the equations of motion in special relativity or Newton's theory take the simplest form in inertial frames. In an inertial frame, a particle on which no external forces act moves with constant velocity. A description of motion in frames which accelerate with respect to inertial frames is complicated by the need to introduce additional forces known as *inertial* (or *pseudo*) forces. For example, an observer in an accelerating or decelerating train experiences forces in addition to the gravitational or electromagnetic forces. Thus, with respect to accelerations, there appears to be an absolute or a preferred frame of reference in which the inertial forces are zero.

Mach's Principle

The inertial frames may be determined by considering the inertial forces on the surface of the earth. For example, the rate of rotation of the earth with respect to the inertial frames may be estimated by the measurement of the centrifugal and coriolis forces on the earth. The rate of rotation thus obtained is found to be approximately the same as the rate of the earth's rotation with respect to distant matter, *e.g.* the distant galaxies. This leads to an important result that *the average motion of distant galaxies with respect to the inertial frame is zero.*

According to Mach (1872), the above result is not an accident. It suggests that the inertial frame is not an absolute frame but is related to the distribution of matter in the universe. In fact, Mach asserted that the concept of inertia (and the inertial frame) can be given meaning only in terms of background stars and galaxies. This is known as *Mach's principle.* It implies that if there were no matter in the universe except for a given body, there would be no inertia or inertial forces and it is meaningless to ask whether it is accelerating with respect to an inertial frame.

A theory of the universe which incorporates Mach's principle cannot include an inertial frame without reference to the distribution of matter in the universe.

Principle of Equivalence

An important input of Einstein's theory of general relativity is the observation that the effect of gravity can be simulated locally by a noninertial frame.

Consider the motion of an object of inertial mass m_I, in a region where the gravitational acceleration g is approximately constant. If there is a nongravitational force $\mathbf{F}$ acting on it, its motion is given by

$$m_I\mathbf{a} = m_g\mathbf{g} + \mathbf{F} \tag{11.1}$$

where the gravitational mass m_g, in principle, may be different from the inertial mass m_I. Alternatively, consider a frame of reference without a gravitational force, but moving with acceleration $-\mathbf{g}$. If the acceleration of mass m_I in this frame is $\mathbf{a}$, the corresponding acceleration in the inertial frame is $\mathbf{a} - \mathbf{g}$ so that the equation of motion is

$$m_I(\mathbf{a} - \mathbf{g}) = \mathbf{F} \tag{11.2}$$

or

$$m_I\mathbf{a} = m_I\mathbf{g} + \mathbf{F} \tag{11.3}$$

Experimentally is observed to a high accuracy (about 1 part in 10^{11}) that $m_I = m_g$, so that Eqs. (11.1) and (11.3) describe the same motion. This result was generalized by Einstein into what is known as the principle of equivalence:

The physical laws are locally the same in an inertial frame with gravitational acceleration **g** *and a noninertial frame with acceleration –* **g** *but no gravity.*

It is important to note that the equivalence is local since the gravitational field tends to zero at large distances whereas the inertial forces in general do not vanish at infinity.

The principle of equivalence gives special importance to freely-falling frames. In these frames, the local effect of gravity is cancelled by the inertial forces so that they form local inertial frames and the considerations of special relativity suffice to describe the physical observations in them. They allow us to deduce two interesting results without going into additional details of the general theory.

1. *Bending of light in a gravitational field:* Consider a beam of light in a gravitational field. Observed from a freely-falling frame which is locally inertial, the beam travels in a straight line. However, the frame with the gravitational field moves with acceleration – **g** with respect to the freely-falling frame so that in this frame the beam appears to bend in the direction of **g**. This is a remarkable result since just the finiteness of the velocity of light implies that light interacts with gravitational fields. The bending of light in a gravitational field is observed in the deflection of light from the stars, moving past the sun, seen during total solar eclipses. However, since **g** is not constant along the path of the beam, quantitative calculations are rather complicated. It may be noted that the predictions of general theory agree with the observations within experimental accuracy.
2. *Gravitational shift of spectral lines:* Consider a photon emitted at $t = 0$, and moving in a direction opposite to the gravitational acceleration **g**. Observed from a freely-falling frame which is at rest with respect to the source at $t = 0$, the frequency of the photon is ν_0. If the photon meets the frame at t, it will have travelled a distance $h = ct$ during this time and the frame at that instant will be moving with a velocity of gt. Hence the frequency ν of the photon observed at a height of h, from a frame at rest with respect to the source (and moving with velocity $- gt$ with respect to the freely-falling frame), is given by

$$\frac{\nu_0 - \nu}{\nu_0} \approx \frac{gt}{c}$$

$$= \frac{gh}{c^2} \tag{11.4}$$

This result, which can also be deduced from energy conservation (see Example 2), has been verified by Pound and Rebka (1960) using Mossbauer effect. They found that a photon falling through a height of 22.6 m shows a shift of $\Delta\lambda/\lambda_0 \approx - gh/c^2 \approx - (2.57 \pm 0.26) \times 10^{-15}$ compared with the predicted value of $- 2.46 \times 10^{-15}$.

The theory of general relativity is based on the principle of equivalence and includes the gravitational effects in terms of the geometry of the space. The theory makes no distinction between gravitational and inertial effects, both being related to the energy and momentum distribution of matter and hence incorporates Mach's principle to some extent (there are some difficulties is the interpretation of boundary conditions in the case of an infinite universe).

11.2 CURVED SPACE-TIME

In the theory of general relativity, the inertial and gravitational effects are described by the geometry of the space. These effects modify the space properties, *e.g.* the surface may become curved, which affect the dynamics of objects. Some of the ideas involved can be illustrated by the following simple example.

Metric Tensor of the Space

Event in inertial frames F and F' are related by the Lorentz transformations

$$x' = \frac{1}{(1 - v^2/c^2)^{1/2}}(x - vt),\ y' = y,\ z' = z \tag{11.5}$$

$$t' = \frac{1}{(1 - v^2/c^2)^{1/2}}\left(t - \frac{v}{c^2}x\right)$$

Trajectories may be characterized by an invariant variable τ, the *proper time*, which gives the invariant interval between two events as

$$(\Delta\tau)^2 = (\Delta t)^2 - \frac{1}{c^2}(\Delta\mathbf{r})^2 \tag{11.6}$$

Now, a gravitational field with acceleration $\mathbf{a}$ in the x-direction can be introduced by going over to a frame with acceleration $-\mathbf{a}$. The required transformations are given approximately by

$$\mathbf{r}' = \mathbf{r} + \frac{1}{2}\mathbf{a}t^2,\ t' = t \tag{11.7}$$

in terms of which the infinitesimal proper time interval is

$$(\Delta\tau)^2 = \left(1 - \frac{a^2t'^2}{c^2}\right)(\Delta t')^2 - \frac{1}{c^2}(\Delta\mathbf{r}')^2 + \frac{2t'}{c^2}\mathbf{a}\,.\,\Delta\mathbf{r}'\,\Delta t' \tag{11.8}$$

Thus, a gravitational (or inertial effect is described by a more complicated measure or metric of $\Delta\tau$ in terms of $\Delta t'$, $\Delta x'$, $\Delta y'$, $\Delta z'$.

The expression of the measure, for a frame in arbitrary motion, can be stated in a compact form by writing

$$(\Delta\tau)^2 = g_{\mu\nu}\, \Delta x^{\mu}\, \Delta x^{\nu} \tag{11.9}$$

where $g_{\mu\nu}$ is called the *metric tensor* of the space. From here onwards, the convention that repeated indices are summed over is used, in this case, $\mu, \nu = 0, 1, 2, 3$ with index 0 standing for the time coordinate and 1, 2, 3 for the three space coordinates. For the inertial frames one has

$$\overset{0}{g}_{00} = 1, \overset{0}{g}_{11} = \overset{0}{g}_{22} = \overset{0}{g}_{33} = -\frac{1}{c^2}, \overset{0}{g}_{\mu\nu} = 0 \text{ for } \mu \neq \nu \tag{11.10}$$

corresponding to flat space. If the coordinates in the second frame are given by the functional relation $x^{\mu} = x^{\mu}(x')$,

$$\Delta x^{\mu} = \frac{\partial x^{\mu}}{\partial x'^{\nu}}\, \Delta x'^{\nu} \tag{11.11}$$

and

$$(\Delta\tau)^2 = g_{\mu\nu}\, \Delta x'^{\mu}\, \Delta x'^{\nu} \tag{11.12}$$

where

$$g_{\mu\nu} = \overset{0}{g}_{\alpha\beta}\, \frac{\partial x^{\alpha}}{\partial x'^{\mu}}\, \frac{\partial x^{\beta}}{\partial x'^{\nu}} \tag{11.13}$$

It should, however, be noted that the metric tensor here is given in terms of only four independent function $x^{\mu}(x')$, though an arbitrary but symmetric $g_{\mu\nu}(x')$ consists of 10 independent functions. The restricted $g_{\mu\nu}$ given in Eq. (11.13) can describe the local gravitational field. Einstein the postulated that the general gravitational field is described by an arbitrary metric $g_{\mu\nu}(x)$ with 10 independent functions. Having defined the means of describing the gravitational and inertial effects, one must now provide (*t*) the framework for the determination of the metric $g_{\mu\nu}(x)$ and (*ii*) the dynamical equations for a given metric.

Elnstein's Field Equations

The metric tensor $g_{\mu\nu}(x)$ is related to the distribution of matter, through the field equations

$$R_{\mu\nu} - \frac{1}{2} g_{\mu\nu} R = -\frac{8\pi G}{c^4} T_{\mu\nu} \tag{11.14}$$

where $T_{\mu\nu}$ is the energy-momentum tensor which acts as the source and $G = 6.67 \times 10^{-11}\, N \cdot \text{m}^2/\text{kg}^2$ is the usual gravitational constant. The quantities $R_{\mu\nu}$ and R are related to the Riemann-Christoffel curvature tensor $R^{\alpha}_{\mu\nu\beta}$ which in turn is related to the metric g_{mv}. The discussion of these relations is beyond the scope of this book (interested reader may refer to Ref. 11). We only note that these field equation determines the metric for a given energy-momentum distribution. In the limiting case of weak gravitational field ϕ, one has $T_{00} = \rho c^2$, $g_{00} = 1 + \frac{2\phi}{c^2}$, $R_{00} = -\frac{1}{2}R = -\frac{1}{c^2}\nabla^2\phi$ with which Eq. (11.14) to first order in ϕ, reduces to the Poisson equation for Newton's gravitational potential, $\nabla^2 f = 4\pi G\rho$.

Geodesics

The path followed by a particle in the presence of gravitational forces is determined by the geometry of the equivalent metric space. To obtain the relation between the path and the metric, it is noted that in 3-dimensional Euclidean space, a particle moves in a straight line, *i.e.* it chooses a path which corresponds to the shortest distance between any two points. However, the ordinary length is not invariant even in special relativity, and is not a suitable quantity for determining the trajectory in the general case.

The proper time τ, defined in Eqs. (11.6) and (11.12), is invariant under general transformations, and the allowed paths may be regarded as corresponding to extrema of τ. It turns out that τ is actually a maximum for the allowed trajectory in the case of special relativity [because of the negative sign of spatial terms in Eq. (11.10)]. For example, the proper time corresponding to points (0, 0) and $(t, 0)$, with metric $g^{\alpha}_{\mu\nu}$ (Eq. 11.10), is

$$\tau_1 = t \tag{11.15}$$

whereas that corresponding to two segments $(0, 0) \rightarrow (t_1, x_1')$ and $(t_1, x_1) \rightarrow (t, 0)$ is

$$\tau_2 = \left(t_1^2 - \frac{1}{c^2}x_1^2\right)^{1/2} + \left[(t - t_1)^2 - \frac{1}{c^2}x_1^2\right]^{1/2} \tag{11.16}$$

which for $x_1 \neq 0$ is less than t (note that the intermediate point has to be taken so that each proper-time interval is real). Thus, the straight line between (0, 0) and $(t, 0)$ corresponds to the maximum proper time. The general situation may be covered by the requirement that the total proper time

$$\tau_{AB} = \int_A^B d\tau \tag{11.17}$$

$$= \int_A^B \left[g_{\mu\nu}\frac{dx^\mu}{ds}\frac{dx^\nu}{ds}\right]^{1/2} ds$$

is an extremum, where Eq. (11.12) has been used, and s is an arbitrary parameter. The extremum paths are called *geodesics*.

The integral condition can be converted into a set of differential equations in the following way. To be specific, let s be the proper time of the geodesic, with values τ_A and τ_B at the end points. Now consider a set of curves $x^\mu(\tau, \varepsilon)$ which connect points A and B, such that

$$x^\mu(\tau, \varepsilon) = x^\mu(\tau, 0) + \varepsilon h^\mu(\tau) \tag{11.18}$$

where $x^\mu(\tau, 0)$ is the geodesic needed, $h^\mu(\tau_A) = h^\mu(\tau_B) = 0$, and ε is a small parameter. Then the proper time is

$$\tau(\varepsilon) = \int_{\tau_A}^{\tau_B} \left[g_{\mu\nu} \frac{dx^\mu}{d\tau} \frac{dx^\nu}{d\tau} \right]^{1/2} d\tau \tag{11.19}$$

$$\equiv \int_{\tau_A}^{\tau_B} f\left(\frac{dx^\mu}{d\tau}, x^\mu \right) d\tau$$

To first order in ε, this expression reduces to

$$\tau(\varepsilon) = \tau(0) + \varepsilon \int_{\tau_A}^{\tau_B} \left[\frac{\partial f}{\left(\dfrac{dx^\mu}{d\tau} \right)} \frac{dh^\mu}{d\tau} + \frac{\partial f}{\partial x^\mu} h^\mu \right] d\tau \tag{11.20}$$

which on integration by parts (and using $h^\mu(\tau_A) = h^\mu(\tau_B) = 0$) gives

$$\tau(\varepsilon) = \tau(0) - \varepsilon \int_{\tau_A}^{\tau_B} \left[\frac{d}{d\tau} \frac{\partial f}{\partial \left(\dfrac{dx^\mu}{d\tau} \right)} - \frac{\partial f}{\partial x^\mu} \right] h^\mu \, d\tau \tag{11.21}$$

Since $\tau(\varepsilon)$ is an extremum at $\varepsilon = 0$, the second term should vanish for all $h^\mu(\tau)$ which implies that

$$\frac{d}{d\tau} \frac{\partial f}{\partial \left(\dfrac{dx^\mu}{d\tau} \right)} - \frac{\partial f}{\partial x^\mu} = 0 \tag{11.22}$$

Equation (11.22) gives four equations, for $\mu = 0, 1, 2, 3$. One of these can be replaced by using the relation

$$g_{\mu\nu} \frac{dx^\mu}{d\tau} \frac{dx^\nu}{d\tau} = 1 \tag{11.23}$$

which follows from Eq. (11.12). For the special case of $\frac{\partial f}{\partial x^{\mu}} = 0$, Eq. (11.22) simplifies to

$$\frac{\partial f}{\partial\left(\frac{dx^{\mu}}{d\tau}\right)} = \text{constant.} \tag{11.24}$$

These differential equations, *i.e.* Eq. (11.22) or Eq. (11.24), with Eq. (11.23), determine the geodesics and hence the dynamics of a particle.

As an illustration, the metric tensor in Eq. (11.8) is considered for the case of acceleration a in the x-direction,

$$g_{00} = 1 - \frac{a^2t^2}{c^2},\ g_{11} = -\frac{1}{c^2},\ g_{01} = g_{10} = \frac{at}{c^2} \tag{11.25}$$

which is equivalent to a space with gravitational field characterized by acceleration a. The corresponding function f in 2-dimensions is

$$f = \left[\left(1 - \frac{a^2t^2}{c^2}\right)\left(\frac{dt}{d\tau}\right)^2 - \frac{1}{c^2}\left(\frac{dx}{d\tau}\right)^2 + \frac{2at}{c^2}\frac{dt}{d\tau}\frac{dx}{d\tau}\right]^{1/2} \tag{11.26}$$

Using Eq. (11.23) and Eq. (11.24) for $\mu = 1$,

$$-\frac{dx}{d\tau} + at\frac{dt}{d\tau} = A \tag{11.27}$$

$$\left(\frac{dt}{d\tau}\right)^2 = 1 + \frac{A^2}{c^2}$$

which lead to $\frac{dx}{dt} = at$ + constant, and therefore reproduce the usual equations for motion in a constant gravitational field.

Curvature of Space

It is clear from the above example that the geodesics in an arbitrary metric space are, in general, curved lines. The space is then said to be *curved* (in contrast to the flat spaces of inertial frames). A measure of the curvature of the space is given by what is known as the *curvature tensor*. Only the simple case of 2-dimensions is considered here. The curvature of a surface in two dimensions, as given by Rindler, is the following: Draw the geodesics starting from a point P, and consider the circle formed by the locus of points which are at a distance

a from P, along the geodesics. If the circumference of the circle is of length l, the curvature is given by

$$K = \frac{3}{\pi} \lim_{a \to 0} \left(\frac{2\pi a - l}{a^3} \right) \tag{11.28}$$

In the simple case of a spherical surface, the distance dl between two neighbouring points (see Fig. 11.1) is given by

$$(\Delta l)^2 = \frac{(\Delta r)^2}{1 - r^2/R^2} + r^2\,(\Delta\phi)^2 \tag{11.29}$$

where R is the radius of the sphere, and r is the distance from the axis. The distance along the geodesic from P, is given by

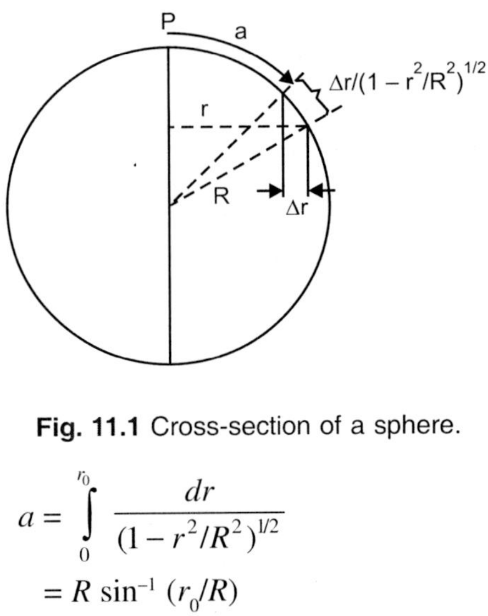

Fig. 11.1 Cross-section of a sphere.

$$a = \int_0^{r_0} \frac{dr}{(1 - r^2/R^2)^{1/2}} \tag{11.30}$$
$$= R \sin^{-1} (r_0/R)$$

while the length of the circumference of the circle is

$$l = 2\pi r_0$$
$$= 2\pi R \sin (a/\mathrm{R}) \tag{11.31}$$

Hence, the curvature is

$$K = \frac{3}{\pi} \lim_{a \to 0} \frac{2\,\pi a - 2\,\pi R \sin (a/R)}{a^3}$$
$$= \frac{1}{R^2} \tag{11.32}$$

For a flat surface ($R \to \infty$), the curvature is zero. In some cases, such as at a saddle point, it can be negative.

An interesting point which emerges from our example is that a is multi-valued (corresponding to going around the sphere an arbitrary number of times), and $r_0 = R \sin(a/R)$ is bounded by the value R. This is an elementary example of a finite universe.

11.3 SCHWARZSCHILD METRIC

As the first application of curved spaces, we analyse the space-time near a point mass M. This would simulate the situation in the neighbourhood of a massive object. In the limit of mass $M \to 0$, the flat space is described by the element

$$(\Delta\tau)^2 = (\Delta t)^2 - \frac{1}{c^2}[(\Delta r)^2 + r^2 (\Delta\theta)^2 + r^2 \sin^2\theta (\Delta\phi)^2] \tag{11.33}$$

expressed in terms of spherical coordinates. With mass $M \neq 0$, the distance from the origin is no longer given by r, through the surface area of the sphere is still $4\pi r^2$ and isotropy is maintained. The solutions of the field equations, Eq. (11.14), were obtained by Schwarzschild (1916) and give the metric

$$(\Delta\tau)^2 = e(r)(\Delta t)^2 - \frac{1}{c^2}[f(r)(\Delta r)^2 + r^2 (\Delta\theta)^2 + r^2 \sin^2\theta (\Delta\phi)^2] \tag{11.34}$$

with
$$e(r) = \frac{1}{f(r)} = 1 - \frac{2GM}{c^2 r} \tag{11.35}$$

This is known as the *Schwarzschild metric*. It may be observed that for $r \to \infty$, the Schwarzschild metric reduces to the metric of the flat space as it should.

The interpretation of the different variables in Eq. (11.33) should be carefully noted. The variable t is the coordinate which, in the absence of any gravitational potential, would represent the time variable. The proper time interval $\Delta\tau$, on the other hand, corresponds to the rate at which local clocks are running. Similarly, the interpretation of r is that the distance measurements for $\Delta t = 0$ give a value $[f(r)(\Delta r)^2 + r^2 (\Delta\theta)^2 + r^2 \sin^2\theta (\Delta\phi)^2]^{1/2}$.

The Schwarzschild metric can be used to explain several important observations. A few of the applications are discussed here.

Rate of Clocks

Consider an atomic clock in the presence of a gravitational field due to mass M. The time interval it shows is

$$\Delta\tau_0 = \left(1 - \frac{2GM}{c^2 r}\right)^{1/2} \Delta t \tag{11.36}$$

where Δt is the displacement in the t-coordinate. The corresponding interval shown by a clock at infinity, is

$$\Delta\tau = \Delta t \tag{11.37}$$

Therefore,

$$\Delta\tau_0 = \left(1 - \frac{2GM}{c^2 r}\right)^{1/2} \Delta\tau \tag{11.38}$$

which means that a clock in a gravitational field runs at a slower rate. This is usually stated as implying that atoms (and human beings) in a gravitational field live longer. Since frequency is inversely proportional to the time interval this also gives the result that the frequency of radiation coming out of the field is

$$\nu = \left(1 - \frac{2GM}{c^2 r}\right)^{1/2} \nu_0 \tag{11.39}$$

and in the weak field limit, *i.e.* for small GM/c^2r,

$$\frac{\nu_0 - \nu}{\nu_0} \approx \frac{GM}{c^2 r} \tag{11.40}$$

This is essentially the gravitational red shift deduced earlier, in Eq. (11.4), from the equivalence principle.

Shift of the Perihelion

The equations of motion in a gravitational field of mass M are obtained from Eq. (11.24) for $x^\mu = t$, ϕ, and Eq. (11.23) with f given by ($\Delta\theta = 0$, $\theta = \pi/2$)

$$f = \left[e(r)\left(\frac{dt}{d\tau}\right)^2 - \frac{f(r)}{c^2}\left(\frac{dr}{d\tau}\right)^2 - \frac{r^2}{c^2}\left(\frac{d\phi}{d\tau}\right)^2\right]^{1/2} \tag{11.41}$$

They lead to

$$\begin{aligned} r^2 \frac{d\phi}{d\tau} &= A \\ e(r)\frac{dt}{d\tau} &= B \\ e(r)\left(\frac{dt}{d\tau}\right)^2 - \frac{f(r)}{c^2}\left(\frac{dr}{d\tau}\right)^2 - \frac{r^2}{c^2}\left(\frac{d\phi}{d\tau}\right)^2 &= 1 \end{aligned} \tag{11.42}$$

Using the first two equations and Eq. (11.35), the last equation simplifies to:

$$\left[\frac{1}{r^2}\left(\frac{dr}{d\phi}\right)^2+\left(1-\frac{2GM}{c^2 r}\right)\right]\frac{A^2}{2r^2}-\frac{GM}{r}=\frac{1}{2}c^2\,(B^2-1) \tag{11.43}$$

where the constant on the right-hand side may be identified with the energy. Compared to the corresponding Newton's equation, this equation has the extra term $-\left(\frac{GMA^2}{c^2 r^3}\right)$. With this additional effective interaction, the planetary orbits are no longer closed ellipses but may be simulated by slowly rotating ellipses. This gives rise to a shift of the perihelion of planets (perihelion is the point on the orbit nearest to the sun). Compared with the leading potential $-\frac{GM}{r}$, the additional term is small,

$$\frac{GMA^2/c^2r^3}{GM/r}\approx r^2/c^2 \tag{11.44}$$

which for Mercury is about 10^{-7}-10^{-8}. The quantitative effect of this term can be calculated from perturbation theory. It gives rise to a rotation of the perihelion of Mercury by about 43″ per century, in good agreement with the observed rotation.

Bending of Light

The equations for the trajectory of light in the Schwarzschild metric can be deduced from Eqs. (11.42). It should however be noted that since $\Delta\tau = 0$ for the propagation of light, the constants A and B are infinite, though A/B is finite. Dividing the last two equations by the first equation in Eqs. (11.42),

$$\frac{e(r)}{r^2}\frac{dt}{d\phi}=\frac{B}{A}$$

$$e(r)\left(\frac{dt}{d\phi}\right)^2-\frac{f(r)}{c^2}\left(\frac{dr}{d\phi}\right)^2-\frac{r^2}{c^2}=0 \tag{11.45}$$

which in terms of x, $= 1/r$ lead to

$$\left(\frac{dx}{d\phi}\right)^2+x^2-\frac{2GM}{c^2}x^3=D \tag{11.46}$$

where $D = (cB/A)^2$. For $\frac{GM}{c^2} = 0$, $x = D^{1/2} \cos \phi$. Treating the gravitational term as a small perturbation, we find to first order in GM/c^2

$$x = D^{1/2} \cos \phi + \frac{GMD}{c^2}(2 - \cos^2 \phi) \tag{11.47}$$

The bending of light is then deduced by obtaining the angles f for $r \to \infty$ or $x \to 0$, which to first order in GM/c^2 satisfy the relation

$$\cos \phi \approx -\frac{2GMD^{1/2}}{c^2} \tag{11.48}$$

or

$$\phi_{\pm} = \pm\left(\frac{\pi}{2} + \frac{2GMD^{1/2}}{c^2}\right) \tag{11.49}$$

Noting that $D^{1/2} \approx 1/r_{\min}$, the deflection of light comes out to be

$$\Delta\phi \equiv (\phi_+ - \phi_- - \pi), \tag{11.50}$$

$$\approx \frac{4GM}{c^2 r_{\min}}$$

For light just grazing the sun ($M \approx 2 \times 10^{30}$ kg, $r_{\min} \approx 7 \times 10^8$ m), this has a value of $\Delta\phi \approx 1.75''$. The bending of starlight grazing the sun during an eclipse (so as to minimize glare), is found to be about $1.89''$ which agrees well with Einstein's prediction. The corresponding prediction of Newton's theory (particles with velocity c accelerated by gravity) is half of Einstein's prediction, *i.e.* about $0.875''$.

Black Holes

Going back to the Schwarzschild metric in Eqs. (11.34), (11.35), it is seen that the metric is singular at $r = r_s$,

$$r_s = \frac{2GM}{c^2} \tag{11.51}$$

where r_s is known as the *Schwarzschild radius*. In most cases, the Schwarzschild radius is quite small, it is about 3 km for the sun. However, the metric is applicable only outside the mass distribution. Inside the distribution, the metric is modified to a nonsingular form. Therefore, in cases where r_s is much smaller than the radius of the mass distribution, the singularity is not relevant. On the other hand, in the case of very massive stars ($m > 3m_{\text{sun}}$), it is expected that the stars may ultimately collapse to size smaller than their Schwarzschild radius. These stars whose radius is smaller than their Schwarzschild radius, are known as *black holes* and the metric singularity gives rise to some unusual behaviour for them.

For radial motion ($d\phi/d\tau = 0$), one obtains from Eqs. (11.42),

$$B^2 - \frac{1}{c^2}\left(\frac{dr}{d\tau}\right)^2 = 1 - \frac{2GM}{c^2 r} \tag{11.52}$$

If the particle starts at large r, with zero velocity, $B^2 = 1$ and the equation of motion reduces to

$$\frac{dr}{d\tau} = -\left(\frac{2GM}{r}\right)^{1/2} \tag{11.53}$$

The solution to this equation is

$$r(\tau) = (2GM)^{1/3}\left(d - \frac{8}{2}\tau\right)^{2/3} \tag{11.54}$$

where d is a constant. It shows that r, as a function of the proper time, does not show any singular behaviour at $r = r_s$, and therefore the fall through $r = r_s$ is smooth.

The behaviour as seen by an observer outside the field, on the other hand, is different. Since $B = 1$ for the case under consideration, it follows from Eqs. (11.42) and (11.53), that the time interval dt, as seen by the observer, is

$$dt = \frac{d\tau}{1 - r_s/r}$$

$$= \frac{r^{1/2}\, dr}{-(2GM)^{1/2}\,(1 - r_s/r)} \tag{11.55}$$

Clearly the time interval $\Delta t \to \infty$ as $r \to r_s$. Physically, this means that with respect to an outside observer, a black hole is frozen at $r = r_s$.

It is easy to show that no light can escape from a black hole. For a light signal one has $\Delta\tau = 0$. Therefore, from the last equation in Eqs. (11.42)

$$dt = \frac{dr}{c\,|1 - r_s/r|} \tag{11.56}$$

which implies that the time taken by the signal to escape from the black hole is infinite:

$$t_{12} = \int_{r_1 < r_s}^{r_2} \frac{dr}{c\,|1 - r_s/r|} \tag{11.57}$$

$$\lim_{r_2 \to r_s} t_{12} \to \infty$$

This explains the name 'black hole' given to the object. It should be mentioned that this result does not take quantum effects into account. It has been shown by Hawking that quantum effects do allow some radiation to come out from the black holes so that, strictly speaking, a black hole is not a black hole.

11.4 KINEMATICS OF THE UNIVERSE

In this section, the metric space of the universe is considered.

It may be expected that the large-scale properties of the universe are not affected by local variations in the distribution of matter. This assumption, stated in the form of a principle, greatly simplifies the analysis. The *cosmological principle* states that *the universe must appear the same to all observers who are at rest with respect to matter in the neighbourhood. The universe must also be isotropic.* This means that clustering of galaxies is a local irregularity and the cosmological scales are must larger than the sizes of the galaxies. Observationally, the distribution of galaxies does appear to be homogeneous and isotropic though there may be some small deviations. The concept of an observer who is at rest with respect to local matter, known as a *fundamental observer*, is very useful in the determination of the kinematics of the universe.

The history of each fundamental observer being the same, the time coordinate can be linked with a local property, say the density, and may be taken to be the proper time of the fundamental observer. It was shown by Robertson and Walker, that the metric of the universe satisfying the cosmological principle, is of the form

$$(\Delta\tau)^2 = (\Delta t)^2 - \frac{1}{c^2}\left[\frac{(\Delta r)^2}{1 - K r^2} + r^2 (\Delta\theta)^2 + r^2 \sin^2\theta\, (\Delta\phi)^2\right] \tag{11.58}$$

where K is the curvature of space. The spatial part of the metric is similar to the metric of a spherical surface given in Eq. (11.29). The curvature of the space may be written as

$$K = \frac{k}{R^2(t)} \tag{11.59}$$

where $R(t)$ is called the *comic scale factor*, with $k = 1$ for positive curvature, $k = -1$ for negative curvature and $k = 0$ for flat space. The metric may be written in a more convenient form in terms of the dimensionless variable σ,

$$\sigma = r/R(t) \tag{11.60}$$

in terms of which the separation between fundamental observers does not change with time. In terms of the co-moving coordinate σ, the metric is

$$(\Delta\tau)^2 = (\Delta t)^2 - \frac{R^2(t)}{c^2}\left[\frac{(\Delta\sigma)^2}{1-k\sigma^2} + \sigma^2\,(\Delta\theta)^2 + \sigma^2\sin^2\theta\,(\Delta\phi)^2\right] \tag{11.61}$$

which is known as the Robertson-Walker metric.

Distances

Distances in the Robertson-Walker metric, with $\Delta\theta = \Delta\phi = 0$, are given by

$$D(t) = R(t)\int_0^a \frac{d\sigma}{(1-k\sigma^2)^{1/2}} \tag{11.62}$$

$$= \begin{cases} R(t)\sin^{-1}\sigma & \text{for } k=1 \\ R(t)\,\sigma & \text{for } k=0 \\ R(t)\sinh^{-1}\sigma & \text{for } k=-1 \end{cases} \tag{11.63}$$

As might have been expected the distances between fundamental observers are proportional to the scale factor. In the $k = 1$ case (positive curvature), the distance $D(t)$ is ambiguous to the extent of $2\pi nR(t)$ corresponding to going around the closed universe n number of times. The surface area of the sphere is given by $4\pi r^2$ or

$$S = 4\pi R^2(t)\sin^2[D(t)/R(t)] \tag{11.64}$$

which is bounded. This is analogous to the length of the circumference of a circle in the two dimensional case [see Eqs. (11.31) and (11.30)].

Velocities

The relative velocities of fundamental observers, are obtained from Eq. (11.62),

$$v = \frac{d\,D(t)}{dt} \tag{11.65}$$

$$= \frac{R(t)}{R(t)}D(t)$$

This allows us to identify the constant of proportionality in *Hubble's law*. It is observed that distant galaxies appear to be moving away with speeds proportional to their distances. This is described by Hubble's law $v = Hr$ where H is called Hubble's constant. Comparison of this relation with Eq. (11.65) leads to

$$H(t) = \frac{\dot{R}(t)}{R(t)} \tag{11.66}$$

Thus, Hubble's constant, in general, is a function of time. At present it has a value of about 1.8×10^{-18} s^{-1}.

Red Shifts

The Robertson-Walker metric provides the proper framework for the description of cosmological red shifts. Since the proper time for the propagation of radiation is zero, $\Delta\tau = 0$, the time interval for the propagation is given by

$$\Delta t = \frac{R(t)}{c} \frac{\Delta\sigma}{(1 - k\sigma^2)^{1/2}} \tag{11.67}$$

Consider now two crests emitted from a galaxy at times t_e and $t_e + \Delta t_e$, which are received by another galaxy at times t_0 and $t_0 + \Delta t_0$ respectively,

$$\int_{t_e}^{t_0} \frac{d(t)}{R(t)} = \frac{1}{c} \int_{\sigma_0}^{\sigma_e} \frac{d\sigma}{(1 - k\sigma^2)^{1/2}} \tag{11.68}$$

$$\int_{t_e + \Delta t_e}^{t_0 + \Delta t_0} \frac{dt}{R(t)} = \frac{1}{c} \int_{\sigma_0}^{\sigma_e} \frac{d\sigma}{(1 - k\sigma^2)^{1/2}} \tag{11.69}$$

From these two relations, it follows that

$$\int_{t_0}^{t_0 + \Delta t_0} \frac{dt}{R(t)} = \int_{t_e}^{t_e + \Delta t_e} \frac{dt}{R(t)} \tag{11.70}$$

For the short intervals under consideration,

$$\frac{\Delta t_0}{\Delta t_e} = \frac{R(t_0)}{R(t_e)} \tag{11.71}$$

and since wavelengths are proportional to the time intervals,

$$\frac{\lambda_0}{\lambda_e} = \frac{R(t_0)}{R(t_e)} \tag{11.72}$$

In the case of an expanding universe, $R(t_0) > R(t_e)$, which would explain the observed red shifts. Expanding $R(t_e)$ at t_0, the red shift z, is

$$z = \frac{\lambda_0}{\lambda_e} - 1$$

$$= \frac{R(t_0)}{R(t_0)\left[1 + H_0\,(t_e - t_0) - \frac{1}{2} q_0\, H_0^2\,(t_e - t_0)^2 + \ldots\right]} - 1$$

$$= (t_0 - t_e) H_0 + \left(\frac{1}{2} q_0 + 1\right) H_0^2 (t_0 - t_e)^2 + ... \qquad (11.73)$$

The parameter q_0 known as the *deceleration paramenter*, is important in the determination of the nature of the universe. For example, a positive q_0 implies a slowing down of the expansion of the universe. It is possible to estimate the value of $(t_0 - t_e)$ from the study of the apparent brightness (essentially the radiation received) of galaxies which together with a knowledge of the red shifts would give an estimate of q_0. Through the experimental uncertainties are too large at present to yield a reliable value for q_0 there are some indications that it is positive (see Ref. 10).

11.5 DYNAMICS OF THE UNIVERSE

The evolution of the universe is determined by the time-dependence of $R(t)$ which is governed by the field equations. However, if the pressure, as a source of gravity, can be neglected (this is certainly reasonable in the present era), most of the general results can be obtained from Newtonian theory of gravity. The discussion here will be based primarily on this simpler approach.

Consider a particle on the surface of a small sphere with its centre at the origin. Then its equation of motion is

$$ma = -\frac{Gm\left(\frac{4\pi}{3} r^3\right)\rho(t)}{r^2} \qquad (11.74)$$

which in terms of $R(t)$ [see Eq. (11.60)] reads as

$$\ddot{R}(t) = -\frac{4\pi}{3} G\rho(t) R(t) \qquad (11.75)$$

Originally, Einstein had considered an additional repulsive term $\frac{1}{3}\Lambda R(t)$ to counteract the attraction. Such a term is ignored in our simplified discussion. Using $\rho(t) = \rho(t_0) R^3(t_0)/R^3(t)$, Eq. (11.75) is integrated after multiplying by $2\dot{R}$ to get

$$\dot{R}^2 = \frac{8\pi}{3} G\rho(t_0)\frac{R^3(t_0)}{R(t)} - kc^2 \qquad (11.76)$$

Here, the constant of integration $- kc^2$, is a measure of the total energy of the particle, and is related to the curvature index $k(k = \pm 1, 0)$ by the solutions to the field equations. The relation suggests that the universe is closed for $k = 1$, *i.e.* $\dot{R}(t)$ becomes zero for sufficiently large $R(t)$ and changes its sign, but open for $k = -1$ or 0, *i.e.* $\dot{R}(t) \neq 0$. This can be compared with what happens to a body thrown up from the surface of the earth. If the initial velocity v_{in} is less than the escape velocity v_{esc} ($E_{tot} < 0$), the body will reach a maximum height and return back to the earth, corresponding to the closed universe. If $v_{in} > v_{esc}$ ($E_{tot} > 0$), the body will go on forever corresponding to the open universe. When $v_{in} = v_{esc}$ ($E_{tot} = 0$), the body just manage to escape from the earth.

In all the three cases of Eq. (11.76), $\dot{R}$ is positive now, and was large at earlier times. Hence the universe described by Eq. (11.76) had a big-bang origin. The solution for $R(t)$ can be obtained by integrating Eq. (11.76) in the usual way:

$$t = \left(\frac{R_m c}{c}\right) \cdot \begin{cases} \sin^{-1}(x^{1/2}) - x^{1/2}(1-x)^{1/2} & \text{for } k = 1, \\ \frac{2}{3}x^{3/2} & \text{for } k = 0, \\ x^{1/2}(1+x)^{1/2} - \sinh^{-1}(x^{1/2}) & \text{for } k = -1, \end{cases} \tag{11.77}$$

$$R_m = \frac{8\pi G\,\rho(t_0)\,R^3(t_0)}{3c^2} \tag{11.78}$$

where $x = R(t)/R_m$. The three cases are considered separately.

(*i*) $k = 0$, *the Einstein-de Sitter model:* In this case the three dimensional space is flat. This gives a permanently expanding universe (see Fig. 11.2) with $R(t)$ given by

$$R(t) = R_m \left(\frac{3ct}{2R_m}\right)^{2/3} \tag{11.79}$$

Since $\dot{R} = RH$ where H is Hubble's constant, one gets from Eq. (11.79), $t = 2/3\ H$. The age of the universe is then estimated to be

$$t_0 \approx 1.2 \times 10^{10} \text{ years} \tag{11.80}$$

(*ii*) $k = 1$, *oscillating universe:* In this case the universe expands to a maximum value of R_m at some time $t_{1/2}$ and collapses back to the original state ($R \to 0$) at $2t_{1/2}$ (see Fig. 11.2). Using Eqs. (11.66) and (11.76), $R(t_0)$ can be obtained in terms of $\rho(t_0)$. The value of R_m then is estimated to be [$\rho(t_0) \approx 1.2 \times 10^{-26}$ kg/m^3]

$$R_m = \frac{8\pi G\,\rho(t_0)\,R^3(t_0)}{3c^2}$$

$$\approx 1.3 \times 10^{10} \text{ parsecs} \tag{11.81}$$

where 1 parses $\approx 3 \times 10^{13}$ km ≈ 3.26 light years, 1 light year being the distance travelled by light in 1 year. The collapse period is

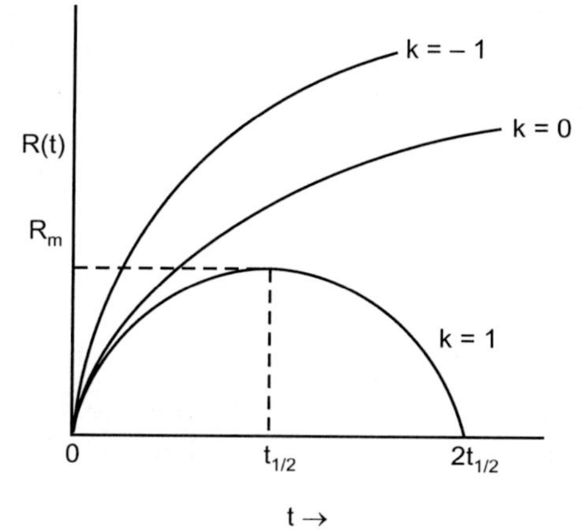

Fig. 11.2 The scale factor $R(t)$ of the universe as a function of time t.

$$2t_{1/2} = \frac{\pi R_m}{c}$$
$$\approx 1.2 \times 10^{11} \text{ years} \tag{11.82}$$

The present age of the universe is estimated from Eq. (11.77) to be

$$t_0 \approx 10^{10} \text{ years} \tag{11.83}$$

Since the universe is closed in this case, there are no problems of boundary conditions at infinity, and Mach's principle is incorporated in the theory.

(*iii*) $k = -1$, *ever-expanding universe:* In this case, $R(t)$ increases as $t^{2/3}$ for small t but increases as t for large t. In this model, following the same steps as for $k = 1$, the present age of the universe comes out to be somewhat larger, [$\rho(t_0) \approx 3 \times 10^{-28}$ kg/m^3]

$$t_0 \approx 1.8 \times 10^{10} \text{ years} \tag{11.84}$$

Though the evidence from various sources, *e.g.* from the estimates of deceleration parameter, is not definitive, it does favour a closed universe. This means that not only did the universe start with a big bang, it will also collapse to the original dense state. What it will do beyond that point is not very clear—it may end at that point or start again giving an oscillating universe.

11.6 THE EARLY UNIVERSE

The big bang thorias discussed above imply that the universe must have started as a very dense, very hot object. This has certain interesting implications for the present-day universe. Two specific results are discussed here.

Background Black-Body Radiation

At the beginning, the universe consisted of radiation and matter in equilibrium at very high temperatures, confined to a small region. At these high temperatures, matter must have been in an ionized state. Since matter in ionized state has much greater interaction with radiation than does matter in atomic state, the radiation had an equilibrium black-body spectrum characterized by temperature T. In this condition, the spectrum of photons, *i.e.* number of photons in the range ν and $\nu + d\nu$, is deduced from the Planck expression [Eq. (2.12)],

$$dN = \frac{8\pi\nu^2 \, V \, d\nu}{c^3 \, [\exp(h\nu/kT) - 1]} \tag{11.85}$$

As the universe expands, the temperature falls. It can be shown that the photon spectrum maintains its black-body characteristics, but with a lower temperature. As a result of the expansion, V changes to V',

$$V' = \frac{R'^3}{R^3} V \tag{11.86}$$

Furthermore, the frequency gets red-shifted (see Eq. (11.72)), the red-shift is present even for reflection by a body moving away) and is given by

$$\nu' = \frac{R}{R'} \nu \tag{11.87}$$

Hence the spectrum is now given by

$$dN = \frac{8\pi(R'\nu'/R)^2 \, (R^3V'/R'^3) \, (R'd\nu'/R)}{c^3[\exp(h\nu'R'/kRT) - 1]}$$

$$= \frac{8\pi\nu'^2 \, V' \, d\nu'}{c^3[\exp(h\nu'/kT') - 1]} \tag{11.88}$$

where $T' = \dfrac{R}{R'} T$. Furthermore, the energy density is given by

$$\varepsilon'_{rad} = \frac{4}{c} \sigma T'^4$$

$$= \frac{4\sigma R^4 T^4}{cR'^4} \tag{11.89}$$

where σ is the Stefan-Boltzmann constant, $\sigma = 5.67 \times 10^{-8}$ in mks units. Such a radiation was observed by Penzias and Wilson (1965), with a characteristic temperature $T' = 2.7$ K. This provides strong support to the big bang theory.

The equivalent mass density is

$$\rho'_{rad} = \frac{4}{c^3} \sigma T'^4$$

$$\approx 4.5 \times 10^{-31} \text{ kg/m}^3 \tag{11.90}$$

which is must smaller than the estimated galactic matter density of about 3×10^{-28} kg/m^3. The present era is therefore known as the *matter-dominated era*. It may be noted that unlike the radiation density which is proportional to $1/R'^4$, the matter density varies as $1/R'^3$.

$$\rho'_{\text{mat}} = \left(\frac{R}{R'}\right)^3 \rho_{mat} \tag{11.91}$$

which implies that

$$\frac{\rho'_{\text{mat}}}{\rho'_{\text{rad}}} = \left(\frac{R'}{R}\right)\frac{\rho_{\text{mat}}}{\rho_{\text{rad}}} \tag{11.92}$$

Clearly at early times ρ_{rad} dominated over ρ_{mat} and that period is known as the *radiation-dominated era*. At the transition between radiation and matter dominated eras, $\rho_{\text{mat}} \approx \rho_{\text{rad}}$ so that

$$\frac{R'}{R} = \frac{\rho'_{\text{mat}}}{\rho'_{\text{rad}}},$$

$$\approx 670 \tag{11.93}$$

which is the present ratio of matter and radiation densities. The transition must have taken place at

$$T = \frac{R'}{R}T'$$

$$\approx 1810 \text{ K} \tag{11.94}$$

At around this temperature (actually at a somewhat higher temperature), hydrogen is in an ionized state, so that at transition, there was enough matter density to produce black-body spectrum for the radiation.

It can be shown that, at early times, the temperature was proportional to $t^{-1/2}$. $T_{\text{rad}} \sim 1/R(t)$. In the radiation-dominated era, solutions in Eq. (11.77) get modified and $R(t) \sim t^{1/2}$ so that $T_{\text{rad}} \sim t^{-1/2}$. General relativity gives the precise relation as

$$T_{\text{rad}} = \frac{1.5 \times 10^{10}}{t^{1/2}} \text{K} \tag{11.95}$$

The temperatures at $t \sim 1$ s were high enough to have created electron-positron pairs. At lower temperatures, T around 10^8 K, fusion reactions must have been present. This is a possible explanation for the large amount of He present, about 1 He atom for every 12 hydrogen atoms (the fusion reactions going on in stars can account for only about 10% of this). It is only at later stages that the condensation into galaxies took place, giving us essentially the present universe.

Radio Source Counting

Radio telescopes can detect sources at enormous distances and hence can be used to obtain informations about the number of ratio sources with apparent brightness (related to the radiation energy received) $l > l_0$. Now, for a homogeneous distribution of sources, since $l = a/r^2$, the number of sources with $l > l_0$ is related to l_0 by

$$N = br_0^3$$
$$= b'/l_0^{3/2} \tag{11.96}$$

where $l_0 = a/r_0^2$, and a, b, b' are constant. This implies that

$$\ln N = -\frac{3}{2} \ln l_0 + d \tag{11.97}$$

Observationally, the number of faint sources appears to be larger,

$$\ln N \sim -2.0 \ln l_0 + d' \tag{11.98}$$

This is an interesting result. In the big-bang theory, it only means that since the radiation from far away sources was emitted earlier, there must have been (*i*) more numerous radio sources and/or (*ii*) brighter radio sources, at early times.

Particle Physics and Cosmology

Finally, we note the important role that our knowledge of particle properties, is playing in our understanding of the universe. Two recent, though tentative, developments are mentioned here.

The preponderance of matter over antimatter has been one of the puzzles in cosmology. The solution to this puzzle may be in the difference in the properties of baryons and antibaryons. The recent attempts at grand unification of strong, weak and electromagnetic interactions (see Sec. 10.5), allow for differences in the properties of matter and antimatter. Estimations based on these theories, and also in rough agreement with the observed ratio of 10^{-9} for the number of baryons (essentially protons and neutrons) to the number of photons.

There has been a question about the missing mass in the universe (apart from the mass of the galaxies). Some recent experiments appear to suggest that neutrinos may have nonzero mass. If this is confirmed, the neutrinos of the universe may contribute a substantial amount of mass to the galaxies and the universe.

A striking characteristic of recent developments in modern physic is that observations in such diverse fields as elementary particles, nuclear physics, atoms and molecules, solid state physics and astrophysics, are interrelated and allow for a unified approach. In a sense, the distinctions between the different domains are becoming blurred and ultimately there may be only the core of basic laws of physics in terms of which all observations can be explained.

11.7 EXAMPLES

In this section, some examples related to the ideas discussed earlier are given.

Example 1

Olbers argued that in an infinite, static universe, every part of the sky must have a brightness comparable to that of the sun.

In an infinite, static universe, every part of the sky will be covered by a star. Let the visible part of one such star subtend a solid angle $\Delta\Omega$ at the earth. Now, the intensity of radiation received at the earth, due to this star at distance d, is proportional to the exposed are $d^2\Delta\omega$ of the star, and decreases as $1/d^2$. Therefore it is proportional to $d^2\Delta\Omega\ (1/d^2) \sim \Delta\Omega$ which is independent of the distance of the star. Therefore, an equivalent area ($d^2_{\text{sun}}\ \Delta\Omega$) of the star might as well be at a distance equal to that of the sun. If we make the reasonable assumption that the stars in general have approximately the same inherent brightness as that of the sun, we would then expect every part of the sky to have the brightness of the sun, day or night. Any attempt at an explanation in terms of absorption does not succeed since at equilibrium the absorbing material must emit as much energy as it absorbs.

Example 2

The red shift of radiation in the presence of a gravitational field, to the leading order in the field, can be shown to follow from energy conservation.

Consider an atom of mass m_2 which goes to a state of lower mass m_1, with the emission of a photon of frequency ν_0. Then energy conservation implies

$$m_2c^2 = m_1c^2 + h\nu_0 \tag{11.99}$$

If the atom is placed in a gravitational potential ϕ, its initial energy is $m_2a^2 + m_2\phi$ while the final energy is $m_1c^2 + m_1\phi$. Then the frequency ν of the photon, which comes out of the potential, is given by

$$\begin{aligned} h\nu &= (m_2c^2 + m_2\phi) - (m_1c^2 + m_1\phi) \\ &= (m_2c^2 - m_1c^2)\ (1 + \phi/c^2) \end{aligned} \tag{11.100}$$

Using Eq. (11.99)

$$\nu = \nu_0\ (1 + \phi c^2) \tag{11.101}$$

which agrees with the general expression in Eq. (11.39) to the leading order. For emission from the sun, $\phi/c^2 \approx -2 \times 10^{-6}$.

Example 3

When two clocks accelerate with respect to each other, they show different proper times. This provides a solution to the twin paradox.

The metric near the surface of the earth, in the local inertial frame (at rest with respect to distant galaxies) is the Schwarzschild metric in Eq. (11.34). Consider two clocks, 1 and 2 which go around with angular velocities ω_1 and ω_2. If they are together at the beginning and again at the end, the proper times shown by them are

$$\tau_i = \int_0^t \left[1 - \frac{2GM}{c^2 r} - \frac{r^2}{c^2}\omega_i^2\right]^{1/2} dt$$

$$= \left(1 - \frac{2GM}{c^2 r} - \frac{r^2}{c^2}\omega_i^2\right)^{1/2} t \tag{11.102}$$

where t is the time coordinate. Therefore

$$\frac{\tau_1 - \tau_2}{\tau_1} \approx \frac{r^2(\omega_2^2 - \omega_1^2)}{2c^2} \tag{11.103}$$

This relation was verified by keeping clock 1 at rest on Earth, $\omega_1 = 2\pi$ rad/day, and taking clock 2 around the earth with velocity v, $\omega_2 = \omega_1 \pm \frac{v}{r}$. For $v \approx 800$ km/h,

$$\frac{\tau_1 - \tau_2}{\tau_1} \approx 1.42 \times 10^{-12} \text{ for } \omega_2 = \omega_1 + \frac{v}{r},$$

$$\approx -0.87 \times 10^{-12} \text{ for } \omega_2 = \omega_1 - \frac{v}{r} \tag{11.104}$$

It is important to note that a clock with greater acceleration shows smaller time, which explains the longer lifetime observed for particles going around in accelerators.

Example 4

An interesting idea in cosmology is what is called the *object horizon*. This is the value σ_{oh} of the farthest object which is visible to us. The signal reaching us, from this object, must have been emitted at the beginning of the universe, *i.e.* at $t = 0$. Since $\Delta\tau = 0$ for the propagation of light, one has [Eq. (11.61)]

$$\int_0^{t_0} \frac{dt}{R(t)} = \frac{1}{c}\int_0^{\sigma_{oh}} \frac{d\sigma}{(1 - k\sigma^2)^{1/2}} \tag{11.105}$$

For the special case of $k = 1$,

$$\sigma_{\text{oh}} = \sin\left(c \int_0^{t_0} \frac{dt}{R(t)}\right) \tag{11.106}$$

The distance of the horizon is [Eq. (11.62)].

$$\begin{aligned} d_{\text{oh}} &= R(t_0) \int_0^{\sigma_{oh}} \frac{d\sigma}{(1 - k\sigma^2)^{1/2}} \\ &= R(t_0) \int_0^{t_0} \frac{c\, dt}{R(t)} \end{aligned} \tag{11.107}$$

Using the solutions in Eq. (11.77) it has been estimated that

$$\begin{aligned} d_{\text{oh}} &\approx \frac{\pi}{2} R(t_0) \text{ for } k = 1, \\ &\approx 10^{10} \text{ parsecs} \end{aligned} \tag{11.108}$$

PROBLEMS

1. A photon is moving horizontally on the surface of the earth. What is the height through which falls in travelling 100 m?
2. Starting from the flat-space metric, obtain the metric for the frame which rotates with angular velocity ω along the z-direction. Write down the equations for geodesics in this frame. Exhibit the coriolis and centrifugal forces in the nonrelativistic approximation.
3. For a particle going around an accelerator, show that the lifetime is given by $\tau = \tau_0 (1 - \omega^2 r^2/c^2)^{-1/2}$, where ω is the angular velocity, and r is the radius of the orbit. What is the expression for τ if ω is changing but r remains a constant?
4. Curvature of a surface may be defined in terms of area also. Show that curvature of a spherical surface is given by

$$K = \frac{12}{\pi} \lim_{a \to 0} \left(\frac{\pi a^2 - A}{a^4}\right)$$

   where A is the area of the surface and a is the distance of any point on the circumference of the circle from the centre of the surface.
5. What is the Schwarzschild radius of the earth?
6. Show that for circular motion in the Schwarzschild metric

$$\left(\frac{d\phi}{d\tau}\right)^2 \left(1 - \frac{3GM}{rc^2}\right) = \left(\frac{GM}{r^3}\right)$$

In the nonrelativistic limit, this relations tends to the usual relation $\omega^2 = GM/r^3$. (It is simpler to start with the geodesic equation for r.)

7. A photon may be bound in a closed orbit by the potential of a black hole. Show that for a circular orbit

$$r = 3GM/c^2 \text{ and } \frac{d\phi}{dt} = c/(3^{1/2}\, r).$$

8. Consider the Robertson-Walker metric with $k = 0$. If a signal is emitted at t and received at t_0,

   (*a*) show that

$$\frac{D(t)}{R(t)} = \sigma$$

$$= \left(\frac{12c}{R_m}\right)^{1/3} (t_0^{1/3} - t^{1/3})$$

   (*b*) show that

$$v(t) = \frac{2}{3t} D(t)$$

   (*c*) show that

$$D(t) = \frac{3}{2} t_0 \frac{v}{(1 + v/2c)^3}.$$

9. In the steady state theory of the universe, the decrease in the density of matter due to expansion of the universe is compensated by continuous creation of matter. Using continuity equation show that the rate of creation is given by

$$\frac{d\rho_c}{dt} = 3\rho_0 H$$

Given that $\rho_0 \approx 3 \times 10^{-28}$ kg/m^3, estimate the rate of creation in terms of protons/m^3/s. Argue that the steady state theory does not imply a sky with a uniform brightness equal to that of the sun.

References

General Books on Modern Physics

1. Leighton R.B., *Principles of Modern Physics*, McGraw-Hill, New York, 1959.
2. Richtmyer F.K., E.H. Kennard and J.N. Copper, *Introduction to Modern Physics*, McGraw-Hill, New York, 1969.
3. French A.P., *Principles of Modern Physics*, John Wiley, London, 1958.
4. Weidner R.T. and R.L. Sells, *Elementary Modern Physics*, Allyn and Bacon, Boston, 1980.
5. Sproull R.L. and W.A. Phillips, *Modern Physics*, John Wiley, New York, 1980.
6. Savelyev I.V., *Physics, a General Course*, vol. III, Mir, Moscow, 1981.
7. Beiser A., *Perspectives of Modern Physics*, McGraw-Hill, New York, 1973.

Special and General Relativity

8. Bergmann P.G., *Introduction to the Theory of Relativity*, Prentice-Hall, Englewood Cliffs, 1942.
9. Rindler W., *Essential Relativity*, Van Nostrand Reinhold, New York, 1969.
10. Berry M., *Principles of Cosmology and Gravitation*, Cambridge University Press, Cambridge, 1976.
11. Landau L. and E. Lifshitz, *The Classical Theory of Fields*, Addison-Wesley, Reading, 1951.
12. Sard R.D., *Relativistic Mechanics*, W.A. Benjamin, New York, 1970.
13. Weinberg S., *Gravitation and Cosmology*, John Wiely, New York, 1972.

Quantum Mechanics

14. Pauling L. and E.B. Wilson, *Introduction to Quantum Mechanics*, McGraw-Hill, New York, 1935.
15. Fermi E., *Notes an Quantum Mechanics*, University of Chicago Press, Chicago, 1961.

16. Landau L. and E. Lifshitz, *Quantum Mechanics–Nonrelativistic Theory*, Pergamon, London, 1958.
17. Schiff L.I., *Quantum Mechanics*, McGraw-Hill, New York, 1955.
18. Merzbacher E., *Quantum Mechanics*, John Wiley, New York, 1961.

Atomic and Molecular Physics

19. Peaslee D.C., *Elements of Atomic Physics*, Prentice-Hall, Englewood Cliffs, 1955.
20. Cagnac B. and J.C. Pebay-Peyroula, *Modern Atomic Physics: Fundamental Principles*, Macmillan, London, 1975.
21. Herzberg G., *Spectra of Diatomic Molecules*, D. Van Nostrand, Princeton, 1950.
22. Cagnac B. and J.C. Pebay-Peyroula, *Modern Atomic Physics: Quantum Theory and its Applications,* Macmillan, London, 1975.
23. Dunford H.B., *Elements of Diatomic Molecular Spectra*, Addison-Wesley, Reading, 1968.
24. Karplus M. and R.N. Porter, *Atoms and Molecules*, Benjamin, Boston, 1970.
25. Fano U. and L. Fano, *Physics of Atoms and Molecules*, University of Chicago Press, Chicago, 1970.

Quantum Statistics and Solid State Physics

26. Mandl F., *Statistical Physics*, John Wiely, London, 1971.
27. Gopal E.S.R., *Statistical Mechanics and Properties of Matter*, Ellis Horwood, Westergate, 1974.
28. Landau L. and E. Lifshitz, *Statistical Physics*, Pergamon, London, 1959.
29. Hart-Davis A., *Solids*, McGraw-Hill, London, 1975.
30. Rosenberg M., *The Solid State*, Clarendon Press, Oxford, 1975.
31. Hall H.E., *Solid State Physics*, John Wiley, London, 1974.
32. Rudden M.N. and J. Wilson, *Elements of Solid State Physics*, John Wiley, Chichester, 1980.
33. Ashcroft N.W. and N.D. Mermin, *Solid State Physics*, Holt, Rinehart and Winston, New York, 1976.
34. Kittel C., *Introduction to Solid State Physics*, John Wiley, New York, 1976.
35. Ali Omar M., *Elementary Solid State Physics*, Addison-Wesley, Reading, 1975.

Nuclear Physics

36. Elton L.R.B., *Introductory Nuclear Theory*, Interscience, New York, 1959.
37. Segre E., *Nuclei and Particles*, Benjamin, New York, 1965.
38. Enge H.A., *Introduction to Nuclear Physics*, Addison-Wesley, Reading, 1966.
39. Bethe H.A. and P. Morrison, *Elementary Nuclear Theory*, John Wiley, New York, 1956.
40. Preston M.A., *Physics of the Nucleus*, Addison-Wesley, Reading, 1962.
41. Murray R.L., *Nuclear Energy*, Pergamon, New York, 1975.

Elementary Particles

42. Longo M.J., *Fundamentals of Elementary Particles*, McGraw-Hill, New York, 1973.
43. Yang C.N., *Elementary Particles*, Princeton University Press, Princeton, 1962.
44. Livigston M.S., *Particle Physics*, McGraw-Hill, New York, 1968.

Answers to Problems

Chapter 1

1. Time period is $(1 + v/c)/(1 - \beta^2)^{1/2}$
5. 14%; 10 km
7. $0.9974c$
9. $(M^2 - m^2)\, c^2/2M$; 224.6 MeV, 3.4 eV
11. $0.875c$, $0.999994c$
12. 287 km/s; 6556.7 Å

Chapter 2

1. 5800 K; for significant number of hydrogen atoms to be in the excited states, $kT \sim 10$ eV
2. 7.13×10^3 J/s; 0.019 J/s
4. 19; 4×10^{17} m^{-2} s^{-1}
5. $[2mc^2\,(h\nu - \varepsilon) + h^2\nu^2 - 2h\nu\,(2mc^2\,(h\nu - \varepsilon))^{1/2} \cos\phi]\; 2\, Mc^2$
5. $\theta = 180°$; $2h^2 \nu_0^2 / mc^2$
7. 0.1484 Å, 0.1 Å; $\theta = 54°$
8. $[(m^2c^4 + h^2\nu^2)^{1/2} - mc^2]/h\nu$; $h\nu/2mc^2$; 10 keV
9. 1.2 Å; yes
10. $n = 1$, $d = 0.5$ Å
12. 0.7 MeV; 40 MeV
13. $n\hbar\omega$
14. 0.5×10^{-13} m; 25.3 MeV
16. 1/1836
17. 2
18. $n\hbar\omega$; $g\left(\dfrac{s}{2} - 1\right)\left(\dfrac{mgs}{\hbar^2 n^2}\right)^{s/(2-s)}$
19. $\dfrac{e\hbar\, Bn}{2m}(1 \pm 1)$ for parallel and anti-parallel cases
20. $n\hbar\omega$
22. 1216 Å, 1026 Å
23. 6000 K
24. 14 K

Chapter 3

1. $\dfrac{1}{\pi}\left(\dfrac{h}{2a}\right)^{1/2} \dfrac{\sin\,(ka/\hbar)}{k}$
2. $T = 4r/(1 + r^2)$, $r = (E + V_0)^{1/2}/E^{1/2}$, $R = 1 - T$
3. $\nu = hn/4ml^2 = v/2l$
4. $A = (\pi a^3)^{-1/2}$, $a = 4\pi\varepsilon_0 h^2/me^2$, $E = -\,\hbar^2/2ma^2$; $P = 13\, e^{-4}$
5. $a = (km)^{1/2}/2\hbar$. $A = \dfrac{(8/3)^{1/2}\,(2a)^{5/4}}{\pi^{1/4}}$

6. $\tan qa = \alpha/q$, $\cot qa = -\alpha/q$, $\alpha = (-2mE)^{1/2}/\hbar$, $q = [2m(V_0 + E)]^{1/2}/\hbar$

7. $\sigma(x) = l\left(\frac{1}{12} - \frac{1}{2\pi^2 n^2}\right)^{1/2}$, $\sigma(p) = \pi\hbar n/l$

8. $\cot qa = -\alpha/q$

9. $\langle L_z \rangle = \langle p_z \rangle = \langle p_x \rangle = 0$, $\langle p_z^2 \rangle = 3a\hbar^2$

10. $3\lambda/4\alpha^4$

11. $\Delta E_0 = 0$, $\Delta E_1 = \pm \lambda l^2 (16/9\pi^2)^2$

Chapter 4

1. $2E_1, -E_1$
2. $13e^{-4}, 0$
3. $E = E_1/4$
4. $a_1, 4e^{-2}/a_1$
5. $10^{-10} Z^2 E_1$
7. $E_\mu = bE_e$, $r_\mu = r_e/b$, $b = m_\mu(m_p + m_e)/m_e(m_p + m_\mu)$
8. $\nu_1 = 2.462 \times 10^{15}\left(1 + \frac{11}{48}\alpha^2\right)\text{s}^{-1}$,

   $\nu_2 = 2.462 \times 10^{15}\left(1 + \frac{5}{16}\alpha^2\right)\text{s}^{-1}$
9. 7 lines
10. $S_0, S_1, P_0, P_1, P_1, P_2; P_1 \to S_1, P_1 \to S_0$,
    $P_0 \to S_1, P_2 \to S_1, P_1 \to S_1, P_1 \to S_0$

Chapter 5

3. 1, 9
6. ${}^2S_{1/2}$, ${}^1S_0$, ${}^2D_{3/2}$, ${}^3F_2$, ${}^4F_{3/2}$, ${}^7S_3$, ${}^6S_{5/2}$, ${}^5D_4$, ${}^4F_{9/2}$, ${}^3F_4$, ${}^2S_{1/2}$, ${}^1S_0$
7. $C_{LS}\hbar^2 = 0.6$ eV
8. 2.1 eV
9. 25.6 kV, 3.8 kV; 0.57 Å; 0.48 Å, 3.3 Å
10. 40; 11.1 keV
11. 4.13×10^{-15}
12. 6.8 keV
13. 3.8 eV
14. 470 N/m, 1.27 Å
15. 0.059

Chapter 6

1. $\Delta E = \pm\frac{e\hbar B}{2m}$, 0; ± 0.00160 Å, 0; ± 0.208 Å, 0
2. $\Delta\nu = \frac{eB}{4\pi m}\left(\frac{4}{5}M_J - \frac{4}{3}M_J'\right)$; max $\Delta\lambda$ is 0.5 Å

3. $\Delta \nu = \dfrac{eB}{4\pi m}\left(\dfrac{4}{5}M_J - \dfrac{2}{3}M_J{}'\right)$

4. $g = 3/2$; 4.2×10^{10} s^{-1} 6. $g = 0.40$

7. 0.025

8. $2 \exp(-E_2/kT) = \exp(-E_1/kT) + \exp(-E_3/kT)$

9. 4.55×10^{-4} % 10. $\Delta\lambda = 3.98$ Å, 6.64 Å

Chapter 7

1. Allowed occupation numbers are $A = (1, 0, 0, 2)$,
$B = (0, 1, 1, 1)$, $C = (0, 0, 3, 0)$
(*a*) 3, 12, 8 (*b*) 1, 2, 4 (*c*) 0, 1, 0 are the numbers of arrangements for *A*, *B* , *C*

2. $\dfrac{1}{2}kT$; 1.35×10^{-6} 3. 7.5×10^{-7} %

4. 11.7 %, 1.6 % 5. 2.8 R

6. $E = 4N_0\, kT\left(\dfrac{T}{\theta}\right)^2 \displaystyle\int_0^{\theta/T} \frac{x^2\,dx}{(e^x - 1)}$ 7. 350 K

8. R mol^{-1} K^{-1} 9. $4\pi V(2v_t^{-3} + v_l^{-3})\left(\dfrac{kT}{h}\right)^3 \displaystyle\int_0^{\theta/T} \frac{x^2\,dx}{(e^x - 1)}$

11. 5.5 eV, 3.3 eV; $1 - 1.8 \times 10^{-5}$, $1 - 5.1 \times 10^{-5}$

12. 0.018 *R*, 2.8 *R* 13. 4.25×10^4 K

14. 1.97×10^{24} m^{-3} 15. 2.6 J/ms K

16. 7×10^{11} s^{-1} 17. 1.96×10^{-5} V

Chapter 8

1. 6.7 eV, 0.89 eV 3. $C = 8.7$ keV, $a = 0.31$ Å

4. 6.0×10^{23} mol^{-1} 5. $\sin\theta = 0.50, 0.71, 0.87$; no; no

6. $R(3^{1/2} - 1)$; $R(2/3^{1/2} - 1)$; $R(3^{1/2}/2^{1/2} - 1)$

8. 13.6 Ω^{-1} m^{-1}; -1.1×10^{-3} $V\,m^3/A\,W$

9. 3 10. 0.73 eV

11. 0. 014 K

12. 6×10^{-10} Ω^{-1} m^{-1}; 1.15×10^2 Ω^{-1} m^{-1}

13. 0.22 A, 4.6×10^{-8} A 14. 9.15×10^{-7} m

16. 0.83×10^{-2} Ω m 18. 2.9×10^{25} m^{-3}

19. 0.85 V

21. $E = \dfrac{Ne\hbar g\ B}{4m}(e^{-y} - e^{y})/(e^{-y} + e^{y})\ y = \dfrac{e\hbar g\ B}{4mkT}$

22. $3T$

Chapter 9

2. $\Delta E \approx 0.72\ A^{2/3}$ MeV 3. 0.32

6. 4.79 $e\hbar/2m_p$ for $l = 2$

7. $s_{1/2}, p_{1/2}, s_{1/2}, h_{9/2}$; – 1.91, – 0.26, 2.79, 2.62 in units of $e\hbar/2m_p$

8. 3.9×10^{-54} kg. m^2, $I_{tot} = 8.7 \times 10^{-54}$ kg. m^2,
   E = 0.336 Mev

9. 0.033 MeV and 0.076 MeV

10. Ru $\xrightarrow[\beta]{}$ Rh $\xrightarrow[\beta]{}$ Pd, Ag $\xrightarrow[\beta^+]{}$ Pd, Cd $\xrightarrow[\beta^+]{}$ Ag;

    Ag and Cd decay also by election capture.

11. Ni $\xrightarrow[\beta]{}$ Cu, Zn $\longrightarrow$ Cu by electron capture

12. 27.9 MeV, 4.23 MeV, 0.07 MeV

13. 8.36 Mev, 0.076 Mev

14. 31. 9739 mu

15. 5.5×10^{-5} MeV, 4.6×10^{-22} kg m/s

16. 8; 6; 1.54×10^{-10} gm; 1.45×10^9 years

17. 2.05 MeV 18. 1.93×10^{-26} m^2

19. 53 g

20. Fusion would require 4.6% change and fission 30% change in 5×10^9 years

Chapter 10

2. $h\nu = mc^2 \dfrac{(E + mc^2)^{1/2}}{(E + mc^2)^{1/2} - (E - mc^2)^{1/2} \cos\theta}$

3. $(m_K^2 - 3m\pi^2)/2m_K$

4. $\Sigma^0 \rightarrow \Lambda 0 + \gamma$ decay is due to electromagnetic interaction, whereas the decay of $\Sigma^{\pm}$ is due to weak interaction

5. About 2×10^{-18} m 6. 4×10^3 s

7. 0.82 Wb/m^2

Chapter 11

1. 5.5×10^{-13} m

2. $g_{00} = 1 - \frac{\omega^2}{c^2}(x^2 + y^2)$, $g_{0x} = \frac{\omega}{c^2}\, y$, $g_{0y} = -\frac{\omega}{c^2}\, x$

3. $\tau_0 = \int_0^{\tau} \left(1 - \frac{r^2}{c^2}\omega^2\right)^{1/2} dt$

5. 9×10^{-3} ms

9. 2×10^{-18} m^{-3} s^{-1}

Index

F

G

H

I

J

K

L

M

N

O

P

Q

R

S

T

U

V

W

X

Y

Z